COLLECTED STUDIES SERIES

Arabic Mathematical Sciences

Also in the Variorum Collected Studies Series:

GEORGE MOLLAND
Mathematics and the Medieval Ancestry of Physics

NATHAN SIVIN
Science in Ancient China: Researches and Reflections

NATHAN SIVIN
Medicine, Philosophy and Religion in Ancient China
Researches and Reflections

ANNE TIHON
Etudes d'astronomie byzantine

JULIO SAMSÓ
Islamic Astronomy and Medieval Spain

A.I. SABRA
Optics, Astronomy and Logic
Studies in Arabic Science and Philosophy

PAUL KUNITZSCH
The Arabs and the Stars: Texts and Traditions on the Fixed Stars and their Influence in Medieval Europe

DAVID A. KING
Astronomy in the Service of Islam

DAVID A. KING
Islamic Astronomical Instruments

DAVID A. KING
Islamic Mathematical Astronomy

ROSHDI RASHED
Optics et mathématiques: Recherches sur l'histoire de la pensée scientifique en arabe

FRANZ ROSENTHAL
Science and Medicine in Islam : A Collection of Essays

BRUCE S. EASTWOOD
Astronomy and Optics From Pliny to Descartes
Texts, Diagrams and Conceptual Structures

Richard Lorch

Arabic Mathematical Sciences

Instruments, Texts, Transmission

VARIORUM
1995

Published by VARIORUM
Ashgate Publishing Limited
Gower House, Croft Road,
Aldershot, Hampshire GU11 3HR
Great Britain

Ashgate Publishing Company
Old Post Road,
Brookfield, Vermont 05036
USA

ISBN 0-86078-555-6

British Library CIP Data
Lorch, Richard.
Arabic Mathematical Sciences: Instruments, Texts, Transmission.
(Variorum Collected Studies Series; CS517).
I. Title. II. Series.
510. 956

The paper used in this publication meets the minimum requirements of the American National Standard for Information Sciences – Permanence of Paper for Printed Library Materials, ANSI Z39.48-1984. ∞ ™

Printed by Galliard (Printers) Ltd, Great Yarmouth, Norfolk, Great Britain

COLLECTED STUDIES SERIES CS517

CONTENTS

This volume contains xii + 354 pages

PUBLISHER'S NOTE

The articles in this volume, as in all others in the Collected Studies Series, have not been given a new, continuous pagination. In order to avoid confusion, and to facilitate their use where these same studies have been referred to elsewhere, the original pagination has been maintained wherever possible.

Each article has been given a Roman number in order of appearance, as listed in the Contents. This number is repeated on each page and quoted in the index entries.

PREFACE

The story of the transmission of the sciences from the ancients to the Arabic-speaking peoples and from them to the Latins has often been told. Until recently, however, little detailed work has been done. At least in the mathematical sciences the historian is beset with difficulties: typically, for one text there is a multiplicity of translations, there are revisions and redactions – which are often hard to distinguish from translations –, and of none of these is there a printed text, let alone a critical edition. In the articles reprinted here some elementary facts about a few texts are presented.

Of the first four items the third makes a simple comparison of the Greek and Arabic texts of passages from three Greek mathematical texts, one of which is treated in more detail in item I. In item II there are examples of perhaps the most difficult type of transmission: here there are no regular translations, but just the puzzling appearance and re-appearance of theorems in fragments and in quotations by later writers, sometimes with an attribution to an author, in this case Archimedes.

Items V–VIII grew out of my doctoral dissertation, "Jābir ibn Aflaḥ and his influence in the West" (Manchester 1971), which was mainly on the establishment of trigonometry – especially spherical trigonometry – in the Latin West through the translation by Gerard of Cremona of Jābir's (12c.) commentary on the *Almagest*.

Item IX is an edition of the Latin translation of the algebraic treatment of the pentagon by Abū Kāmil (9c.). Luckily, the translation is fairly literal and, though we have only one Arabic manuscript and one Latin manuscript, it was possible to add a little information on comparative vocabulary.

Even with the articles on instruments (XI–XVIII), problems of transmission arise time and again. Perhaps the most startling case is article XVI: two different universal instruments are found in the work of Jābir b. Aflaḥ, one in Arabic and the other in the Latin translation – the latter with diagrams for the instrument described in Arabic. This is perhaps a fitting place to acknowledge the previous publication of Dr. Sevim Tekeli on the instruments, including an edition of the Arabic text[1]. My article, written without knowledge of her work, is reproduced in this book since it includes English translations of both the Arabic and Latin passages and also a detailed discussion of the "Latin" instrument. A model of the "Arabic" instrument was made for the World of Islam Festival in 1976 and was shown at the exhibition in the Science Museum, London. Because of the practical difficulties encountered – beginning with the size of the principal armilla (six *ashbār*) and ending with the weight of the quadrant, which had to be kept still during observations – we may suppose that the instrument was a theoretical construct.

Items XI–XIII are devoted to the *sphaera solida*, a rotatable model of the heavenly sphere set in a fixed horizon-ring, usually with other rings and quadrants. This instrument was perhaps the ancestor of the plane astrolabe and had many of the same applications. It was known in antiquity, but the details of its transmission to the medieval Islamic world are unknown. Equally, how it became the subject of a description "Totius astrologie speculationis radix ... ", with its rehearsal of the applications, is unknown.

The *sphaera solida* of item XI is mechanically rotated to imitate the daily revolution of the heavens. This text was used to make a model of the device for the Institut für Geschichte der Arabisch-Islamischen Wissenschaften in Frankfurt. In the self-rotating sphere may be seen many of the elements of the early mechanical clock: drive by falling weight, toothed gears, and resulting rotation once in 24 hours. In article XI and in the other excursion into mechanical technology, article XV, the mechanics are traced back to Hellenistic times.

Item XIV, on al-Khāzinī's *qibla* table may be considered as an appendix to item XI, which has a short section on finding the *qibla*. The formula found to be underlying this table was subsequently found also to underlie the "Abbasid *qibla* table"; and the table attributed to al-Khāzinī was identified by David King (Frankfurt) as a "hopelessly corrupt"version of this[2]. He interprets the formula by reference to the "method of the *zījes*".

[1]S. Tekeli, "Nasirüddin, Takiyüddin ve Tycho Brahe'nin rasat aletlerinin mukayesesi", *Ankara Üniversitesi, Dil ve Tarih-Coğrafya Fakültesi Dergisi*, 16, no. 3–4 (1958). The section on Jābir's instrument and the torquetum is on pp. 383–393, the last four pages being an edition from the Berlin manuscript.

[2]D. King, "The Earliest Islamic Mathematical Methods and Tables for Finding

In article XVII it is suggested that the common form of the horary quadrant embodied the graphical solution of a trigonometric formula. This formula underlies other instruments and also several sets of tables for finding the time of day from the height of the Sun[3].

In his treatise on a generalized kind of stereographic projection (last item), al-Ṣaghānī (late 10c.) adapted standard graphical methods to finding the characteristics of the conic sections generated by his projection. Even here questions of transmission arise, for al-Ṣaghānī quotes Apollonius in a translation apparently different from that of the Banū Mūsā recently edited by Toomer[4].

I am grateful to the various journals and other publications for permission to reproduce the articles in this book, to the present publishers for their patience and to Menso Folkerts and Paul Kunitzsch for allowing joint publications to be reprinted.

I thank Paul Kunitzsch for constant help in the last seventeen years in reading manuscripts, rooting out numerous mistakes and interpreting texts. It is also a pleasure to acknowledge help and encouragement from Menso Folkerts, Ted Kennedy, David King, the late Donald Hill and the late Haskell Isaacs, who gave me so much of his time and expertise in Manchester.

RICHARD LORCH

Munich

July 1995

---

the Direction of Mecca", *Zeitschrift für Geschichte der Arabisch-Islamischen Wissenschaften*, 3 (1986) 82–149. The "Abbasid" table is considered on pp. 118–129 and the "al-Khāzinī" table on pp. 134–138.

[3]See D. Girke and D. A. King, *An Approximate Trigonometric Formula for Astronomical Timekeeping and Related Tables, Sundials and Quadrants from 800 to 1800*, Johann Wolfgang Goethe-Universität, Institut für Geschichte der Naturwissenschaften, Preprint Series No. 1, 1988.

[4]*Apollonius'* Conics *Books V to VII. The Arabic Translation of the Lost Greek Original in the Version of the Banū Mūsā*, ed. ... by G. J. Toomer, 2 vol., N.Y. etc. 1990.

ACKNOWLEDGEMENTS

Grateful acknowledgement is made to the following for kindly permitting the reproduction of articles originally published by them: the Institut für Geschichte der Arabisch-Islamischen Wissenschaften, Frankfurt am Main (I, XIII); Casa Editrice Leo S. Olschki, Florence (II); the Institut Supérieur de l'Éducation de la Formation Continue, Université de Tunis (III); the Institute for the History of Arabic Science, Aleppo (IV, XI, XIV, XVII); the Instituto de Cooperación con el Mundo Arabe, Madrid (part of VIII); Munksgaard International Publishers Ltd., Copenhagen (VI, XII, XVI); Martinus Nijhoff Publishers, Dordrecht (X); the Istituto della Enciclopedia Italiana, Rome (XV); The New York Academy of Sciences (XVIII). Item V is reprinted with kind permission from Birkhäuser Verlag Basel (Switzerland) from Demidov S.S./Folkerts M. et al. (Eds.); *Amphora. Festschrift for Hans Wussing on the Occasion of his 65th Birthday*, 1992. Item IX is a chapter in *Vestigia mathematica* (copyright: Editions Rodopi B.V., Amsterdam - Atlanta, GA 1993); it is reprinted by permission of Kluwer Academic Publishers.

I

THE ARABIC TRANSMISSION OF ARCHIMEDES' *SPHERE AND CYLINDER* AND EUTOCIUS' COMMENTARY*

The Greek text[1]

Our present knowledge of the Greek text of Archimedes' works depends upon three manuscripts, two of them no longer extant, together with an autograph of a medieval Latin translation made direct from Greek. First there is the ninth- or tenth-century ancestor (called "A" by HEIBERG in his second edition), now lost, of the Renaissance manuscripts that were used by HEIBERG in his first edition of the works of Archimedes. "A", which was probably copied at the instance of the ninth-century mathematician Leo in Constantinople,[2] was once owned by George Valla. Also lost is a collection of the mechanical and optical works, which was used as a supplement to "A" by William of Moerbeke in 1269 to prepare his literal translations into Latin. These translations, in the autograph in MS Vatican Ottob. Lat. 1850, are HEIBERG's "B". Finally, the famous palimpsest (originally kept at Jerusalem) discovered by HEIBERG in Istanbul, containing the *Method*, previously known only through citations, also contains large parts or fragments of other works by Archimedes. This manuscript is of the tenth century and was used by HEIBERG (MS "C") for his second edition.

* Acknowledgement: it is a pleasure to thank Prof. Dr. Paul KUNITZSCH for help with the decipherment and translation of the Arabic texts considered in this paper, especially the passages in Apendix 1.

[1] The information in this paragraph has been taken from the Prolegomena in HEIBERG I and III and from the useful summary in DIJKSTERHUIS 36–44.

[2] HEIBERG III xxii–xxiii.

For the *Sphere and Cylinder* the Greek text depends upon "A", as attested by various Renaissance copies and by "B", and the palimpsest "C"; and for Eutocius' commentary we have only "A" and "B".[3] Heiberg points out[4] that the text of the *Sphere and Cylinder* and the *Measurement of the Circle* has been translated from the Doric dialect and otherwise altered by scribes, who added and omitted material as they saw fit. It is possible that in some cases the Arabic texts preserve in translation readings nearer to the original than the Greek manuscripts.

The Arabic translations of the Sphere and Cylinder[5]

In the preface to his *Taḥrīr* Naṣīr al-Dīn al-Ṭūsī (6/13th century) reports[6] that there were two translations of the *Sphere and Cylinder* into Arabic: one improved by Thābit b. Qurra and the other translated by Isḥāq b. Ḥunayn. The Thābit text, with which he started, left out some axioms (*muṣādarāt*) and shows other signs of incompleteness. Most of his comments mentioning Isḥāq or Thābit are entirely concerned with the division into theorems. For instance, to the alternative proof I 7 he says:

> "Thābit makes this another theorem. In Isḥāq's text it and what precedes is one theorem."

Further, near the end of the preface, he says that Thābit had 48 theorems in the first book and Isḥāq 43. We are reminded of similar remarks he made about the two translations of Euclid available to him.[7] But we may ask how fixed the numeration was in the early days. Of the Greek citations of the *Sphere and Cylinder* mentioned by HEIBERG[8] – by Hero, Pappus, Proclus and Simplicius – only two, by Pappus, give the proposition number;

[3] HEIBERG III iii.

[4] HEIBERG I 9.

[5] Only translations are considered here. For the "indirect transmission", e. g. by the Banū Mūsā, see Abū Ja'far 155–156 and the references given there.

[6] Ṭūsī 2. Unless otherwise indicated the proposition-numbers of the Greek text have been used.

[7] See THAER.

[8] *Quaestiones Archimedeae* 15.

and these correspond to no known numeration of the propositions.[9]

MS Istanbul Fatih 3414 contains, among other works by Archimedes, a version of the *Sphere and Cylinder*,[10] which lacks definitions (ἀξιώματα) 1 to 4 and all the assumptions. Of the preliminary matter only the prefatory letter to Dositheus (somewhat changed)[11] and definitions 5 and 6, of the solid sector and solid rhombus, remain. Where the missing definitions should be is a note to say that the translator into Syriac had left some difficult parts untranslated.[12] Further, propositions I 7 ii and 31 are treated as separate theorems, as Ṭūsī says was done by Thābit and not by Isḥāq, and so are I 23, 28 and 43, which Ṭūsī says Isḥāq did not do. So far, the text represented by Fatih 3414 would seem to be the one that Ṭūsī said was improved by Thābit. There are 49 propositions in book I, but, despite Ṭūsī's careful recording of such matters, this is probably of small moment. On f 8r there is a framed title ascribing the translation to Qusṭā b. Lūqā, without mentioning Thābit. Qusṭā is better known as a writer and translator of medical works.[13] The translation of several mathematical works is ascribed to him,[14] including Theodosius' *Spherics* up to III 5[15], but a full discussion of such ascriptions will only be possible when comparative vocabulary and styles of translation have been studied.[16] For convenience the

[9] *Collection* V 24 and 33 quote *Sphere and Cylinder* I 16 and 14 as I 17 and 15 respectively. See Ver Eecke 285, 300. For the numbering in the Arabic tradition see Appendix 2.

[10] Ff 8v–59v, 676 H. I am most grateful to Professor Sezgin, Director of the Institut für Geschichte der Arabisch-Islamischen Wissenschaften in Frankfurt, for sending me a copy of the Archimedes material in this manuscript. Other manuscripts (see Sezgin V 129) are: Istanbul, Aya Sofya 2758/4, 71v–72v, 8c. H (fragment); Bursa, Haraççı 1174, 101v–144v, 6c. H; Teheran, private library Muʿtamid.

[11] For a translation see Appendix 1.

[12] Fatih 3414, f 9r, last line, to f 9v, line 2. For the scribe's comment on this note see below and Appendix 1.

[13] See Sezgin GAS III 270–274.

[14] See e.g. Sezgin GAS V 145, 155; VI 75, 152, 153; Diophantus (Sesiano), 8–9; Diophantus (Rashed), xxv ff.

[15] According to Ṭūsī in the preface to his Taḥrīr of the *Spherics*.

[16] Both Professor Sezgin and Professor Kunitzsch have expressed misgivings (private communications) at a too ready acceptance of Qusṭā as transla-

text in MS Fatih 3414 will be referred to as "Quṣṭā-Thābit", but always in inverted commas.

There is a fragment of another Arabic translation (hereafter for convenience called the "second" translation) in Fatih 3414, ff 6r–7r, comprising the prefatory epistle to Dositheus (though about the first half is missing, no doubt because of the loss of a folio), the definitions and assumptions, all in the Greek order. There is a note by the scribe on f 7r[17] to say that in some manuscripts he found this fragment along with the regular text and that he included it to fill in the lacuna in the introductory material. He quotes the remark, probably by the translator from Syriac into Arabic, explaining this lacuna, *verbatim*.

The "second" translation transcribes the name "Eudoxus", which occurs twice in the prefatory epistle to Dositheus, as *afrdqsys*,[18] the "*a*" and "*y*" standing for *alif* and *yā'* respectively; but in the "Quṣṭā-Thābit" translation three folios further on the name does not appear at all, the passages in question not being exactly equivalent to the Greek – though the name is translated in one instance as *ahl al-qawl bi'l-ḥaqq*. In general the "second" translation is fairly close to the Greek – certainly closer than the "Quṣṭā-Thābit" translation –, but, at any rate in our manuscript, Assumption 3, that of surfaces with the same extremities the plane is the least in area, and the second of Eudoxus' theorems quoted in the prefatory epistle are missing, probably by *homoioteleuton*.

Fatih 3414 was written by Muḥammad b. 'Umar b. Aḥmad b. abī Jarāda, to whom Suter attributes a *taḥrīr*, written in the year 691/1292, of Thābit b. Qurra's book on the section of the cylinder.[19] In the colophon he says he finished copying the work on 16th Rabī' I, 676 [17th August 1277]. The manuscript was carefully written, with informative marginal notes, often signed with his name. One of these notes (f 35v) says that the lettering

tor of the *Sphere and Cylinder*. Quṣṭā's treatise on the *sphaera solida* is not so strictly mathematical as the *Sphere and Cylinder*.

[17] For a translation, see Appendix 1. The note was transcribed by Sezgin in GAS V 129.

[18] f 6r, lines 1 and 4.

[19] Suter, section 385, p. 158. Could this possibly be Thābit's translation or revision of the *Sphere and Cylinder*?

of text and diagram did not tally – this is also sometimes true of the Greek manuscripts.[20] Another (40v) comments on the vagaries of copyists or translators. A curious remark on f 62r appears to say that Eutocius used his own reworking of Apollonius and not that of the Banū Mūsā. These comments would repay investigation.

The Taḥrīr of Naṣīr al-Dīn al-Ṭūsī

As we have seen, Ṭūsī had two translations of the *Sphere and Cylinder*, one of them apparently corrected by Thābit. In the preface he says that he patched up this text, which he found faulty both in the copy available to him and in the translation itself, and that he then found a copy of Eutocius' commentary translated by Isḥāq b. Ḥunayn, which also carried a text up to I 14.[21] It is clear, however, from his comments on the text itself that – e. g. to his I 8, 26, 31, 33 and 46 (I 7 part i, 23, 28, 31 and between 43 and 44 in Heiberg) – that he had access to a longer text. Ṭūsī says that he added commentary from Eutocius – or other geometers[22] – where appropriate and that he separates comment from text [no doubt by the use of such introductions as *qāla* and *aqūlu*]. He attaches the *Measurement of the Circle* to the *Sphere and Cylinder*, since it uses the same assumptions.

Since the fragment of the "second" translation is so short, we can only guess whether it is from the translation that Ṭūsī ascribes to Isḥāq and, if so, which text he generally prefers. In definitions 5 and 6, of "solid sector" and "solid rhombus", his style is more similar to that of "Qusṭā-Thābit" – for instance, both give the term *rmys*[23] *mujassam* as an alternative to *mu'ayyan mujassam* whereas the "second" translation does not. On

[20] See HEATH, xxiii.

[21] I.e. book I proposition 14. This abbreviated form of reference will be used throughout the article. A Roman "i" or "ii" after such a proposition number indicate the first or second of two clearly defined parts of the proposition. See Ṭūsī 2 for this passage. For a translation of the preface, see SEZGIN, GAS V 128–129.

[22] An example is the commentary by al-Qūhī to the last theorem of book II. See Ṭūsī 115 foll.

[23] Derived from the Greek ῥόμβος.

the other hand, Eudoxus appears in his own person, if only once. Assumption 3 (left out in the "second" translation) was clearly not in the text that Ṭūsī had in front of him, for he treats this assumption as a special case of the following one and has no separate enunciation for it, introduced by *qāla*, as with the other assumptions.

In Ṭūsī's *Taḥrīr* the numbering of the theorems is taken seriously. Not only does he note where the two translations differ in this respect, but he occasionally quotes theorems in his own text by number – e. g. in the last theorem of book I (his 48th) he quotes his I 35 and 36 (33 and 34 in HEIBERG).

The Hebrew translation

A Hebrew translation of the *Sphere and Cylinder* exists in two Bodleian manuscripts, Laud 93 and Heb. d. 4, and (to the middle of I 16) in Cambridge Add. MS 1220.[24] At the beginning of the text in MSS Laud 93 and Cambridge add. 1220 the Arabic is attributed to Qusṭā b. Lūqā; and in a note at the end of Laud 93, in a different hand, the translator from Arabic into Hebrew is given as Qalonymus b. Qalonymus. Neither name appears in the title or colophon of MS Heb. d. 4, which in general carries the same translation, though differing markedly in some places (e.g. end of book I). The numbering of the theorems in Laud 93 (there is none in Heb. d. 4 and no regular numbering in the Cambridge manuscript) is different from HEIBERG, from Fatih 3414 and from Ṭūsī, though the total for book I is 49, as in Fatih 3414. Appendix 2 is a conspectus of the theorems in the four texts.

Laud 93 agrees well enough with "Qusṭā-Thābit" in several specimens taken,[25] but a revised text must have been used – or the translation may have been revised, influenced by some other text. For the missing axioms are present, but appear so that the assumptions come before the definitions, in contrast to HEIBERG

[24] Laud 93, 1r–28v; Heb. d. 4, 108r–141v; Cambridge add. 1220, 49r–59v. Of the three I have examined only Laud 93 in any detail.

[25] E.g. there are long passages of the last theorem of book I in which the translation is word-for-word.

and Ṭūsī. A comparison of the definitions and assumptions reveals an almost word-for-word translation of the "second" Arabic translation,[26] with the interesting exception of Definitions 5 and 6. Here we have a fairly literal translation of "Quṣṭā-Thābit". Even the alternative terms for "solid rhombus" (*rmys mwgśm*[27]) is given. It would therefore appear that the material missing in "Quṣṭā-Thābit" was supplied by the "second" translation. The name Eudoxus is transcribed in the Hebrew as *audfss* and *audfsus*, the "*a*" standing for *aleph* and the "*u*" for *waw*. These transcriptions are nearer the Greek than those in our manuscript of the "second" Arabic translation.

The Latin fragment

The Latin fragment edited by CLAGETT[28] from MSS Madrid lat. 10010, f 84, and Bodleian Digby 168, f 122r (123r) contains six enunciations found in the prefatory epistles in books I and II. It is an almost word-for-word rendering of Fatih 3414, ff 8v10–9r2, 9r7–9 and 46r4–6. The closeness to the Arabic may be seen from the following short extract, the first quoted proposition of Eudoxus. For comparison, the Greek text and the "second" Arabic translation are also given.[29]

... ὅτι πᾶσα πυραμὶς τρίτον ἐστὶ μέρος πρίσματος τοῦ βάσιν ἔχοντος τὴν αὐτὴν τῇ πυραμίδι καὶ ὕψος ἴσον, ...

(f 6r) . . . أن كل شكل نارىّ فهو ثلث المنشور الذى قاعدته مساوية لقاعدته وارتفاعه مساوٍ لارتفاعه . . .

(f 9r) . . . أن كل مخروط قاعدته شكل مستقيم الخطوط إذا كان هو ومنشور

[26] Specimens taken: Definitions 1–3, assumptions 1, 2 and 4. Assumption 3 is also missing in Hebrew. There are some differences: e.g. a negative is missing from Definition 1 in Hebrew.

[27] The "s" in "rmys" represents the letter samekh. No attempt has been made to vocalize the words.

[28] CLAGETT I.

[29] HEIBERG 4.5–7; MS Fatih 3414, 6r2–3; ibid 9r7–9. I have taken the Latin text from the Madrid manuscript alone (so *simul* for CLAGETT'S *semel*). For most of this passage I could not read the available film of the Digby manuscript, which anyway appears to be inferior.

جميعاً على قاعدة واحدة ويكون ارتفاعهما واحداً فإن ذلك المخروط هو ثلث المنشور

. . .

Omnis piramis, cuius basis est figura rectilinea, aut ipsa et serratile simul sunt super basim unam et est utriusque altitudo una, tunc illa piramis est due tertie serratilis.

With one exception the correspondence is exact, even to a small scribal error in the third proposition: the Arabic has *wa-qāʿidatihā* ("and its base") for *wa-qāʿidatayhā* ("and its two bases"), the *yā'* appearing to have got lost, at least in our manuscript, and the Madrid manuscript reproduces this with "et basis".[30] We may also note in passing that the second quoted proposition by Eudoxus, no. 5 in the Latin, is missing from our manuscript of the "second" Arabic translation.

Clearly, the six enunciations have been taken from the introductions to the two books, the first five from book I and the sixth from book II. For although the first three also appear in the preface to book II, they are not formulated in the same way and do not correspond with such exactness to the Latin; further, the corresponding enunciations in the body of the book are not identical to the wording of the preface.

As to how this compilation was made, two possibilities suggest themselves. First, as CLAGETT supposes,[31] it could have been made from a translation, now lost, of the *Sphere and Cylinder*. In favour of this hypothesis is the presence of the fragment among extracts of other Gerard translations in the Digby manuscript. Because of the literalness of translation, the fragment may well have been translated by Gerard of Cremona, though no separate translation is mentioned in the list of his translations attached to the famous *Vita*. The only suitable title is "Liber Archimedis tractatus I"[32] and this is probably to be associated with the *Measurement of the Circle*, of which we have a full Latin text.[33]

[30] CLAGETT has "duo bases", presumably from Dig. 168. I cannot read the available film at this place. The exception to exact correspondence is *due tertie* for *thulth*.

[31] CLAGETT I 433.

[32] See SUDHOFF, p. 77.

[33] CLAGETT I p. 30 ff.

On the other hand, the list may not be complete or both Archimedes texts may have been taken under one title – on analogy with the Ṭūsī *Taḥrīr*, which, as we have seen, attaches the *Measurement of the Circle* to the *Sphere and Cylinder.*

A second possibility is that the compilation was made in Arabic, before translation. In this context we may note the presence in MS Madrid 10010 of snippets or small treatises that have yet to be investigated.[34] In either language, to make such a collection of propositions was an understandable procedure, since, being Archimedes' end-results on the area and volume of a sphere and its segments, they were of more general interest than the others. Most of the Greek citations of the *Sphere and Cylinder* are of these propositions.[35] Even Hero, who cites propositions from book II as such (though not by number), cites the Archimedean propositions of our fragment simply as being by Archimedes and in fact quotes no other result from book I.[36]

Some mathematical terms in the "Qusṭā-Thābit" version

No systematic attempt has been made to investigate the translation technique or comparative vocabulary of this translation,

[34] E. g. Madrid 10010 ff 50v–51v, "Hec propositiones sunt necessarie in theoremate undecimo partis secunde libri Theodosii de speris . . .", almost certainly a translation of a gloss existing in some Arabic manuscripts of Theodosius, and ff 51v–52v, "Nota in tercia decima primi libri Geber . . .", which is perhaps a purely Latin comment.

[35] Of course, the hypothesis of a compilation in Greek would not help us here, since the "Qusṭā-Thābit" translation is clearly the basis of the Latin text. For Greek citations in general, see HEIBERG, *Quaestiones*, 15. Pappus cites propositions other than those in the fragment in *Collection* V 2, 23, 24, 33, 35, 43.

[36] II 2, 3 and 4 are quoted in the *Metrica* (see Heron III 122.16, 172.11 and 184.27 resp.). The propositions given in the prefaces are quoted in the *Metrica* (see Heron III 2.18, 86.29, 88.11, 88.26, 120.28) and in the *Stereometrica* (see Heron V 2.4, 6.9, 8.6, 116.9). We may speculate that Hero had only book II to hand – it would, for instance, go some way to explaining his vagueness about Archimedes' authorship of these propositions in the preface to *Metrica*, book II (Heron III 92; see DIJKSTERHUIS, p. 35 note 1) –, but this is scarcely to our purpose here. It is enough to note that the propositions of the Latin fragment were eminently collectible.

but a few facts have been collected. We notice in the following specimens that πολύγωνον (polygon) is translated literally as (*shakl*) *kathīr al-zawāyā* and that ἡ ἐκ τοῦ κέντρου (radius) is simplified to *niṣf quṭr*.[37] Again, the rendering of δύναται + accusative or ἴσον δύναται + dative[38] by *yaqwā ʿalā* is in principle a translation.[39]

1. from prop. 16. Greek 70.22–72.4; Arabic 24v1–6.
καὶ ἐπεὶ τὸ ὑπὸ τῶν ΒΑ, ΑΗ ἴσον ἐστὶ τῷ τε ὑπὸ τῶν ΒΔ, ΔΖ καὶ τῷ ὑπὸ τῆς ΑΔ καὶ συναμφοτέρου τῆς ΔΖ, ΑΗ διὰ τὸ παράλληλον εἶναι τὴν ΔΖ τῇ ΑΗ, ἀλλὰ τὸ μὲν ὑπὸ ΑΒ, ΑΗ δύναται ἡ ἐκ τοῦ κέντρου τοῦ Λ κύκλου, τὸ δὲ ὑπõ ΒΔ, ΔΖ δύναται ἡ ἐκ τοῦ κέντρου τοῦ Κ κύκλου, τὸ δὲ ὑπὸ τῆς ΔΑ καὶ συναμφοτέρου τῆς ΔΖ, ΑΗ δύναται ἡ ἐκ τοῦ κέντρου τοῦ Θ, τὸ ἄρα ἀπὸ τῆς ἐκ τοῦ κέντρου τοῦ Λ κύκλου ἴσον ἐστὶ τοῖς ἀπὸ τῶν ἐκ τῶν κέντρων τῶν Κ, Θ κύκλων·

والمجتمع من ضرب ب ا فى اح مساوٍ للمجتمع من ضرب ب د فى دز ومن ضرب ا د فى د ز و اح جميعاً لأن د ز موازٍ لـ اح والمجتمع من ضرب ب ا فى اح يقوى عليه نصف قطر دائرة ل والمجتمع من ضرب ب د فى د ز يقوى عليه نصف قطر دائرة ك والمجتمع من ضرب دا فى د ز و اح جميعاً يقوى عليه نصف قطر دائرة ط فمربع نصف قطر دائرة ل مثل مربعى نصفى قطرى دائرتى ك و ط،

2. from prop. 30. Greek 112.2–6; Arabic 33r13–18.
ἴσον ἄρα ἐστὶν τὸ περιεχόμενον σχῆμα ὑπό τε μιᾶς πλευρᾶς τοῦ πολυγώνου καὶ τῆς ἴσης πάσαις ταῖς ἐπιζευγνυούσαις τὰς γωνίας τοῦ πολυγώνου τῷ περιεχομένῳ ὑπὸ τῶν ΖΘΚ· ὥστε ἡ ἐκ τοῦ κέντρου τοῦ Λ κύκλου ἴσον δύναται τῷ ὑπὸ ΖΘΚ·

فالشكل الذى يحيط به ضلع من أضلاع الكثير الزوايا وخط مساوٍ للخطوط التى تصل بين زوايا الشكل الكثير الزوايا هو مساوٍ لما يحيط به ز ط و ط ك، فيكون نصف قطر دائرة ل فى القوة مساوياً للمجتمع من ضرب ز ط فى ط ك،

[37] This is the regular translation. The only exception I have found is in the introduction (see Appendix 1).
[38] For these terms, see Mugler 150–152.
[39] The form *fī'l-quwwa* in specimen 2 is unusual.

3. from prop. 37. Greek 136.26–138.1; Arabic 39r3–4.
τὸ δὲ ὑπὸ τῆς ΕΘ καὶ τῶν ΕΖ, ΓΔ, ΚΑ δέδεικται ἴσον τῷ ὑπὸ τῶν ΕΛ, ΚΘ περιεχομένῳ·

والمجتمع من ضرب ه ط فى ه ز ج د ا ك كان قد تبين أنه مساوٍ للمجتمع من ضرب ه ل فى ك ط،

But some of the other expressions – for which the specimens were selected – are not so regular. For instance, the expression of the rectangle contained by *AB* and *BG*[40] variously expressed in Greek as (1) τὸ ὑπὸ (τῶν) ΑΒ ΒΓ, (2) τὸ ὑπὸ (τῶν) ΑΒΓ, (3) τὸ περιεχόμενον ὑπὸ τῆς ΑΒ καὶ τῆς ΒΓ, (4) τὸ περιεχόμενον ὑπὸ τῶν ΑΒΓ etc.[41] is usually translated as *al-mujtamiʿ min ḍarb AB fī BG*,[42] but there are other translations, such as the form *mā yuḥīṭ bihi AB wa-BG*, which occurs with *ZT* and *TK* in the middle of specimen 2 or the similar form at the beginning that uses *al-shakl*. A further variation[43] is found with *al-saṭḥ* for *al-shakl*. The Arabic phrase *mujtamiʿ min ḍarb* may indicate a more definitely algebraic interpretation than the Greek expressions, but the alternative forms should make us wary of drawing too hasty conclusions.

The only term to be investigated thoroughly (using Heiberg's index) is τὸ ἀπὸ (τῆς) ΑΒ, which means the square on *AB*. In specimen 1 it is rendered in the form *murabbaʿ AB*, which is the most frequently encountered,[44] but in several places more algebraic-sounding forms occur, such als *al-mujtamiʿ min ḍarb AB fī mithlihi*[45] or *alladhī yakūn min AB fī mithlihi*[46], and once[47] we find *mujtamiʿ min* for *murabbaʿ*.

[40] The expression with *AB* and *BG* is taken as the standard form. In the examples other letters or phrases take their place. The Greek or Arabic words in brackets are sometimes, but not always, found in the expressions in which they occur.

[41] For the περιεχόμενον expressions see Mugler 341–342.

[42] E.g., the Heiberg page and line being separated from the folio and line of Fatih 3414 by an oblique stroke, 70.19/24r19 foll. for Greek expressions 1 and 2; 94.6–8/29v3–4, 94.18/29v9, 110.3/32v19 and 110.12–16/33r5–7 for expression 3; 138.1–2/39r4 for expression 1; and so on.

[43] 146.7–8/41r1–2.

[44] Other examples are 72.2–3/24v5–6, 170.25/46v19 foll., 176.7–8/48r2–3, 178.11/48v5, 178.18/48v9 foll., 184.16–17/50r7.

[45] 138.2/39r4–5, 98.23/30v7–8, 152.16/42r6.

Extra material in the Arabic Sphere and Cylinder[48]

To explain the differences between the Greek text as we have it and the Arabic versions, some idea of the content of the work should be given. In book I Archimedes approximates the surface and volume of a sphere by figures formed by rotating inscribed and circumscribed polygons about their axes of symmetry. Accordingly, he finds the areas of a cone without base (Prop. 7, 8, 14, 15), of a frustrum of a cone without bases (Prop. 16), of the rotated polygon just described, formed from cones and frustra (Prop. 21) and of the sphere (Prop. 33). The same procedure is applied for the area of a segment of the sphere (Prop. 42) and for the volume of the sphere (Prop. 34) and of a segment of a sphere (Prop. 44). The whole is preceded by inequalities and other lemmata (Props. 1, 2, 3, 5, 6) used in the exhaustion proofs; and the other propositions[49] may be considered as intermediary results.

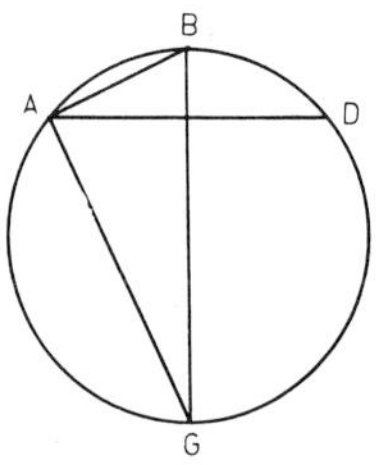

In the Arabic translation two theorems about the areas of prisms inscribed in, and circumscribed about, a cylinder follow strictly analogous theorems (Heiberg 7, 8) about pyramids and cones. There are also additional theorems near the end of the book to generalize the results about the area and volume of a segment of a sphere. In I 42 it is proved that the area of a segment (*ABD* in diagram) less than a hemisphere is equal to the

[46] 148.16/41r9; also in the "extra" theorems not in the extant Greek text, e.g. 43v7.

[47] 96.24/30r4–5.

[48] This section is mostly about the "Qusṭā-Thābit" translation, but, as will be seen, the extra material is also to be found in the Ṭūsī *taḥrīr* and in the Hebrew translation.

[49] Except Prop. 43 – see below.

circle whose radius is the line (*BA*) from vertex to base. Archimedes himself (according to the extant Greek text) generalizes this to apply to segments greater than a hemisphere by saying that the square on *BA* (to use the lettering of the diagram) together with the square on *GA* is equal to the square on *BG*, and that therefore the circle with radius *BA* together with the circle with radius *GA* is equal to the circle with radius *BG*, which (Prop. 33) is equal to the surface of the sphere. Thus the area of segment *AGD* is proved to be equal to the circle with radius *GA*. In the Arabic translation there is another theorem, with the same proof, for the case of a segment equal to a hemisphere. At the end of the book there is a generalization, again on the same principles, of the result of Prop. 44 on the volume of a segment. Curiously, only segments greater than a semicircle are mentioned here explicitly.

Other extra material in the Arabic includes a list of enunciations of book II at the beginning of the book.[50] They are not identical to the enunciations in the body of the book. All these additions appear not only in the "Qusṭā-Thābit" text, but also in Ṭūsī and, with small differences, in the Hebrew translation. One such difference is in the second of the two extra theorems after Prop. 8, where "and the way we deal with this is as before"[51] is substituted for a full proof, as in "Qusṭā-Thābit" and Ṭūsī.

Eutocius' Commentary on the Sphere and Cylinder

Apart from Ṭūsī's *Taḥrīr*, the Arabic manuscripts of Eutocius' commentary, none of them complete, are as follows:

Escurial 960, 22v–42v;
Istanbul, Fatih 3414, 60v–66v, 72r–v[52], 13c AD;
Paris BN 2457, 191v, 10c AD;
Beirut, St. Joseph 223, item 20, pp. 153–157, 15c AD.[53]

50 Fatih 3414 ff 46r8–46v7

51 I am grateful to Dr. H. D. Isaacs for correcting my transcription and translation of this passage.

52 What is now folio 72 clearly got misbound at some stage.

53 This fragment was reproduced in facsimile, transcribed and translated, with introduction, in Eratosthenes.

The Escurial manuscript contains only the parts of the commentary on the second book dealing with finding two mean proportionals between two given lines (for Prop. II 1) and the lemma for II 4, both of which may be treated as cubic equations. Eutocius' reports of the methods by various Greek mathematicians for finding two mean proportionals – equivalent to duplicating the cube – begin on the following folios and lines: Hero 22v8, Philo 23r7, Apollonius 23v22, Diocles 24r7, Pappus 24v23, Sporus 25v27, Menaechmus 26r24 and 26v28, Eratosthenes 27v5, Nicomedes 29r9, Plato 31r14, Archytas 31v9. It will be noticed that the order is different from that in the Greek text: Plato, Hero, Philo, Apollonius, Diocles, Pappus, Sporus, Menaechmus, Archytas, Eratosthenes, Nicomedes. The commentary to II 4 (begins 32r26) contains the methods of proving the lemma that Eutocius ascribes to Archimedes (33r4) and then those of Dionysodorus (37r29) and Diocles (38v15) – this time in the Greek order. In both series of proofs Eutocius' introductory material is included.

The text in the Istanbul manuscript is written in the same hand as the other Archimedes material. After Eutocius' introduction to II 4, "Archimedes'" proof is given, labelled in the margin as three propositions or parts. This is followed by the last part of Eutocius' discussion of the first two postulates of book I, Heiberg 10.4–14.20, the labels in the margin running from "4" to "7". At the end the scribe says that he finished it on 6th *Dhū 'l-Qa'da* 684. Then:

> I confined myself to copying these theorems from the commentary of Eutocius on the *Sphere and Cylinder*, since I did not find in the rest of what I found . . . [illeg.]; and I only found his commentary on the first book, but did not find what he had promised to mention of the method of Dionysodorus and the method of Diocles in the proof of theorem 4 of the second book . . .

No doubt this refers to the switch to book I at the section labelled "4". The part common to Escurial 960 agrees well enough and is from the same translation.

The Paris fragment contains only the introduction to the results on mean proportionals and Hero's method, breaking off shortly before the end (Heiberg 60.23). It is almost word-for-word the same as the corresponding passage in Escurial 960, the only significant difference being that it once has *musāwiyan li-murabba'* where the Escurial manuscript has *mithl murabba'*. In the first few lines the text is ascribed to abū 'l-Ḥasan Thābit b. Qurra.

The Beirut fragment is Eutocius' report of Eratosthenes' method of finding two mean proportionals. It is not the same as the equivalent passage in Escurial 960, though some parts are similar.

As expected, the text of the two Diocles passages is not the same as the Arabic translation of Diocles' *On burning mirrors*. Again, Ṭūsī's *Taḥrīr* in the commentary to II 4, which reproduces Eutocius, has a different wording. We may note in passing that in his commentary to II 1 Ṭūsī gives only one method of finding two mean proportionals and it is not taken direct from Eutocius. The Geniza fragment mentioned by Sezgin[54] contains a discussion of Menaechmus' methods for finding two mean proportionals in the hands of Arabic commentators.

In the Bodleian manuscript Heb. d. 4 ff 195v–206v there is a Hebrew translation of almost all of the extant Greek text of the commentary to book I. Folio 199, which mentions regular solids, is probably displaced from the previous treatise in the codex, on the fourteenth book of Euclid. The folio apparently takes the place of the equivalent of Heiberg 14.7–22.22. According to Renan[55], the translation, which is literal, is probably by Qalonymus b. Qalonymus, and he argues for a date before 1311. The numbers of the propositions commented upon are given in the text. These are included in the table of Appendix 2.

In sum, it is clear that Eutocius' commentary on the first book of Archimedes' *Sphere and Cylinder* and at least parts of the commentary on the second were translated into Arabic, but not all manuscripts contain all this text. We note Ibn abī Jarāda's disappointment with the copy that he had and also Ṭūsī's relative silence on II 1. But the commentaries to II 1 and II 4 were clearly fairly well known and it is these, reformulated and developed as they were by Arabic writers, that play the important role in the solution of the cubic equation.

[54] Sezgin GAS V 130.
[55] Renan, p 438.

APPENDIX 1: LITERAL TRANSLATION OF TWO EXTRACTS FROM MS FATIH 3414

1. *The note on f 7r*

I found the introduction [*ṣadr*] of the book of *The Sphere and Cylinder* which follows this one and which is the one that is at the beginning of the book [and that] he had said in the course of it what the following is a copy of:

> I found in the copy that the translator of this book from Greek into Syriac mentioned that he omitted in this place a small passage which he did not translate from the Greek book because of its difficulty for him.

I found in some copies two introductions, one of them the introduction that follows this one and the other is this introduction. I found their meanings in harmony, but in this one there is an addition – this must be the one of which it was mentioned[1] that the translator of that introduction omitted it. Perhaps another translator translated this introduction and did not leave anything out from it, and so I copied this introduction also.

Muḥammad b. ʿUmar b. Aḥmad b. Hibat Allāh b. Aḥmad b. abī Jarāda wrote it.

2. *The preface on ff 8v–9r*

The beginning of the book.

To Dositheus from Archimedes, a greeting to you. I had earlier written you a letter [*kitāb*] in which I described what I had discovered from the sciences [i.e. of mathematics]: that

> any section of a right-angled cone [i.e. any parabolic segment] which a straight line and a portion [*ṭāʾifa*] of the line of the section contains is one-and-a-third times the triangle whose base is its [sc. the section's] base and whose altitude is as its altitude.

I furnished in it the proof for it. Now I shall put down for you, correct with proofs, what valuable discoveries [*al-ʿulūm*] have occurred to me. It is that

[1] *or* he mentioned

[for] any sphere the surface containing it is four times the greatest circle occurring on this sphere.

The surface of any segment of the sphere is equal to the circle, the line that is drawn from whose centre to its circumference is equal to the straight line that is drawn from the vertex-point of the segment to the circumference-line of the circle of this segment's base.

Any cylinder whose base is equal to the greatest circle occurring on some sphere and whose altitude corresponds [*muwāzin*] to the diameter of that sphere – that cylinder is one-and-a-half times the sphere.

Also, the surface of that cylinder and its base [*sic*] together are one-and-a-half times the surface of the sphere.

Now these properties, which are in the figures that we mentioned, were originally in them in nature; and knowledge of them was not revealed to any of the Ancients who were concerned with geometry before our time – they did not determine the quantity of the area of these figures, some in respect to others. Therefore[2] I thought it right to add what I was able to discover in that [subject] to what those geometers who were before me had set down, whereas some of the people of true doctrine have mentioned that the superfluities of discoveries of the bodies [i.e. of theorems about the sphere, cylinder, etc.] are numerous.[3] According to this:

[for] any pyramid whose base is a rectilinear figure, if it and a prism together are on the same base, and their altitudes are the same, that pyramid is a third of the prism.

Every cylinder-cone [*makhrūṭ usṭuwāna*, i. e. right circular cone] whose base is the base of the cylinder and whose altitude is equal to its [sc. the cylinder's] altitude is a third of it [in volume].

Since everything that we have described of the condition of those figures is in nature, as we have mentioned, and [since]

[2] A superfluous *mā* has been omitted in the translation of this sentence.

[3] This sentence corresponds to HEIBERG I 4.2–5. "Some of the people of true doctrine" is the translation of *qawman min ahl al-qawl bi'l-ḥaqq*. "Eudoxus" was not understood by the translator as a proper name, but as a compound common noun (or possibly phrase). Professor KUNITZSCH calls such a procedure "etymologisierende Übersetzung".

there will be after us – among those esteemed in the science of geometry, however many they might be – those who do not know this and do not arrive at knowledge of it, then it is clear that those who search for knowledge of that need to be judicious and of fine understanding. It is necessary that that was evident and clear in the time of the esteemed Conon [*Qūnūn*]. I am of the opinion that I have reached – from an appreciation of that, a mastery of it and the establishment of argument and proof in it – the end that is necessary in that. Now I am going to explain what I have discovered and established, in its proof walking the path of the mathematical sciences. I premise what I must premise in that, God willing (be He exalted).

APPENDIX 2: THE NUMBERING OF PROPOSITIONS IN BOOK 1

Heiberg	Q	T	Heb.	Eut.	Content
			1		L E in
1	1	1	2		L E in
2	2	2	3	2	L E ei
3	3	3	4	3	L E ei
4	4	4	5		L E ei*
5	5	5	6		L E ei
6 i			7		L E ei*
6 ii					L E i (*)
6 iii	6	6	8	6	L E e (*)
7 i	7	7	9		A C i
7 ii	8	8	–		
8	–	9	–	8	A C e
–	9	10	10		A Z i
–	10	11	[11]		A Z e
9	11	12	12	11	A C in
10	12	13	13	12	A C en
11	13	14	14		A Z in
12	14	15	15		A Z en
13	15	16	16	15	A Z
14	16	17	17	16	A C
15	17	18	18		A C

Heiberg	Q	T	Heb.	Eut.	Content
16	18	19	19	18	A F
lemmata		lemmata	20		V C Z
17	19	20	21		A/V C
18	20	21	22		A/V C
19	21	22	23		A/V C
20	22	23	24		A/V C
21	23	24	25		L i
22	24	25	26	24	L i*
23	25	26	27		A S in
24	26	27	28		A S i
25	27	28	29		A S in
26	28	29	30		V S i
27	29	30	31		V S in
28	30	31	32		V S en
29	31				A S e
30	32	32	33	29	A S en
31	33	33	34		V S en
32	34	34	35	30	AV S ei
33	35	35	36		A S
34	36	36	[37]	32	V S
34 cor.	37		38		AV S
35	39	38	40		A S i* (cf 24)
36	38	37	[39]		A S in* (cf 23)
37	40	39	41	35	A S in (cf 25)
38	41	40	42		V S i* (cf 26 & 27)
39	42	41	43	37	V S en* (cf 28 & 29)
40	43	42	44	38	A S en* (cf 30 & 31)
41	44	43	45	39	AV S ie* (cf 32)
42	45	44	46	40	A S * (cf 33)
43	46	45	47		
–	47	46			
44	48	47	48	42	V S * (cf 34)
–	49	48	49		

The first four columns represent respectively the number given in HEIBERG'S Greek text, the "Qusṭā-Thābit" translation in MS Fatih 3414, the *Taḥrīr* of Naṣīr al-Dīn al-Ṭūsī according to the printed text, the Hebrew translation in MS Laud 93 and (by implication) in the Eutocius commentary in MS Heb. d. 4.

The unnumbered proposition at the beginning is in the introductory material.

The column on the right indicates the general nature of the theorem according to the following conventions:

A	surface area	Z	cylinder
V	volume	S	sphere
L	lemma	i	inscribed figure
C	cone	e	circumscribed figure
F	frustrum of cone	n	inequality

Theorems about segments or sectors are indicated by *. Where possible the HEIBERG number of the corresponding theorem about whole figures is given in brackets.

Bibliography

Abū Ja'far. Richard LORCH, "Abū Ja'far al-Khāzin on Isoperimetry and the Archimedes Tradition", *Zeitschrift für Geschichte der arabisch-islamischen Wissenschaften,* 3 (1986), 150–229.

Marshall CLAGETT, *Archimedes in the Middle Ages,* Vol. I, Maddison 1964.

Diocles. *Diocles on Burning Mirrors,* edited and translated by G. J. TOOMER, New York 1976.

Diophantus (RASHED). *Diophante, Les Arithmétiques,* ed. Roshdi RASHED, Volumes III and IV, Paris 1984.

Diophantus (SESIANO). Jacques SESIANO, *Books IV to VII of Diophantus' Arithmetica in the Arabic Translation Attributed to Qusṭā ibn Lūqā,* New York 1982.

E. J. DIJKSTERHUIS, *Archimedes,* Copenhagen 1956.

Eratosthenes. Amin MUWAFI and A. N. PHILIPPOU, "An Arabic Version of Eratosthenes on Mean Proportionals", *Journal for the History of Arabic Science,* 5 (1981), 147-165.

HEIBERG, Quaestiones. J. L. HEIBERG, *Quaestiones Archimedeae,* Hauniae 1879.

HEIBERG. *Archemedis opera omnia cum commentariis Eutocii,* ed. J. L. HEIBERG, Leipzig. Vol I, 1910; Vol. III 1915.

T. L. HEATH, *The Works of Archimedes,* Cambridge 1897.

Heron. *Heronis Alexandrini opera quae supersunt omnia,* Leipzig. Vol. III, *Rationes dimetiendi et Commentatio dioptrica,* ed. H. Schoene, 1903; Vol. V, *Stereometrica et De mensuris,* ed. J. L. HEIBERG, 1914.

Charles MUGLER, *Dictionnaire historique de la terminologie géométrique des Grecs,* Paris 1958.

Ernest RENAN, "Les écrivains juifs français du XIVe siècle", *Histoire Littéraire,* 31 (1893).

Fuat SEZGIN, *Geschichte des arabischen Schrifttums,* Leiden. Vol. III, 1970; Vol. V, 1974; Vol. VI, 1978.

K. SUDHOFF, "Die kurze 'Vita' und das Verzeichnis der Arbeiten Gerhards v. Cremona . . .", *Archiv für Geschichte der Medizin*, 8 (1915), 73–92.

Heinrich SUTER, "Die Mathematiker und Astronomen der Araber und Ihre Werke", *Abhandlungen zur Geschichte der mathematischen Wissenschaften*, 10 (1900).

C. THAER, "Die Euklid-Überlieferung durch aṭ-Ṭûsî", *Quellen und Studien zur Geschichte der Mathematik, Astronomie und Physik*, Abt. 3: Studien, 3 (1936), 116–21.

Ṭūsī. Naṣīr al-Dīn al-Ṭūsī. *Kitāb al-kura wa 'l-usṭuwāna li-Arshimīdis. Taḥrīr*. Hyderabad 1359/1940–41.

VER EECKE. Pappus d'Alexandrie, *La collection mathématique*, translated with notes by Paul Ver Eecke, Paris 1982.

Note added in proof

In the *Fihrist* (see G. FLÜGEL's edition, p. 266) Ibn al-Nadīm mentions the two books of the *Sphere and Cylinder*, but gives no translator. For the "indirect transmission" (see n. 5) see also Wilbur R. KNORR, "The Medieval Tradition of Archimedes' *Sphere and Cylinder*". I am grateful to the author for letting me see this in typescript.

The Geniza fragment of the Arabic Eutocius (see n. 54) is in two parts, both from works by al-Shannī: from his criticism of Abū 'l-Jūd about the construction of the regular heptagon (see J. P. HOGENDIJK, *Arch. Hist. Exact Sci.*, 30 [1984], 277–78. I am grateful to the author for this reference) and from his treatise on Heron's theorem (see F. SEZGIN, GAS V 352). The Hebrew translation in Bodleian Heb. d. 4 also contains part of the "Archimedes" proof to II 4 – roughly equivalent to HEIBERG III 130.11–144.23, though the content is not identical, the diagram letters in particular frequently not being equivalent. In general, the translation differs in many places from HEIBERG (see Y. T. LANGERMANN, *Italia*, 7 [1988], p. 38 n. 50. I am grateful to the author for pointing this out to me). The text breaks off suddenly in the middle of the last page of the codex. The missing folio replaced by the present f 199 is f 188.

II

SOME GEOMETRICAL THEOREMS ATTRIBUTED TO ARCHIMEDES AND THEIR APPEARANCE IN THE WEST

In late antiquity and the Middle Ages several mathematical theorems were attributed to Archimedes which do not appear in the Greek *corpus* of his writings. Such attributions are to be found not only in the writings of later authors, but also in the titles of collections of lemmas. Of the latter perhaps the best known is that printed by Heiberg in 1913 as a Latin translation from the Arabic under the title of *Liber assumptorum*.[1] The Arabic text appears to exist in only one manuscript, the thirteenth-century codex Istanbul Fatih 3414, which contains other Archimedes material: the *Dimensio circuli*, the *De sphaera et cylindro* and part of Eutocius' commentary to it. The redaction by abū'l-Ḥasan ʿAlī b. Aḥmad al-Nasawī, who reworked the text with commentary, was re-edited in the thirteenth century by Naṣīr al-Dīn al-Ṭūsī and of the result there are numerous manuscripts. In the seventeenth century the Nasawī redaction was twice translated into Latin, by John Gravius, published by Samuel Foster in 1659, and by Abraham Ecchellensis, published by Alfonso Borelli in 1661. Whether the text taken was in the form of Ṭūsī's *Taḥrīr* remains to be investigated. Heiberg took over Borelli's edition, having first removed al-Nasawī's comments. There are commentaries by al-Kūhī, some of whose comments are included by al-Nasawī, and by al-Sijzī (Paris BN ar. 2458). Sezgin mentions another short tract by Sijzī, apparently on the same text, in a Teheran manuscript.[2]

---

1 *Archimedis opera*, II, 509-525.

2 SEZGIN, V 334. It is not absolutely clear that the Archimedes work commented on is the *Lemmata*.

Another collection of lemmas is known in two Arabic manuscripts. In the better of these, Istanbul Aya Sofya 4830, it is ascribed to one Aqāṭun, but in the other, Bankipore 2468, it is called «Archimedes' book on the Elements of Geometry». Following the recent editor and translator, Yvonne Dold-Samplonius, we shall call it the *Book of assumptions*. It should be added that there is another book of the same name by Thābit b. Qurra, which was the subject of a *taḥrīr* by al-Ṭūsī.[3] Also in Bankipore 2468 is *Archimedes' Book on Tangent Circles*, published from the unique manuscript in the Teubner series in 1975. Finally, there is a book *On the construction of the heptagon* in a unique[4] manuscript revised by the scribe, Muṣṭafā Ṣidqī, in 1740. Not all of the text is concerned with the heptagon.

To these sources of lost Archimedean theorems we may add Pappus' *Collection*, extant in Greek, but apparently not translated into Arabic except for the last book. Yet another source is al-Bīrūnī's *Istikhrāj al-awtār*, which has recently been studied by Dr. Abdul-Latif of Amman, who concludes that there are two versions of the text, long and short, both by Bīrūnī.[5] To what extant attributions by Bīrūnī and others depend on the titles of collections of lemmas it is difficult to assess.

A characteristic of our sources for the lemmas is that essentially the same result turns up in different forms. For instance, the first proposition of the *Liber assumptorum* states that if *CD* in figure 1 is parallel to *AB*, then *EDB* is a straight line. The converse appears in Pappus' *Collection* VII 106 and also, with a different proof, in a short series of propositions, probably dependent on Arabic sources, in two Latin manuscripts.[6] More striking is the variety of treatment accorded to what is essentially a single result attributed to Archimedes in Bīrūnī's *Istikhāj*.[7] If, in figure 2, from point *D*, the midpoint of arc *ABG, DE* is drawn perpendicular to *AB*, then $AE = EB + BG$. The first proof, also given as by Archimedes, sets

[3] *See* DOLD-SAMPLONIUS (1976), p. 14.

[4] Cairo, Dār al-kutub, Muṣṭafā Fāḍil Riyāḍa 41m, ff. 105*r*-110*r*.

[5] Private communication from Dr. ABDUL-LATIF, quoting his 1987 publication.

[6] Bonn, Universitätsbibliothek, S 73, ff. 52*v*-53*r*, and Vat. Reg. lat. 1268, ff. 41*r*-*v*. See BUSARD (1985), p. 155.

[7] In Suter's translation, p. 13. Suter follows the text in MS Leiden Or. 513.

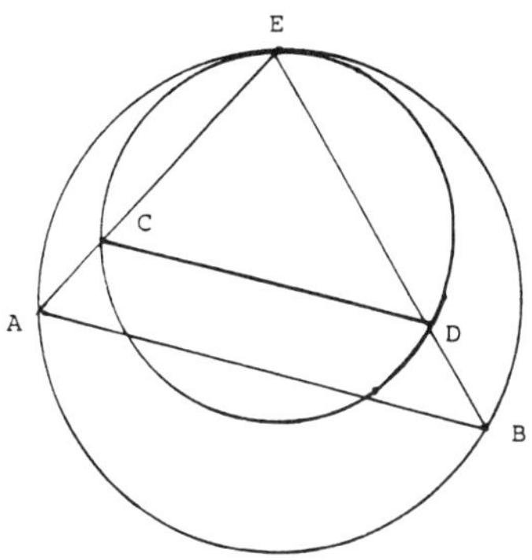

Fig. 1

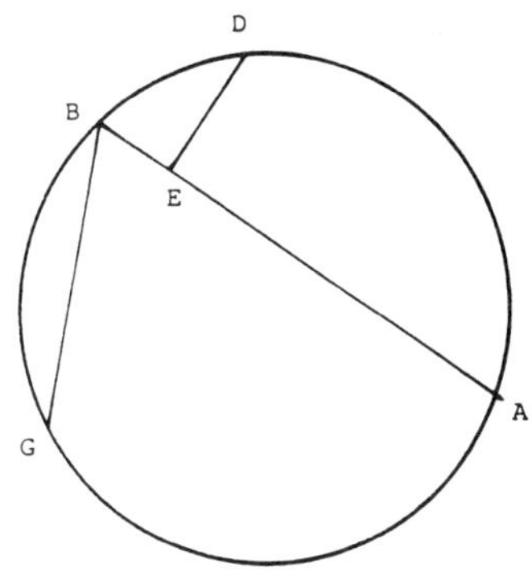

Fig. 2

(figure 3) arc $DL = BD$ and $EM = BE$ and proves that AL, which is equal to BG (of figure 2), is equal to AM. In the *Liber assumptorum* a similar proof is applied, but the line BG does not appear. In the *Heptagon* text[8] there again is no line BG, the segment AB is a semicircle, arc BD is put equal to DL and AM to AL, and the result is effectively a proportion between the lines, possibly to facilitate the calculation of chords. Did Archimedes prove a series of related results or are our sources heavily reworked?

Some of the lemmas in the various collections are elementary. For example, the bisecting line AD in figure 4 is equal to each of the

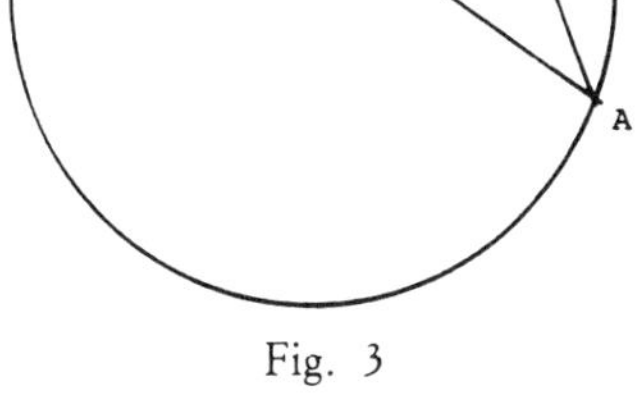

Fig. 3

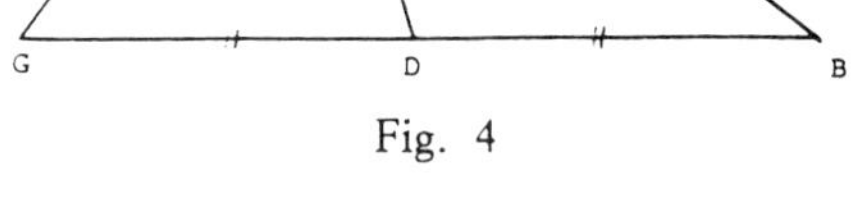

Fig. 4

---

[8] Schoy, p. 81.

halves of the triangle's base if, and only if, angle *BAG* is right. This result is used, for instance, in a solution of the trisection problem, but it must have had dozens of other applications. Again, the difference-of-squares lemma in figure 5, $AB^2 - AG^2 = BD^2 - GD^2$ is used in the Banū Mūsā's proof of the Hero formula for the area of a triangle, but often elsewhere – in an application in Ptolemy's *Planisphaerium*, the result was not even explicitly cited. Both these lemmas are in the *Book of Assumptions.*[9] On the other hand, the collection also contains important and interesting results.

In this paper we should like to present three such results – the trisection of an angle, the construction of the regular heptagon and Hero's formula for the area of a triangle in terms of its sides –, all three the subject of intense investigation in Arabic,[10] and to see how they appear in the medieval Latin West soon after the period of translation from the Arabic.

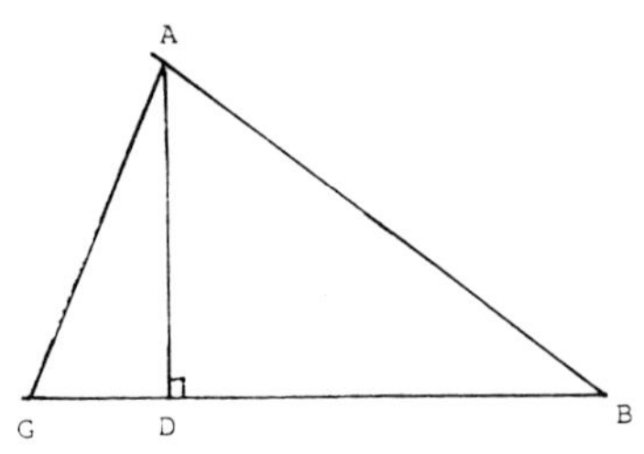

Fig. 5

1. Trisection of an angle

In the eighth proposition of the *Liber assumptorum* a *neusis* (verging) construction is described for solving the problem.[11] In figure 6 *AB* is rotated about *A* until $BC = r$. This makes angle *C* a third of angle *ADE*. For angle *DAB* = angle *DBA* = double angle *C*, and so angle *ADE* = angle *DAB* + angle *C* = triple angle *C*. The solution

[9] Dold-Samplonius (1976), pp. 26-28; Archimedes, *Kitāb al-uṣūl*, pp. 13-14.

[10] For Arabic writings on the trisection problem see Hogendijk (1979); on the heptagon Anbouba, Rashed and Hogendijk (1984); and on Hero's formula Abdul-Latif.

[11] *Archimedis opera*, II 518.

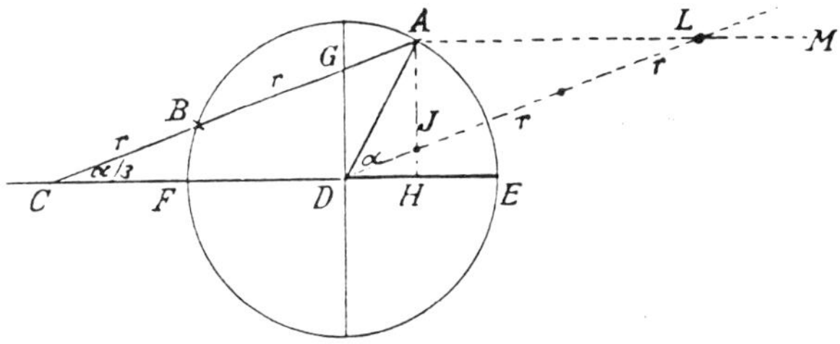

Fig. 6

seems to be generally accepted as Archimedean. A different treatment is to be found in Pappus, including a different *neusis*.[12]

The first solution of the problem in the medieval Latin West is to be found in the *Liber trium fratrum*, the geometrical treatise by the Banū Mūsā translated by Gerard of Cremona in the 12th century. The Banū Mūsā, the sons of Mūsā ibn Shākir, were among a group of scholars in ninth-century Baghdad who acquired manuscripts of Greek scientific works and translated them into Arabic – or reworked existing translations. The oldest of the brothers, Muḥammad, is said to have met Thābit ibn Qurra of Ḥarrān and to have persuaded him to come to Baghdad. From the *Fihrist* and other sources we know that the three brothers, Muḥammad, Aḥmad and Ḥasan, wrote, among other works, treatises on mechanics, on the ellipse, and on the movement of the first sphere.

It is possible that Thābit had something to do with the preparation of the *Liber trium fratrum*, in Arabic entitled *The Book of the Knowledge of the Measurement of Plane and Spherical Figures*. This small, but important, treatise is concerned with the measurement of the areas and volumes of plane and solid curved figures. It not only gave the formulae for the area of the circle and the volume of the sphere, but presented demonstrations of these and other formulae of an Archimedean character. This treatise belongs to the large number of works translated into Latin by Gerard of Cremona in the second half of the twelfth century and, as Marshall Clagett has pointed out, it is principally through this treatise and the *Dimensio circuli* that Archimedean ideas were transmitted to the West.[13] The *Liber trium fratrum*,

---

[12] See TROPFKE, III 123 f.

[13] CLAGETT, I 223-225, 558-559.

or *Verba filiorum,* was known in the 13th century to Leonardo Fibonacci, who used it in his *Practica geometrie* (1220), to Jordanus Nemorarius, who used it in his *Liber philotegni*, and to the unknown mathematician who reworked this treatise into the *Liber de triangulis*, to Roger Bacon and others.

In the *Verba filiorum* there is the first introduction into the West of the problem of finding two mean proportionals between two given quantities, which from Greek antiquity had been used to solve the problem of duplicating the cube. The Banū Mūsā present two solutions, which Eutocius attributed to Archytas and to Plato. It is well known that the problem of the duplication of the cube leads to a cubic equation which the Greeks solved by a *neusis* construction or by equivalent means, such as using conic sections or other curves of higher degree. Another classical geometrical problem was the trisection of an angle, which also leads to a cubic equation and can therefore be solved in a similar way. The Banū Mūsā present the first solution in Latin of this problem, too – a solution which reminds us to some extent of the one found in the *Liber assumptorum*.[14] The crucial point is that the Banū Mūsā give a mechanical solution which reduces the problem to a *neusis* construction. If *ABG* is the given angle, a circle of any radius *r* is drawn, defining the points *D, E* and *L* (see figure 7), *BZ* is perpendicular to *ADL.* The point *T* moves on the circle between *Z* and *L* until *TS* = *r*, where *S* is the intersection of *TE* and *BZ.* The trisection is completed by drawing line *MKB*

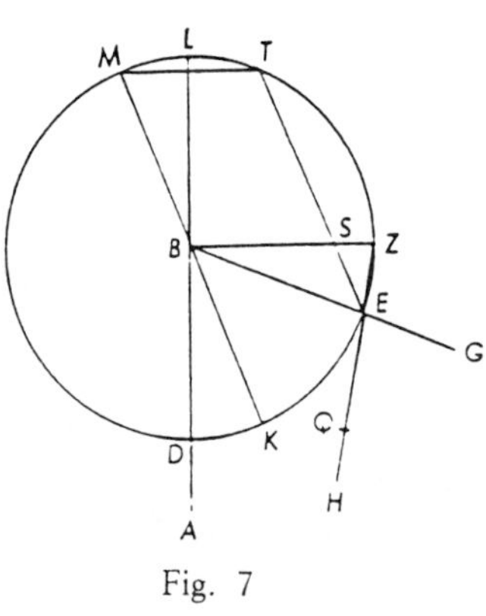

Fig. 7

[14] Prop. 18; ed. CLAGETT, I 344-349.

parallel to *TSE*. Now angle *DBK* = $^1/_3$ angle *ABG*. For since *MB* and *TS* are equal and parallel, so are *MT* and *BS*, and therefore *ML* = *LT*. Hence arc *DK* = arc *ML* = $^1/_2$ arc *MT* = $^1/_2$ arc *KE*, and arc *DK* must be a third of arc *DE*. In modern terms, the point *Q* (where *TQ* = *r*) traces the inner part of a conchoid of a circle which intersects *BZ* in *S*. The idea in the Banū Mūsā text is similar to that in the eighth proposition of the *Liber assumptorum*. The main difference is that in the Archimedean text the line segment equal to the radius is outside the circle, while in the *Liber trium fratrum* it is inside. Thus in the *Liber assumptorum* the intersection of the outer part of the conchoid has to be found.[15]

The problem of trisecting an angle is also treated in the *Liber de triangulis*, which up to the publication of volume 5 of Marshall Clagett's *Archimedes in the Middle Ages* was thought to be by Jordanus de Nemore, who in Professor Clagett's words[16] was «a first-rate and greatly influential mathematician». He lived certainly before 1260, most probably in the first half of the thirteenth century, and may have been master at the university of Paris. Since volume 5 of Clagett's *Archimedes* we know that Jordanus is the author of the so-called *Liber philotegni*, but not of the *Liber de triangulis*, which depends upon the *Liber philotegni* and refers to it. Whereas the *Liber philotegni* contains 64 propositions, most of which have to do with the determination and comparison of triangles and other polygons, the *Liber de triangulis* is somewhat longer. Its unknown author recast many of the proofs of the original work and omitted or added others. It should be noted that he omitted propositions 47-63 of the *Liber philotegni* and added 18 propositions, which are not in the *Liber philotegni*, to book IV of the *Liber de triangulis* (IV.10, 12-28). These additional propositions are of special interest, because most of them have to do with works translated from the Arabic; in Clagett's opinion the author of the *Liber de triangulis* added to his tract, almost intact, a collection of Arabic-based propositions.[17]

One of these additional propositions, IV.20, concerns the trisec-

[15] See the comparison of the methods in TROPFKE, III 124, and the analysis in CLAGETT, I 666-670.

[16] CLAGETT, V 146.

[17] CLAGETT, V 324.

tion of an angle.[18] It is divided into three parts. The first (lines 3-24) is a paraphrase of proposition 18 of the Banū Mūsā. There follows (lines 25-28) a minor modification of part of that solution. More interesting is the last part (lines 29-43) which gives formal geometric authority for the *neusis* used in the proof. As the author expresses it: *The said demonstration concerning the trisection of an angle does not at all suffice for me, for I can find no certainty in it. To make it suffice for me, I demonstrate the same thing as follows.* That authority is a proposition from Alhazen's *Perspectiva*[19] whose proof is based on the use of conic sections. It seems to be clear that the whole proposition as we have it comes from a single author working in the Latin West.[20]

There seems to be a connection between the *Liber de triangulis* and the Euclid reworking attributed to Campanus of Novara. In the first edition of the Euclid (Venice, 1482) there is an addition at the end of book IV which gives a solution of the trisection problem. This addition seems not to be in the Latin text of the Campanus manuscripts, but we should add that most of them have not yet been checked (there are more than a hundred of them). But it has been found in the margin of one late manuscript.[21] Therefore it is not very probable that Campanus was the author of this text.[22] The solution in the Campanus edition is similar to the third part of the solution in the Pseudo-Jordanus *Liber de triangulis*, but this text has a serious geometrical defect, which Copernicus recognized in a note to his personal copy of the *editio princeps*.[23]

Another solution of the trisection problem is given by Johannes Regiomontanus in a letter written to Giovanni Bianchini in 1464.[24] Regiomontanus answers a question posed by Bianchini in a letter from 5 February 1464: in preparing a chord and arc table, Bianchini had

---

18 Ed. Clagett, I 672-677, and V 412-414.

19 Clagett, V 414 (text), 469 (translation) and 323 f. On Ibn al-Haytham and the trisection problem see Hogendijk (1979), pp. 20-22.

20 Clagett, V 594.

21 Naples, VIII.C.21, 15th c.

22 See Clagett, I 678 f., with edition on p. 680 f.

23 See Clagett, I 678.

24 Ed. Curtze, II 258 f. See Clagett, III 351-353.

asked him to find a way for trisecting the angle of 60°.[25] In his answer Regiomontanus solves the problem in exactly the same way as in the Banū Mūsā text. But because Regiomontanus gives a reference to Alhazen's *Perspectiva*, we may assume that he had it from the *Liber de triangulis*. We know from the 1512 catalogue of Regiomontanus' books that he possessed a *liber triangulorum* which might have been the Pseudo-Jordanus text. That Regiomontanus knew this proof from a Campanus manuscript is not very probable; at least the «Campanus» proof is not in the Euclid manuscript in the Campanus version which was in Regiomontanus' possession and is now in the Stadtbibliothek Nuremberg.[26]

From the 16th century on there are some solutions of the trisection problem which may have been influenced by the solution given in the printed Campanus text. We mention here only the efforts of Viète. But that is another story.

2. Heptagon

There are numerous texts in Arabic on the regular heptagon. Much of the basic work of edition and comparison has been done by Professors Anbouba and Rashed and most recently by Dr. Hogendijk. A construction of the side of the regular heptagon is to be found in the *Heptagon* treatise. The solution depends on the following lemma: if in figure 8 the side BA of the square $ABCD$ is produced, say to H, and the line DEZ is rotated about D until triangles AZE and CTD are equal in area, then $AB \cdot KB = AZ^2$ and $ZK \cdot AK = KB^2$. These follow from:[27]

$$CD\ (= AB) : AZ = AE : TL = AZ : LD\ (= KB)$$

$$TL\ (= AK) : KT\ (= KB) = LD\ (= KB) : ZK$$

In the application of this lemma (figure 9) C, D are taken on AB so that $AD \cdot CD = DB^2$ and $CB \cdot DB = AC^2$; $CH = CA$,

[25] Ed. Curtze, II 238.

[26] Cent. VI 13.

[27] We note in passing that only the first result depends on the equality of triangles AZE and CTD.

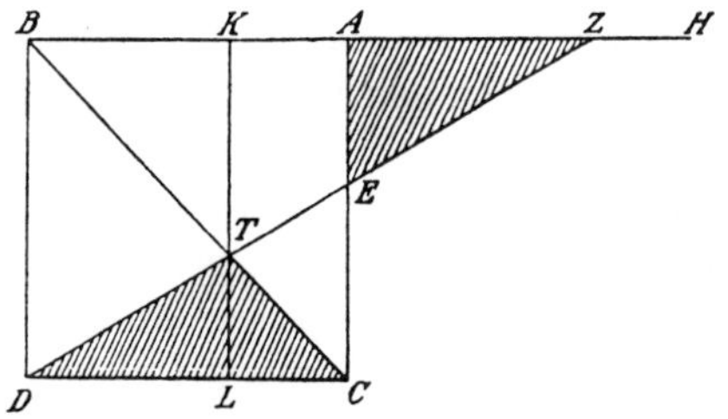

Fig. 8

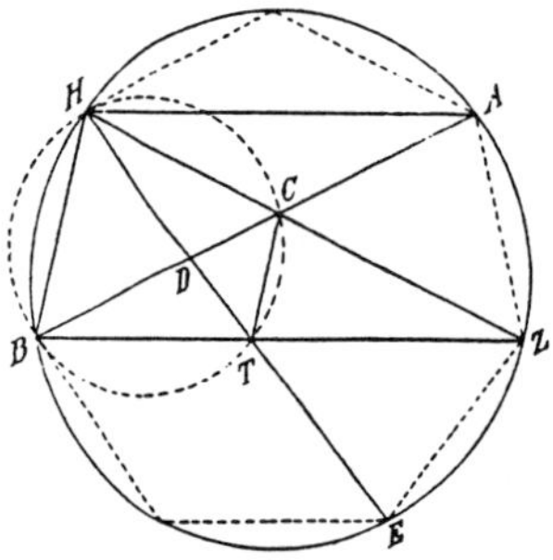

Fig. 9

$DH = DB$; circle AHB is drawn and the diagram completed. Now BH is the side of the regular heptagon in circle AHB. In essence the proof runs:

angles $HAC = AHC$, so arcs $AZ = HB$.

Since $AD \cdot CD = DB^2 = DH^2$, triangles AHD and CHD are similar.

Therefore angles $DAH = CHD$, i.e. arcs $ZE = BH$

B, H, C, T, are concyclic, since angles CHD and TBD, standing on equal arcs in circle AHB, are equal.

Therefore (since $DH = DB$) $CD = DT$,[28] and so $CB = TH$.

But $TH \cdot DH = CB \cdot DB = AC^2 = HC^2$, so triangles THC and CHD are similar and angles $DCH = HTC$.

Therefore angles $DBH = CTH = DCH = 2\ CAH$, and so arc $AH = 2$ arc HB;

And since $DHB = DBH$, arc EB also $= 2$ arc HB.

Since arc AE is again double arc HB, it is clear that the circle has been divided into seven parts, each equal to HB. The question of authorship is unsettled. Professor Rashed considers that the construction, apart from the lemma, is the work of Arab mathematicians.[29]

---

[28] The text is not clear at this point. HOGENDIJK (1984), p. 206-207, suggests that $BD \cdot DC = HD \cdot DT$ (EUCLID, *Elements* III 35) is invoked. We have in any case attempted only a general idea of the proof.

[29] Private communication.

This is in contrast to the opinion of Dr. Hogendijk, who gives reasons for supposing a Greek origin[30] and points out that in Arabic writings the solution is generally ascribed to Archimedes.[31]

The first known text in the Latin West which deals with the heptagon is a long fragment translated by Gerard of Cremona and found in the important manuscript Paris, Bibliothèque Nationale, lat. 9335 and in some other manuscripts. This codex, which contains many Gerard translations, has recently been dated to 1200 or shortly before.[32] The text was not known to Hogendijk and first edited by Clagett.[33]

The text seeks to inscribe a regular heptagon in a circle. The first part is a mechanical solution using a *neusis* construction (figure 10, left): to the diameter *AGB* there are two perpendiculars *HQ* and *GD* with *AH* = *HG*; there is a third line *TZ* perpendicular to *GD* with *DT* = *TG*. Then the line *AD* is moved about *A* as center so that *D* stays on the circumference of the circle. Thus the point *E*, where the line *AD* intersects the fixed perpendicular *QH*, slides along *QH*, and the perpendicular *TZ* moves and intersects *DL* at *K*. These motions continue until *K* falls upon the diameter *AB*. This situation is shown in the diagram on the right, primed letters being used for movable points. In the text of the fragment it is proved that then the

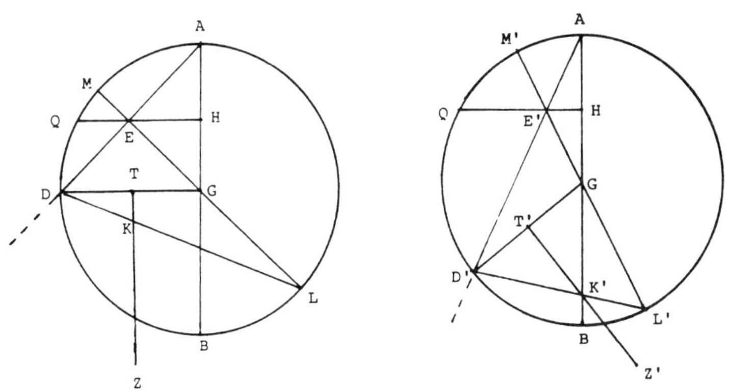

Fig. 10

[30] Hogendijk (1984), 211-212.

[31] *Ibid.*, p. 212.

[32] By Professor Bernhard Bischoff, Munich.

[33] Clagett, V 596-599.

line *D'B* is the side of the inscribed regular heptagon. Since triangles *K'D'G* and *AE'G* are isosceles (*HE'* and *T'K'* being right bisectors of *AG* and *GD'* respectively), angle *E'GD'* is double angle *D'GK'*, which in turn is double *AGE'*. But the sum of these three angles, or $3\,^1/_2$ times angle *D'GB*, is two right angles; and so *D'B* is seen to be the side of the heptagon.

In this text there follows a brief statement that if a line *ADCB* is divided into its three constituent segments so that $CB : AD = AD : (CB + DC)$ and $DC : CB = CB : (CD + AD)$, and if a triangle is constructed with these segments, then its greatest angle is double the middle and the middle is double the smallest (figure 11). From this it follows that, if the triangle is circumscribed by a circle, the side, or chord, that subtends the smallest angle is the side of the regular inscribed heptagon. It can be shown that this passage is very close to the so-called Archimedean solution of the inscribed heptagon.[34]

After this short, but very interesting, note there is a final treatment giving an Indian rule to find an approximation to the side of any regular polygon inscribed in a circle. This formula may have come from Hero's *Metrica*, but not from Archimedes.

This fragment on the heptagon was taken over nearly word for word into the *Liber de triangulis* falsely attributed to Jordanus.[35] This proposition belongs to that part for which there is no equivalent in the genuine *Liber philotegni*.

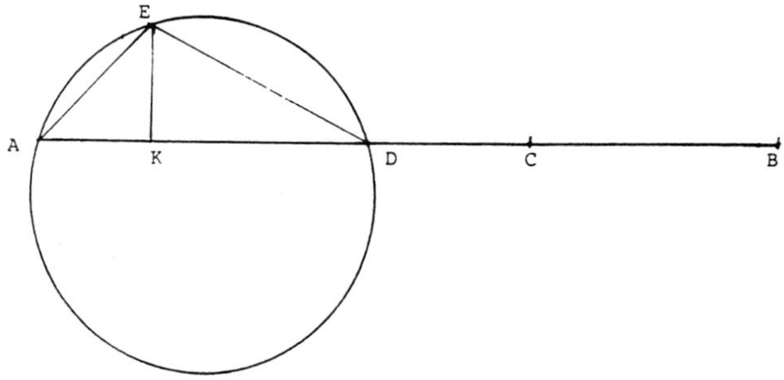

Fig. 11

[34] See CLAGETT, V 327 f., and V 473, note 6.

[35] *Liber de triangulis*, IV 23. Ed. CLAGETT, V 418-423; English translation with notes V 471-474; analysis V 325-328.

3. HERO'S FORMULA FOR THE AREA OF A TRIANGLE

In the *Istikhrāj* Bīrūnī ascribes the enunciation of the Hero theorem to Archimedes, but the proof he gives is by abū ʿAbdallāh al-Shannī,[36] whom Abdul-Latif dates to c. 955-c. 1020, working in or near Baghdad.[37] This proof (see figure 12) need not concern us here, but is connected with the theorem about the broken chord (figure 2). Another proof by al-Shannī is of quite a different nature. Let *ABG* be the triangle and, according to the usual modern convention, let the sides be a, b, g, whose semi-sum is s. In figure 13 let $BW = BH = BZ = BA$ and $BE = BD = BG$. Al-Shannī shows that four times the square of the area of the triangle is $GD \cdot WH \cdot EG \cdot AW$, and the result follows from two applications of Ptolemy's theorem:

$$GD \cdot WH = GD \cdot AZ = AD^2 - AG^2 = (a+g)^2 - b^2 = 4s(s-b)$$
$$EG \cdot AW = AG^2 - GW^2 = b^2 - (a-g)^2 = 4(s-a)(s-g)$$

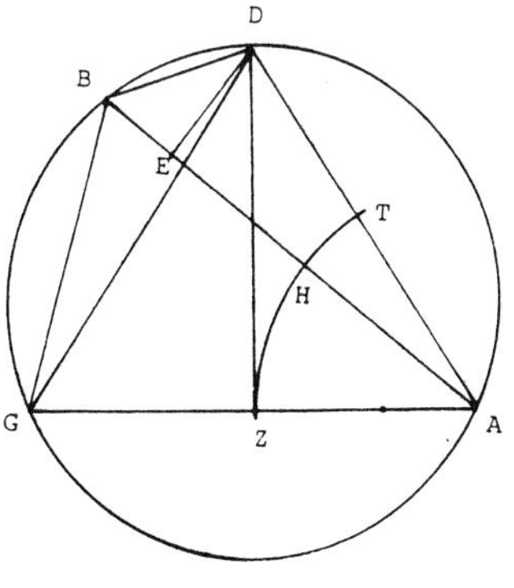

Fig. 12

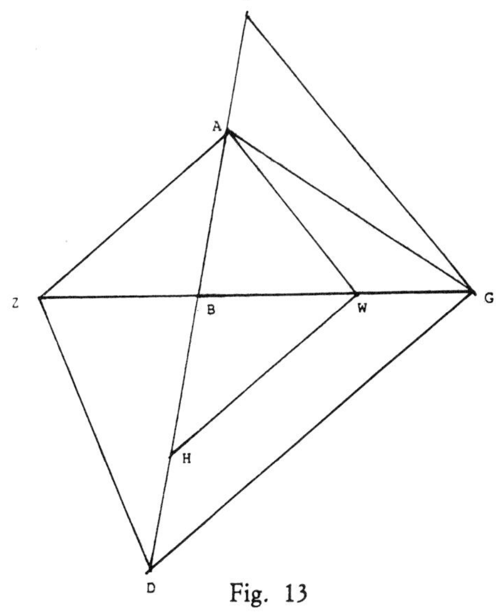

Fig. 13

[36] See SUTER's translation, pp. 39-40.

[37] Private communication. Dr. ABDUL-LATIF included this in a paper presented at the 13th Annual Conference on the History of Arabic Science in Aleppo in May 1989.

There are other proofs in Arabic, for instance by Abū'l-Wafā'.[38]

The result appears in several places in the Hero *corpus*. His proof, in *Metrica* prop. 8 and *Dioptra* prop. 30, may be summarized (see figure 14).

Construct L by making angles GHL and GBL right.

Therefore angles GLB = supplement of $GHB = AHD$.

Therefore triangles AHD and LGB are similar.

Therefore $GB : AD = BL : DH\ (= EH)$.

But $BL : EH = BK : KE$, since triangles BLK and EHK are similar.

Put $BT = AD$. Therefore $GB : BT = BK : KE$.

Therefore $GT : BT = BE : KE = GE \cdot BE : GE \cdot KE = GE \cdot BE : EH^2$.

Therefore $(GT \cdot EH)^2 = GT \cdot BT \cdot GE \cdot BE$.

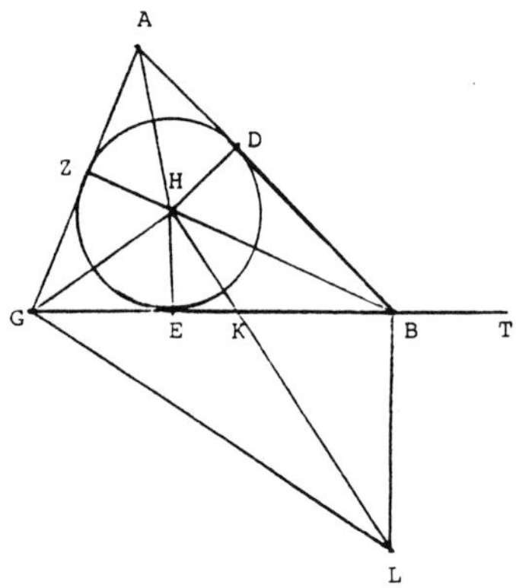

Fig. 14

We may note a series of theorems in the *Heptagon* text on right-angled triangles. The treatment of area involves an inscribed circle and the results involve some of the same quantities as the Hero formula.

Hero's formula for the area of a triangle was known very early in the Latin West. We find it in the *Corpus agrimensorum*, a collection of treatises of the Roman surveyors, some of which originate in late

38 See Kennedy and Mawaldi.

antiquity, others in the early Middle Ages. Hero's formula is given in the so-called *Liber podismi*, a short anonymous treatise which was influenced by Hero of Alexandria, especially by his geometry. Therefore it is not surprising that this formula is used in the *Liber podismi*. Its unknown author does not give a proof of it, and it is strange that he uses it for finding the area of a triangle with the sides 6, 8 and 10, i.e. a right-angled triangle.[39] We can assume that from this source, which was not influenced by Arabic authors, the Latin West learned the formula.

The formula is also present in the *Mishnat ha-middot*, the oldest mathematical work in Hebrew, which might have originated about 150 A.D, but was probably a compilation made after the ninth-century translations into Arabic. The treatise, which is of a practical nature, gives Hero's formula in IV,9 without proof.

The Hispano-Jewish philosopher, mathematician and astronomer Abraham bar Ḥiyya, better known as Savasorda, who lived in Barcelona and died in or after 1136, wrote in 1116 a treatise on practical geometry, which was translated from Hebrew into Latin in 1145 by Plato of Tivoli under the title *Liber embadorum*. In the second part of this treatise, which deals with the measurement of triangles, Savasorda treats Hero's formula for the area of a triangle.[40] The author gives the general formula and then an example with the same lengths of the sides as in the *Liber podismi*, i.e. 6, 8 and 10. He does not give a proof, but ends with the words: *Haec quidem in geometriae demonstrationibus est intricata, quapropter tibi leviter explanari posse non existimo.*

The first *demonstration* of Hero's formula was made available in the West by Gerard of Cremona's translation of the *Liber trium fratrum* in the 12th century. It was this proof, proposition 7 in the Banū Mūsā text, which was known to Leonardo of Pisa, Luca Pacioli and others and taken over into their works.

Compared with Hero's proof, that of the Banū Mūsā is somewhat unwieldy.[41] In figure 15 AB, AG and AH are produced; AW and AY are put equal to s (in the conventional notation), so that $BW = ZG$

---

[39] Ed. Bubnov, p. 515, problem 5.

[40] Curtze, I 72-74: no. 33.

[41] Ed. Clagett, I 278-289.

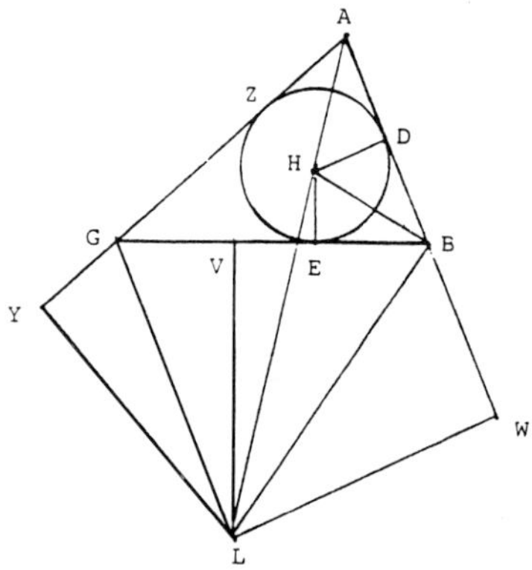

Fig. 15

and $GY = BE$; YL is drawn perpendicular to AY, so that $YL = WL$ (since triangles AYL and AWL are congruent). The steps of the proof are: LV (where $BV = BW$) is perpendicular to BG; triangles LBV and LBW are congruent; triangles LBW and BHD are similar, and so $HD \cdot WL = DB \cdot WB$; $HD^2 : HD \cdot WL = HD : WL = AD : AW$. The first step, which is rather complicated, is proved by showing that $BL^2 - LG^2 = BV^2 - VG^2$.

Savasorda's *Liber embadorum* was the model for the *Practica geometrie* written by Leonardo Fibonacci in 1220.[42] In contrast to Savasorda, Leonardo gives a proof of Hero's theorem[43] similar to that of the Banū Mūsā and apparently borrowed from it.

In the Latin West there is another proof of Hero's theorem of the area of a triangle, a proof somewhat different from that by the Banū Mūsa and much closer to the original proof given by Hero. This proof, which existed in two versions, has been associated with the name of Jordanus, although it is neither part of his *Liber philotegni* nor of the *Liber de triangulis*, but was transmitted as a special text in some manuscripts. Clagett has discussed the problem of the authorship[44] without being able to decide whether it is by Jordanus or not. But it seems to have come from an Arabic source, because

---

[42] Ed. Boncompagni, II 1-224.

[43] Boncompagni, II 40, line 7, - 41 bottom.

[44] Clagett, I 636-639.

it is stated: *This rule is said to have been written in Arabic* (version 1, beginning) and *This rule for a triangle was written in Arabic* (version 2, end). It should be noted that the version 1 text on Hero's formula was taken over into a short work entitled *Liber de triangulis datis* which is transmitted in at least three manuscripts from the 13th century.[45] This treatise has three propositions. The first two are equivalent to two propositions of the *Liber philotegni*, and the third is version 1 of Hero's theorem for the area of a triangle.

The version 1 proof[46] is similar to that in Hero's *Metrica* and *Dioptra*, but it is – in Professor Clagett's words – not very economical: there is an excessive number of steps to draw conclusions which from our point of view are obvious. On the other side, some geometrical steps are lacking. These steps are added by the author of the reworked version 2, which has come to us in only one late manuscript.[47] In this reworking it is also stated that there is another method, by means of the same demonstration, to find the area of a triangle. This method uses the fact that the product of the radius of the inscribed circle and the semiperimeter is equal to the area of the triangle, and informs how to find the radius from the given sides.

Other witnesses to Hero's theorem in the West give it without proof: it is in the geometrical treatise of Abū Bakr translated by Gerard of Cremona into Latin,[48] in John de Muris' *De arte mensurandi*, in many Italian treatises on practical geometry, and in Chuquet's *Geometry*.[49] We also find the formula in Leonardo Mainardi's treatise on practical geometry written in the second half of the 15th century in Italian. Mainardi says that he has it from a book on mechanics (*in alcune libro de mechanici*).[50] Another author is Johannes Widman who published in 1489 one of the earliest printed German *Rechenbücher*.

Luca Pacioli, who in 1494 published his well-known *Summa de arithmetica geometria proportioni et proportionalità*, brings, in the eighth chapter of the first part (*distinctio*) of the geometrical section, Hero's

---

[45] See CLAGETT, V 585.

[46] Ed. CLAGETT, I 642-647.

[47] Ed. CLAGETT, I 648-657.

[48] Ed. BUSARD (1968), p. 112.

[49] Ed. l'HUILLIER, p. 119, with note.

[50] CURTZE, III 396.

theorem together with a proof which is substantially the same as that in Leonardo's *Practica geometrie*.[51]

Any one of our examples could have been treated in greater detail. Other examples could have been taken. We are at the beginning of the task of tracing the transmission of Archimedean fragments.

BIBLIOGRAPHY

ALI I. ABDUL-LATIFF, *Muᶜādalat* Heron *ᶜibra 'l-ᶜuṣūr*, «Journal of the Institute of Arabic Manuscripts», 31, 1987, 59-145.

A. ANBOUBA, *Qaḍīya handasīya wa-muhandisūn fī 'l-qarn al-rābiᶜ al-Hijrī: tasbīᶜ al-dā'ira*, «Journal for the History of Arabic Science», 1, 1977, 384-352 (Arabic) and 2, 1978, 264-269 (French summary).

[ARCHIMEDES] *Archimedis liber assumptorum interprete Thebit Ben-Cora exponente Almochtasso ... Abrahamus Ecchellensis Latine vertit. Jo. Alfonsus Borellus Notis Illustravit* [appended to an edition of Apollonius' *Conica*], Florence, 1661.

[ARCHIMEDES] *Lemmata Archimedis apud Graecos et Latinos jam pridem desiderata ... a Johanne Gravio traducta et nunc primum cum Arabum scholiis publicata, revisa ... a Samuele Foster*, London, 1659.

Archimedis opera omnia cum commentariis Eutocii, ed. J. L. HEIBERG, Leipzig. Vol. I, 1910; Vol. II, 1913; Vol. III, 1915; Vol. V (*Archimedes, Über einander berührende Kreise*, ed. Yvonne Dold-Samplonius, Heinrich Hermelink & Matthias Schramm), Stuttgart, 1975.

[ARCHIMEDES] *Kitāb fī 'l-Uṣūl al-handasīya li-Arshimīdis naqalahu ... Thābit b. Qurra*, Hyderabad, 1948.

[ARCHIMEDES] *Kitāb fī 'l-dawā'ir al-mutamāssa li-Arshimīdis*, Hyderabad, 1948.

BALDASSARRE BONCOMPAGNI, *Scritti di Leonardo Pisano*, 2 vols., Rome, 1857-1862.

[AL-BĪRŪNĪ] HEINRICH SUTER, *Das Buch der Auffindung der Sehnen im Kreise von Abū'l-Raiḥān Muḥ. el-Bīrūnī*, «Bibliotheca Mathematica», 3.F., 11, 1910-11, 11-78.

NICOLAUS BUBNOV, *Gerberti postea Silvestri II papae Opera Mathematica (972-1003)*, Berlin, 1899.

H. L. L. BUSARD, *Der codex orientalis 162 der Leidener Universitätsbibliothek*, XIIᵉ Congrès International d'Histoire des Sciences, *Actes*, Tome III A, Paris, 1971, pp. 25-31.

H. L. L. BUSARD, *L'algèbre au moyen âge: Le «Liber mensurationum» d'Abû Bekr*, «Journal des Savants», 1968, 65-124.

H. L. L. BUSARD, *Some Early Adaptations of Euclid's Elements and the Use of its Translations*, in *Mathemata. Festschrift*

[51] See HULTSCH, pp. 242-246.

für Helmuth Gericke, Stuttgart, 1985, pp. 129-164.

MARSHALL CLAGETT, *Archimedes in the Middle Ages*. Vol. I, Madison, 1964; vol. II-V, Philadelphia, 1976-1984.

MAXIMILIAN CURTZE, *Urkunden zur Geschichte der Mathematik im Mittelalter und der Renaissance*. Part I: *Der «Liber embadorum» des Savasorda in der Übersetzung des Plato von Tivoli*. Part II: *Der Briefwechsel Regiomontan's mit Giovanni Bianchini, Jacob von Speier und Christian Roder*. Part III: *Die «Practica geometriae» des Leonardo Mainardi aus Cremona*, Leipzig, 1902.

YVONNE DOLD-SAMPLONIUS, *Kitāb al-mafrūdāt li-Aqāṭun: Book of Assumptions by Aqāṭun*, Heidelberg, 1976.

YVONNE DOLD-SAMPLONIUS, *Archimedes: Einander berührender Kreise*, «Sudhoffs Archiv», 57, 1973, 15-40.

T. L. HEATH, *Archimedes, The Works of Archimedes*, Cambridge, 1897.

J. L. HEIBERG, *Quaestiones Archimedeae*, Hauniae, 1879.

JAN P. HOGENDIJK, *Greek and Arabic Constructions of the Regular Heptagon*, «Archive for History of Exact Sciences», 30, 1984, 197-330.

JAN P. HOGENDIJK, *On the trisection of the angle and the construction of a regular nonagon by means of conic sections in medieval Islamic geometry*, University of Utrecht, Department of Mathematics, Preprint no. 113, 1979.

HERVÉ L'HUILLIER (ed.), *Nicolas Chuquet, La Géométrie*, Paris, 1979.

FRIEDRICH HULTSCH, *Der Heronische Lehrsatz über die Fläche des Dreieckes als Function der drei Seiten*, «Zeitschrift für Mathematik und Physik», 9, 1864, 225-249.

YUSUF ID & E. S. KENNEDY, *A Medieval Proof of Heron's theorem*, «The Mathematics Teacher», XVII, no. 7 (November 1969), 585-587. Reprinted in E. S. KENNEDY, *Colleagues and Former Students, Studies in the Islamic Exact Sciences*, Beirut, 1983, 492-494.

E. S. KENNEDY and MUSTAFA MAWALDI, *Abū al-Wafā' and the Heron Theorems*, «Journal for the History of Arabic Science», 3, 1979, 19-30.

[Mishnat ha-middot] SOLOMON GANDZ, *The Mishnat ha-Middot, the first Hebrew Geometry of about 150 C.E. and the Geometry of Muhammad ibn Musa ...*, «Quellen und Studien zur Geschichte der Mathematik, Astronomie und Physik», Abt. A, 2, 1932, 1-96.

NAṢĪR AL-DĪN AL-ṬŪSĪ, *Kitāb ma'khūdhāt li-Arshimīdis, taḥrīr*, Hyderabad, 1940.

Pappi Alexandrini Collectionis quae supersunt, ed. FRIDERICUS HULTSCH, Berlin, 1876-1878.

ROSHDI RASHED, *La construction de l'heptagone régulier par Ibn al-Haytham*, «Journal for the History of Arabic Science», 3, 1979, 309-386.

CARL SCHOY, *Die trigonometrischen Lehren des persischen Astronomen Abu'l-Raiḥân Muḥammad ibn Aḥmad al-Bîrûnî*, Hannover, 1927.

FUAT SEZGIN, *Geschichte des arabischen Schrifttums*, Leiden. Vol. V (Mathematik), 1974.

JOHANNES TROPFKE, *Geschichte der Elementar-Mathematik*, 3. Asuflage, Berlin & Leipzig: Bd. III, *Proportionen, Gleichungen*, 1937; Bd. IV, *Ebene Geometrie*, 1940.

J. VERNET & M. A. CATALÁ, *Dos tratados del Arquímedes árabe: Tratado de los circulos tangentes y el libro de los triángulos*, «Memorias de la Real Academia de Buenas Letras», 13, 1972, 33-80. Reprinted in JUAN VERNET, *Estudios sobre historia de la ciencia medieval*, Barcelona, 1979, 97-144.

III

Remarks on Greek Mathematical Texts in Arabic

Despite the plentiful material, the Arabic translations of Greek mathematical works have been little studied. (An exception is Diophantus' *Arithmetic*, but the extant parts in Greek and Arabic do not match.) In the present paper it is suggested that these translations can tell us something of the fortunes of the Greek texts in late antiquity and might sometimes even transmit a better reading than the extant Greek manuscripts. To this end three examples are taken of substantive differences between Greek and Arabic.

The first example is taken from Euclid's *Elements*, whose complicated transmission into Arabic and Latin is described in this conference by Professor Folkerts. Suffice it to say here that there are available many manuscripts attributing the translation they carry to Isḥāq b. Ḥunayn and its improvement to Thābit b. Qurra, and that for book XI we have at least one supposed witness to the Ḥajjāj translation, for the scribe of MS Copenhagen 81 said he had to use it for want of the other. In theorem 2 Euclid shows that any two straight lines that intersect lie in one plane and that every triangle lies in one plane. In the Arabic it is proved in this way (see fig. 1):

> AB, GD intersect at E. We show that they, and also triangle ZEH (formed by taking points Z and H on DE and BE respectively) lie in the same plane.
>
> For, if it were otherwise, a part of triangle ZEH would be in the plane and a part more elevated (*fī 'l-sumk*).
>
> But this would mean that parts of the lines ZE, EH are in the plane and parts more elevated – which contradicts what was proved in theorem 1.
>
> Therefore triangle ZEH lies in the one plane.

Therefore the two lines *AB*, *GD* are also in one plane, since they are both in the same plane as triangle *ZEH*.

In the Greek text the argument is basically the same, but extra points are introduced on the diagram (see fig. 2) to specify the various parts to be considered as in the same plane or not. It seems at least as likely that these points were added by a commentator as that they were removed by one. In this particular instance, therefore, the possibility that the Arabic translation represents a text nearer the original should at least be taken seriously; and this may also be true for other passages. A few remarks may be added here:

1. In this theorem there is no difference, apart from those due to the usual scribal errors, between the "Isḥāq-Thābit" manuscripts – of which MSS Oxford Thurston 11, Cambridge add. 1075 and Tehran Majlis 200 have been consulted – and the "Ḥajjāj" manuscript Copenhagen 81. In MS Escorial 907 we find more serious divergences from the common reading, as well as a relatively large number of trivial differences, but not enough to qualify for a different translation – and certainly no extra points.

2. The Latin translations attributed to Adelard of Bath ("Adelard I") and Gerard of Cremona both carry the same argument as in the Arabic texts; and very similar arguments are to be found in the redaction of Naṣīr al-Dīn al-Ṭūsī (in MS Munich 848) and in the "pseudo-Ṭūsī" redaction printed in 1594. Certainly no extra points are introduced.

3. As Sir Thomas Heath remarks in his English translation of the *Elements*, the first three propositions of book XI are unsatisfactory. In all three there are substantive differences in the Arabic.

The second example is taken from Theodosius' *Spherics*. Here Ṭūsī reports that Qusṭā b. Lūqā translated the text to the fifth proposition of book III, the rest being translated by someone else, and that the whole was revised by Thābit b. Qurra. Apart from the thirteenth-century redactions by Muḥyī al-Dīn Yaḥyā b. Muḥammad b. abī al-Shukr and by Ṭūsī himself, there are at least three Arabic texts of the *Spherics*, one being attributed to Qusṭā b. Lūqā as translator, one to Ḥunayn b. Isḥāq, and one with no attribution. This last was the basis of the Latin translation by Gerard of Cremona. Between definitions 5 and 6 of the Greek text of book I there are in the Arabic two extra definitions: on distances of circles on a sphere from the centre and on planes inclined, but not perpendicular, to each other. Further, the Greek definition 6, on equally inclined planes, appears in Arabic in a more general form. The two extra definitions are obvious enough and it seems likely that they were added by some commentator. But there is no reason to suppose that they were added in Arabic, for they appear in all three putative translations and in the Ṭūsī redaction. Now Ibn abī al-Shukr omits the extra definitions and substitutes for the Greek definition 6 a definition of inclination of one plane upon another – perhaps taken from Euclid, and perhaps from one of the Theodosius translations – and adds the generalization of the three putative translations. Whatever sources Ibn abī al-Shukr had, it seems reasonable to suppose that the three translations – or as many of them as *were* translations – were taken from a Greek text containing the extra material. Perhaps this brings us little nearer the original text – though Theodosius' treatment of inclination of planes is not entirely satisfactory in the Greek –, but it does tell us something about what happened to the text in late antiquity.

The third example consists of some extra material in the Arabic translations of Archimedes' *Sphere and Cylinder*. According to the titlepage of a manuscript (Istanbul, Fatih 3414) written in 676/1277 by Muḥammad b. ᶜUmar b. Aḥmad b. abī Jarāda, it was trans lated by Qusṭā b. Lūqā. In the text there is a note to say that a passage is left out because the translator from Greek into Syriac had not understood it. This lacuna and the division into theorems square well with a text described by Ṭūsī in his *Taḥrīr* as having been revised by Thābit. For the present we may perhaps assume that Thābit revised Qusṭā's translation from the Syriac. In the same Istanbul codex there is a fragment of another translation, which may or may not be the same as the second translation reported by Ṭūsī, which he ascribes to Isḥāq b. Ḥunayn. Now since Naṣīr al-Dīn, according to his own preface, patched up the "Thābit" translation and only later compared it with the "Isḥāq" translation, we may doubt the extent to which the latter has been transmitted to us. On the other hand, Ṭūsī gives us details of the division of the text into propositions according to the two translators: it would be highly unlikely that he failed to report the total absence of a theorem.

There are several extra theorems in the "Qusṭā-Thābit" translation of book I: a pair between propositions 8 and 9 in the Heiberg numbering and two near the end. The pair of theorems concerns the surface area of prisms inscribed in and circumscribed about a cylinder. We will consider only the first, the second being strictly analogous. In the diagram (fig. 3) the bases of the prism, $ABGD$ and $EZHT$, are equilateral and the rectangle MNS has

NS equal to the perimeter of base $EZHT$ and MN equal to BZ. Now

$GB.BZ$ = rectangle BH.

Similarly $GD.BZ = GT$; $DA.BZ = DE$; $AB.BZ = AZ$.

Therefore perimeter $EZHT.BZ$ = area of prism without bases.

But perimeter $EZHT = NS$ and $BZ = MN$:

Therefore $MN.NS$ = area of prism without bases.

But $MN.NS$ = area of surface MS.

Therefore surface MS = area of prism without bases.

As may be seen even from this shortened account, the proof procedes at a leisurely pace; also the enunciation contains the unnecessary assumption that the base of the prism be equilateral. The proposition was probably added to the text on analogy with theorem 7, a similar result about a pyramid inscribed in a cone. In this theorem the assumption that the bases be equilateral is required, since the equivalent of line BZ is the perpendicular from the vertex of the cone to one of the sides and must be the same for each. Archimedes drops the assumption in theorem 8, about a pyramid circumscribed about a cone, no doubt because it is unnecessary.

There are also additional theorems near the end of the book to generalise the results about the area and volume of a segment of a sphere. In Proposition 42 it is proved that the area of a segment (ABD in fig. 4), less than a hemisphere, is equal to the circle whose radius is the line (BA) from vertex to base. Archimedes himself (according to the extant Greek text) generalises this to apply to segments greater than a hemisphere by saying that the square on BA together with the square on GA is equal to the square on BG, and that therefore the circle with radius BA together with the circle with radius GA is equal to the circle with radius BG, which (Prop. 33) is equal to the surface of the sphere. Thus the area of segment AGD is proved to be equal to the circle with radius GA. In the Arabic translation there is another theorem, with the same proof, for the case of a segment *equal* to a hemisphere. At the end of the book there is a generalization, again on the same principles, of the result of Prop. 44 on the volume of a segment. Curiously, only segments *greater* than a semicircle are mentioned here explicitly. *Prima facie* it would seem reasonable to assume that these theorems, like those about inscribed and circumscribed cylinders were constructed on analogy with one in the text, namely Prop. 43, and were added by a commentator. Certainly the generalization of the theorem about surface area to segments *equal* to a hemisphere seems to be too trivial for Archimedes to have mentioned separately. One might even be inclined to doubt the authenticity of the generalization to a segment *greater than* a hemisphere. For according to the editor, Heiberg, the Greek has been translated from its original Doric dialect and otherwise altered by scribes, who added and subtracted material as they saw fit.

Naṣīr al-Dīn says that the first of these generalizations was not counted as a separate theorem in the "Isḥāq" translation; and since he does not say that the others are missing, we may assume that they were present. Two different translators would scarcely include such extra material it if were not in the original. This leads us to speculate about other ways in which this original might have differed from the text as we have it. In particular, does the Arabic terminology reflect late Greek usage? For instance we find that the expression for "the [rectangle] contained by AB, BG" usually appears as *al- mujtami*ᶜ *min ḍarb AB fī BG*, though sometimes by the closer *mā yuḥīṭ bihi AB wa BG*. The matter is complicated by the existence of intermediate Syriac versions.

Our three examples are of three different kinds of deviation in the Arabic translation from the traditional Greek text. In the Euclid example the translation probably reflects a reading nearer the original; the extra theorems in the Archimedes text are almost certainly added, and even induce doubt about the genuineness of other propositions; and the Theodosius definitions were probably mostly added, but might in part preserve some genuine material. But in all we have a window on the state of mathematical texts in the ninth century – the date of the earliest Greek manuscripts, or a little earlier.

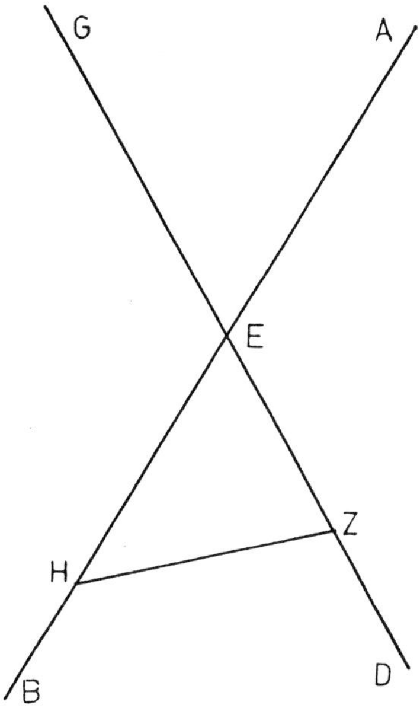

Fig. 1

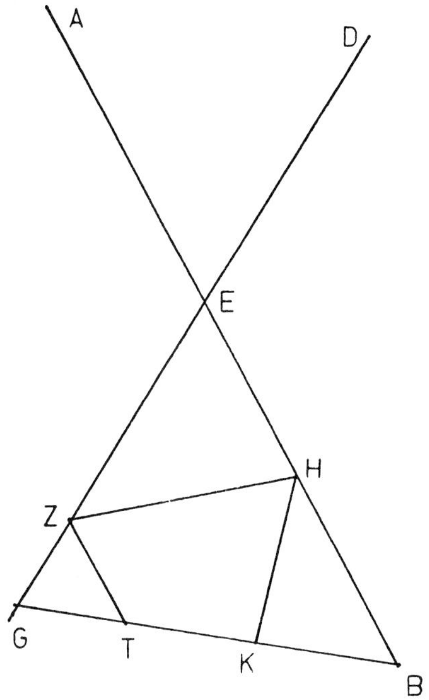

Fig. 2

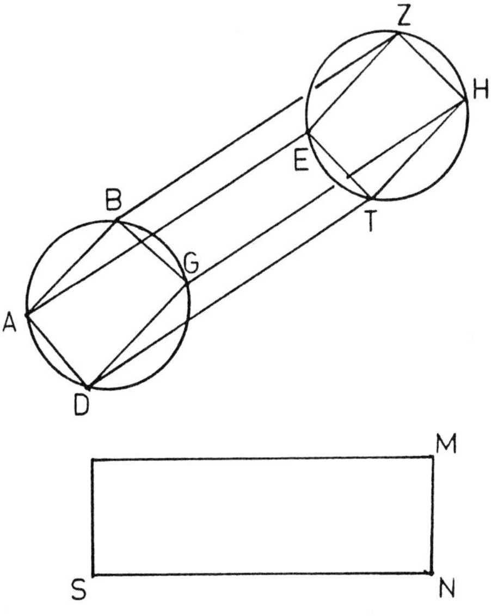

Fig. 3

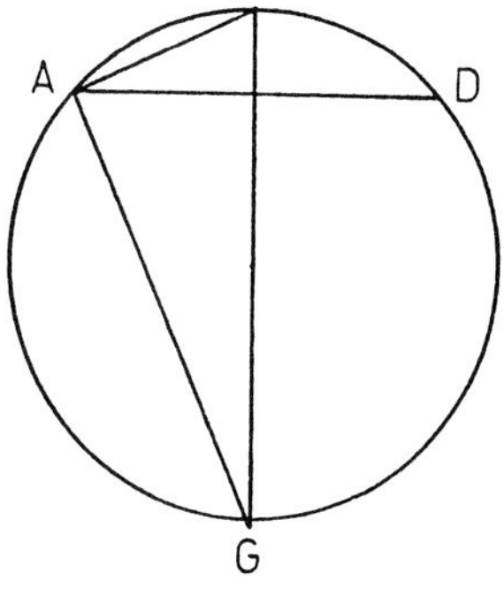

Fig. 4

IV

A Note on the Technical Vocabulary in Eratosthenes' Tract on Mean Proportionals

The table below is organized alphabetically by the roots of the Arabic words, or, in the case of phrases, of the principal Arabic words involved. Apart from such changes as the occasional removal of the Arabic article or inseparable preposition, the words have not been reduced to standard from, such as nominative singular for Greek nouns. By this means it is hoped that a spurious generality will be avoided, for it may be that a Greek word is translated in a given way only when it appears in a certain form or forms. In addition, the forms given serve as a reminder of the syntactical contexts from which they have been taken, contexts often different in the two languages. Greek third-person imperatives are normally rendered in Arabic by the first person plural of some form of the imperfect. Participles are normally rendered by clauses.

References by pairs of numbers divided by colons are to page and line of the Greek text (see reference 3 in the article above) for pages 102-114, and to page and line of the Arabic MS reproduced above for pages 153-157.

This *Note* is appended as a small contribution to the study of Greek mathematical and mechanical texts in Arabic translation.

It is a pleasure to thank Professor Paul Kunitzsch (Munich) for looking over this appendix and pointing out several errors in it. He considers the suggestion of the last paragraph speculative.

English equivalent of Greek	*Greek*	*Arabic*
half-cylinder	ἡμικυλίνδρων 106:4	نصف اسطوانة ١٥٤ : ٤
in the instrument	ἐν τῷ ὀργάνῳ 110:22	في الآلة ١٥٦ : ١٢
with an instrument	ὀργανικῶς 110:2	بآلة ١٥٦ : ١
a way ... by using an instrument	τις ... ὀργανικὴ λῆψις 106:8-9	عمل .... يعمل بآلة ١٥٤ : ٧
has been shown	ἀποδέδεικται 110:2	بر[هنا] ١٥٥ : ١٧
remaining	μένοντος 108:5	وليبقى (correctly وليبق) ١٥٥ : ٦
problem	πρόβλημα 104:9	باب ١٥٣ : ١٢
we shall show	δείξομεν 112:3	نبين ١٥٦ : ١٦
below	ὑποκάτω 108:6	من تحت ١٥٥ : ٧
in grooves	ἐν χολέδραις 110:6	(فيجريان) في مجاري لها ١٥٦ : ٤ (وانظر دفع)
solid	στερεόν 104:6	مجسم ١٥٣ : ١٠
I move	συνάγω 110:22	ولنحرك ١٥٦ : ١٢
Let them be drawn (pl.)	ἤχθωσαν 108:3	نخرج ١٥٥ : ٥
let there be drawn through points	διήχθω διὰ τῶν ... σημείων 108:8 – 9	نخرج من نقط ... ١٥٥ : ٨
wooden	ξύλινον 110:3	من خشب ١٥٦ : ٢
line	εὐθεῖα 108:9	خطاً ١٥٥ : ٨ (انظر علم ، نسب)
two means	δύο μέσας 106:2	خطين ١٥٤ : ٣ (انظر نسب)
	106:10	خط[ين] وسطين ١٥٤ : ٨
	110:3	خطين متوسطين ١٥٦ : ٢
lines	γραμμάς 110:9	خطوط ١٥٦ : ٥ (انظر عطف)
with no gap [between them]	ἄσχαστα 110:11	ولم يكن بينهما خلل ولا فرجة ١٥٦ : ٦–٧
let it approach, be brought together	συνωσθήτω 108:5-6	وندفع ١٥٥ : ٧
are capable of being pushed forward	ἐπωστοί εἰσιν 110:6	وليكن... يدفعان فيجريان ١٥٦ : ٣–٤
surface	ἐπίπεδον 104:5	سطح ١٥٣ : ٩ (انظر وزى)
equal	ἴσους 110:4	متساوية ١٥٦ : ٣
unequal	ἄνισοι 106:28	غير متساويين ١٥٥ : ٤
in a straight line	κατ' εὐθεῖαν 110:22–23, 108.8	على [مستوى] واحد ١٥٦ : ١٢–١٣
		متصلا على [استواء] ١٥٥ : ٨
bronze, brass	χαλκοῦν 110:4	من شبه ١٥٦ : ٢
putting together, assembly	συνάγεσθαι 110:10	صنعة ١٥٦ : ٦
figure	σχήματος 108:7	صورة ١٥٥ : ٨
has been drawn	γέγραπται 110:18	صورت ١٥٦ : ١٠
doubled	διπλασιασθεισῶν 104:4-5	اذا ضعفت ١٥٣ : ٩ (انظر كعب)

double	διπλάσιος 104:2	ضعف ١٥٣ : ٧
eightfold. eight times	ὀκταπλάσιον 104:6	ثمانية اضعاف ٤ ١٥٣ : ١٠
(the cube) will be doubled	διπλασιασθήσεται (ὁ κύβος) 104:15	ضعف (المكعب) ١٥٣ : ١٤
duplication (of the cube)	(κύβου) διπλασιασμός 104:9	اضعاف (المكعب) ١٥٣ : ١٢
sides	πλευρῶν 104:4	اضلاع ١٥٣ : ٩
by the lines called curved	διὰ τῶν καλουμένων καμπύλων γραμμῶν 106:4-5	بالخطوط التي تسمى المنعطفة ١٥٤ : ٤–٥
two given lines	δύο τῶν δοθεισῶν εὐθειῶν 110:20	خطين معلومين ١٥٦ : ١١
two given [lines]	δύο τῶν δοθεισῶν 106:2	الخطين المعلومين ١٥٤ : ٣
ivory	ἐλεφάντινον 110:4	من عاج ١٥٦ : ٢
above	ἐπάνω 108:6	من فوق ١٥٥ : ٧
a hundred feet	ἑκατόμπεδος 102:24	مائة قدم ١٥٣ : ٦
diagonals in them [rectangles]	διάμετροι ἐν αὐτοῖς 108:3-4	اقطارها ١٥٥ : ٦
let there be erected	συνεστάτω 108:2	نقيم ١٥٥ : ٤
cube	κύβου 104:9	المكعب ١٥٣ : ١٢
base, frame, board (in the shape of a brick. The Arabic repeats the Greek etymologically)	πλινθίον 110:3	لبنة ١٥٦ : ٢
is fixed in	ἐνήρμοσται 110:5	وليكن ... قد ألصق ١٥٦ : ٣
has been attached with solder	καθήρμοσται .. προσμεμολυβδοχοημένον 110:13-14	وقد ألصق ... برصاص ١٥٦ . ٨
let it meet	συμπιπτέτω 108:9	حتى يلقى ١٥٥ : ٩
small, thin tablets, panels, plates	πινακίσκους 110:4,10	ألواح صغار ١٥٦ : ٣ ألواح ١٥٦ : ٦
tablets, panels, plates,	πίνακας 110:22	الالواح ١٥٦ : ١٢
lying alongside one another evenly	ὁμαλῶς συναπτόμενα ἀλλήλοις 110:11-12	تكون ماسّها بعضها لبعض باستواء ١٥٦ : ٧
proportional	ἀνάλογον 112:4	متناسبة ١٥٦ : ١٧
$ZK : KH = BZ : GH$	ὡς ... ἡ ZK πρὸς KH, ἡ BZ πρὸς ΓH 108:21	نسبة زك الى كح كنسبة بز الى جح ١٥٥ : ١٥
two means in continued proportion	δύο μέσας ἀνάλογον ... ἐν συνεχεῖ ἀναλογίᾳ 106:28-9	خطين مناسبين (لهما) على الولاء ١٥٥ : ٣
	the same 104:14-15	خطين .... مناسبين لهما حتى تتوالى النسب ١٥٣ : ١٣ – ١٤
produced	ἐκβληθείσῃ 108:9-10	نفذ ١٥٥ : ٩
at [point] K	κατὰ τὸ K 108:10	على نقطة ك ١٥٥ : ٩
geometers	τοὺς .... γεωμέτρας 104:19-20	المهندسين ١٥٤ : ١
in terms of geometric surfaces	ἐπὶ τῶν γεωμετρουμένων ἐπιφανειῶν 110:1	بسبيل السطوح الهندسية ١٥٦ : ١

parallel (to each other)	παράλληλα 110:11	متوازية ١٥٦ : ٦
in the parallels ... ... (i. e. because ... is parallel to ...)	ἐν ... ταῖς .... παραλλήλοις 108:12	من قبل موازاة .... لخط .... ١٥٥ : ٩ — ١٠
parallelograms	παραλληλόγραμμα 108:2	سطوح متوازية الاضلاع ١٥٥ : ٥
the middle parallelogram	τοῦ μέσου παραλληλογράμμου 108:5	سطح ... الاوسط المتوازي الاضلاع ١٥٥ : ٦ — ٧
middle	ὁ ... μέσος 110:5	الاوسط ١٥٦ : ٣
successively	ἐφεξῆς 108:3	متوالية ١٥٥ : ٥ (انظر نسب)

Technical terms are sometimes translated by descriptive phrases. Thus καταπαλτικὰ καὶ λιθοβόλα ὄργανα (106:20-21), "catapults and stone-throwing implements", becomes (154:15.) الآلات التي تستعمل في الحروب لترسل على الحيطان الحجارة (in the MS there is a و wrongly inserted before the last word). ἄσχαστα (110:11), "without a gap", becomes (156: 6-7) ولم يكن بينها خلل ولا فرجه For the liquid measure μετρητής (see 106:17) the description is preceded by a transcription (154:13) مطريتيس وهو مما يكال به الأشياء الرطبة

An example of difference of form without real change in meaning is the expansion of εὑρεῖν (110:21), "to find", into (156:11) وبنتيجته نريد ان نبين كيف نجد But distinct differences in meaning do occur, though it is seldom clear that they are intentional. For instance, the Arabic in 153:13-14 omits the condition, foun din the Greek of 104:12-15, that one of the lines mentioned is double the other. Again, in 154:12-14 of the Arabic there appears to be no mention of a cube and its side, as in the Greek of 106:16-19 – though the whole passage, 106:8-19/154: 8-14 is not very clear in either language. τρεῖς πινακίσκους ἴσους ὡς λεπτοτάτους (110:4-5) "three equal panels as thin as possible", becomes (156:3) ألواح صغار دقاق متساوية (Escurial 960 is nearer the Greek in this passage). Other differences are to be found in 104:15-16/153:15-16 and 110:14-16/156:8-9. There is a subtle difference between τὸ δοθὲν στερεὸν παραλληλογράμμοις περιεχόμενον, "the given solid bounded by parallelograms" (106:12-13) and جسم معلوم متوازي الأضلاع (154:9). A final example: in the votive offering - ἀνάθημα, translated as قائم مربع (see 110:12/156:7-8 & foll.) - the instrument is placed ὑπ' αὐτὴν τὴν στεφάνην τῆς στήλης "under the very crown of the stele", but على رأس ذلك القائم . The Arabic here is a simplified translation, "on top of the stele", thus avoiding the difficult technical terms of Greek temple architecture. Parts of the Greek may have given the translator some trouble. For several words – e.g. ἐπωστός, χολέδρα, προσμολυβδοχέω (see 110:6,6,14) – Liddell & Scott's *Greek-English Lexicon* gives only this text of Eratosthenes. Of course we have no guarantee that the Arabic is a straight translation and has not been reworked.

Of some interest is the obtrusion of an unknown "Titanus" (154:6) in front of "Menaechmus" (106:8). The correspondence of text and translation is by no means clear in this place, but it is possible that طيطانس بن (the reading بن is uncertain) is a corruption of τι τοῦ, which precedes "Menaechmus" in the Greek. If this is so, some at least of the corruption probably occurred in Greek, since one of the manuscripts reads επιβραχυτητι, apparently for the whole phrase ἐπὶ βραχύ τι τοῦ. It is worth noting that طيطانس بن does not occur in MS Escurial 960.

V

Some Remarks on the *Almagestum parvum*

Contents

1 General description of the text

In the *Speculum astronomiae* attributed to ALBERTUS MAGNUS (d. 1280) an astronomical work based on PTOLEMY's *Almagest* and AL-BATTĀNĪ's astronomical work beginning "Inter universa ..." is reported:[1]

> Ex hiis quoque duobus libris collegit quidam vir librum secundum stilum Euclidis, cuius commentarium continet sententiam utriusque, Ptolemaei scilicet atque Albategni, qui sic incipit: *Omnium recte philosophantium etc.*

This treatise, which goes under many names, among them *Almagestum parvum*, is a compendium of the first six books of the *Almagest* with additional material from AL-BATTĀNĪ and others. It is written in the form of proposition and proof, as in EUCLID's *Elements*. Where appropriate, a rule for calculation is added, introduced by *regula* or *Unde manifestum est* or similar formula. The "Euclidean style" is also to be seen in the numbering of the propositions. Although in some manuscripts the numeration is incomplete, it is probably original, since there are internal references to it. Examples of this are "... restat ex kata disiuncta et nona bis et undecima semel assumpta propositum colligere" at the end of I 13 (the first form of

[1] [ALBERTUS MAGNUS], p. 9.

Reprinted with kind permission from Birkhäuser Verlag Basel (Switzerland) from Demidov S.S. / Folkerts M. et al. (Eds.); Amphora, Festschrift for Hans Wussing on the occaision of his 65th Birthday, 1992.

Menelaus' theorem for spherical arcs),[2] "... ex xxx^{am} presentis ..." near the end of II 34 (on the intersection of ecliptic and altitude circles), and "... ut ex premissa et 20^{a} habetur collige et ... ex 19^{a} et 21^{a} et 22^{a} attende" at the beginning of the proof of III 24 (on the inequality of solar days).[3]

There are no tables in the *Almagestum parvum*. The treatise must be assumed to have been intended as a theoretical introduction to astronomy rather than a manual for practical astronomers or astrologers.

In the main, the lettering of the diagrams, at least in the earlier books, agrees with that of the translation of the *Almagest* itself by GERARD OF CREMONA.[4] The diagrams checked for this purpose are those up to II 22, or *Almagest* II 10, and the many-lettered diagram for determining the solar eccentricity and apogee, III 11, or *Almagest* III 4. Of the manuscripts compared, MS Paris BN lat. 16657, Vatican, lat. 1261 and Berlin Staatsbibl. lat. Q 510, the third usually has the best diagrams, the first two often having mistakes in the drawing as well as missing or wrong letters. At the beginning of the work we find two exceptions: *h* appears instead of *e* in the diagram for the very first theorem, I 1, on finding the side of a regular decagon etc., and again in I 5, on the chord of the sum of two arcs whose chords are known. In this there is a parallel to the translation of the *Elements* by ADELARD OF BATH and the compilation of the *Elements* known as *Version II*.[5] The diagrams that appear in the *Almagestum parvum* but not in the *Almagest* itself have not been fully investigated. Suffice it to say in this preliminary survey that the diagram for IV 1, which was added to clarify PTOLEMY's statement about lunar observations and parallax, has the letters A, B, C, D, E, F, G, K.

In book II the numerical details of the characteristics of the various parallels (*Almagest* II 6) are replaced by a general discussion, in the manner of THEODOSIUS' *De habitationibus* and similar texts, of the corresponding general results. For instance, the first part of II 9 states that for a parallel further away from the equator the inequality of the days and nights is greater.

The proofs and explanations are of variable length. Here is an example of a proposition (I 9) with a particularly short proof (see fig. 1):[6]

> *Si in circulo continui arcus sumantur et uterque minor semycirculo, dyametrus producta a communi eorum termino lineam*

---

[2]MS Paris BN lat. 16657, f 86r. The following proposition has a similar remark.

[3]MS Paris BN lat. 16657, f. 95v and 103r. All the internal references quoted here appear in other manuscripts.

[4]Manuscript used: Vatican, lat. 2057. For a discussion of the possible dependence of the *Almagestum parvum* on this translation, see below. For the diagram letters in Latin translation from the Arabic see [KUNITZSCH 1991].

[5]See [EUCLID, ADELARD] *passim* and [EUCLID Version II], e.g. p. 163 for V 5.

[6]MS Paris BN lat. 16657, f. 85r.

rectam reliquos eorundem terminos continuantem secabit secundum proportionem corde dupli arcus unius ad cordam dupli arcus alterius.

Fiat ergo linea perpendicularis super semydyametrum BD, et sit medietas corde arcus duplicantis arcum GB. Item sit AZ perpendicularis super eandem dyametrum, et sit sinus arcus AB. Quare fient trianguli GEH et AEZ similes.

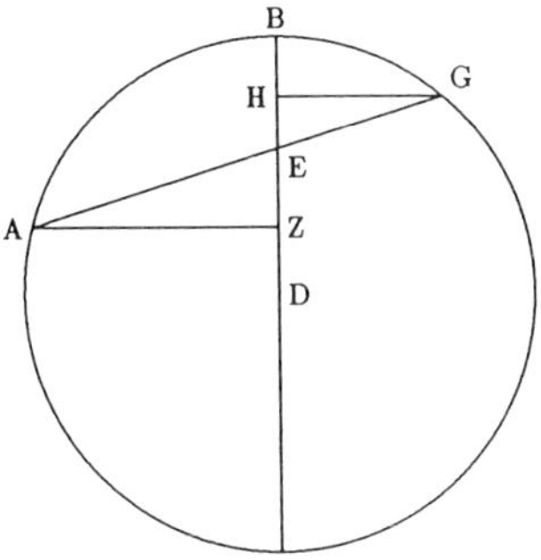

Fig. 1

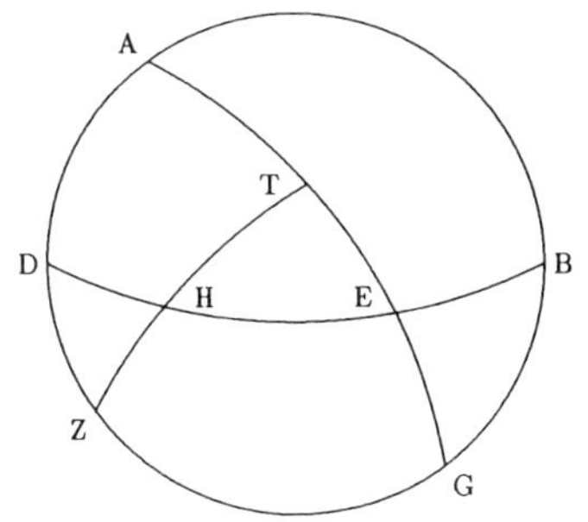

Fig. 2

MS Paris BN lat. 16657 adds, perhaps in the same hand, "ex hoc habebis propositum cum adiutoris xv sexti". One is reminded of the proofs, or rather sketches of proofs, to be found in the Latin text of EUCLID's *Elements* now known as *Version II* and attributed to ROBERT OF CHESTER.[7] Further, the internal citations of propositions by number, three of which are quoted above, are similar to those of *Version II*. For instance, the proof of Pythagoras' theorem consists of the construction of three lines and "Age ergo ex XIIII[a] bis assumpta et ex IIII[a] bis et iterum bis et XLI[a] bis et iterum bis argumentum elice".[8] On the other hand, some passages are long-winded. An example is the passage on finding the chords of various particular arcs — after the theorems have been given and the method made clear.

There are numerous references to AL-BATTĀNĪ (as "albategni" etc.), particularly for his values of astronomical parameters. In VI 4, for instance, on the limits of lunar eclipses, he is cited six times, alternately with PTOLEMY.[9] Other authorities mentioned in the text include THEODOSIUS for his *Sphaerica*, THĀBIT B. QURRA (as "tebit Benchorat" etc.),

[7]See [EUCLID Version II], *passim*. For a discussion of the proofs see pp. 26ff.
[8][EUCLID Version II], p. 130.
[9]MS Paris BN lat. 16657, ff. 124v–125r.

AL-ZARQĀLLU (as "arzacel" etc.) and the Toledan Tables.[10] Perhaps the most interesting of such passages is the reference at the end of III 1 to the Toledan Tables as very recently drawn up by AL-ZARQĀLLU:[11]

> ... et super hoc arzacel tabulas motuum toleti novissime composuit.

At the end of III 11 there is a reference to the theory of trepidation:[12]

> Huius forsan diversitatis causa ex parte esse potuit error in instrumento et ex parte motus octave spere ante et retro.

It is clear that the *Almagestum parvum* is a compilation. Perhaps this accounts for the varying terminology in the treatise. We find, for example, that though *circulus* (or *orbis*) *declivis* and *circulus* (or *orbis*) *signorum* are the usual terms for the ecliptic, *zodiacus* and *circulus zodiacus* are also often used. Again, *elevatio* and *ascensio* are both common for ascension. As for trigonometrical functions, *sinus* and *sinus perfectionis* are used for sine and cosine respectively, and *corda dupli* (or *duplicantis* or *duplantis*) *arcus* for the chord of double the arc. Finally, though the vocabulary is on the whole common for medieval astronomy, that is, largely based on the terminology of the Arabic-Latin translations of the twelfth century, there are some striking examples of words of Greek origin: thus *applani* for heavens in the preface and *amphytritis* in the definitions introducing book II (see below). In book I we find *apodixis* for demonstration alongside *teoreuma*, *demonstratio* and *speculatio*:[13]

> ... ex prime speculationis ratione ... ex eodem teoreumate ...
> ex [s]equentium demonstratione ... ex sequenti apodixi ...

Against "speculationis" MS Paris BN lat. 16657 has "id est propositionis"; and it is clear from the context that each of the four words means theorem or proposition. Perhaps *speculatio* is connected with θεώρημα.

2 Spherical astronomy in the *Almagestum parvum*

For the second form of Menelaus' theorem for spherical arcs (*Almagestum parvum* I 14), a separate diagram, lettered differently to I 13, the first form of the theorem, is supplied. The proof of I 14 is parallel to that of I 13: both prove the collinearity of three points and then say that the result follows

[10]E.g. MS Paris BN lat. 16657, f. 89 for THEODOSIUS, f. 96v (III 1) for THĀBIT, f. 100 (III 11) for AL-ZARQĀLLU, f. 119r (IV 14) for the Toledan tables.

[11]At the end of III 1, MS Paris BN lat. 16657, f. 97r. BIRKENMAJER noticed this passage in his account of the text, [BIRKENMAJER], p. 144.

[12]MS Paris BN lat. 16657, f. 100r.

[13]In the passage on the value of particular chords, MS Paris BN lat. 16657, ff. 83v–84r. For Amphitrite see [HERMANN], p. 335 (under 77rE). For HERMANN's uses of the term see also p. 220 (last word of 77vH).

from previous theorems (see quotation above). PTOLEMY himself simply states the result in the second case for the figure of the first, saying that it follows similarly. In several manuscripts the two diagrams for I 7 and 8, which are the plane cases of I 14 and I 13 and form the bases of their proofs, are labelled respectively "alkata coniuncta" and "alkata disiuncta" (with various spellings).[14] Not only are I 13 and I 14 proved "ex kata disiuncta" and "per kata coniunctam", but the compound proportion in I 13 is described as "sicut in kata disiuncta".[15] After this the *kata* terms are transferred to I 13 and I 14 themselves and the principal theorems in spherical astronomy are proved "per kata coniunctam" or "per kata disiunctam". An example is II 3, in which the altitude of the pole is found (see fig. 2):[16]

> *Altitudinem poli per arcum diei minime notum presto indagare.*
>
> *Regula: si sinum differentie medie diei minimi ad equinoctialem diem ducas in sinum perfectionis quarte orizontis, productumque dividatur per sinum arcus*[1] *orizontis qui est inter ortum tropyci et equinoctialem, atque quod exierit ducatur in sinum quadrantis, productumque dividatur per sinum arcus medii diei, exibit altitudinis poli.*
>
> Supraposita figura denuo assumpta quantitatem arcus ZB, que est altitudo poli, querimus. Igitur per kata disiunctam proportio[2] sinus ET ad sinum arcus AT componitur ex proportione sinus EH ad HB sinum et proportione sinus ZB ad sinum ZA. Quare si ducas[3] primum in quartum et productum dividas per tertium, exibit quiddam quod sic se habebit ad secundum sicut quintum ad sextum. Sed tria nota, duo enim per ypotesim, tercium quia est quarta circuli: ergo quartum notum est, quod intendebamus. Posito ergo arcum diei minimi habere horas rectas ix et dimidiam invenies altitudinem poli esse fere xxxvi graduum.
>
> 1. medii *add.* MS; 2. .1. *add.* MS; 3. ducas] dividas MS.

The term *kata* is to be derived from the Arabic *qaṭṭāʿ* (cutting), the usual term for Menelaus' theorem being *al-shakl al-qaṭṭāʿ*. For this derivation it suffices to compare two translations of the opening words of THĀBIT B. QURRA's treatise on the theorem: where GERARD has[17]

[14]For the spelling we here follow MS Paris BN lat. 16657, f. 84v, and also for the spelling, "kata", when the theorems are quoted (e.g. 87v).

[15]MS Paris BN lat. 16657, f. 84v and 85r. In this manuscript the last-mentioned "disiuncta" is misspelt "diiuncta".

[16]MS Paris BN lat. 16657, ff. 88r–v. The emendations are made with the support of MS Berlin, lat. Q 510, f. 120.

[17][THĀBIT B. QURRA], pp. 159 and 161, the first being quoted from BJÖRNBO's edition and the second from MS Paris Maz. 3642 , 9r–11v (14c) and three other manuscripts.

Quod de figura que nominatur sector dixisti ... intellexi ...

another translation runs

Intellexi quod dixisti super figuram que nominatur catha ...

In fact it is probably through this translation that the indeclinable noun *catha* (or *kata* etc.) was introduced into Latin. Accordingly, the occurrence of similar-sounding terms in THEON's commentary on the *Almagest*, κατὰ διαίρεσιν and κατὰ σύνθεσιν (in each case preposition + noun)[18] appears to be a coincidence.

The usual trigonometrical functions are sine and cosine, but the *kata* theorems and their associated lemmata are expressed in chords: The proof of I 9, given above, is an exception. At the end of the demonstration of I 16, the first application of a *kata* theorem for spherical arcs, both chords and sines are used; the author adds "sinum noto medietatem corde dupli arcus".[19] As in the *Almagest* itself, the tangent functions are not used.

3 The introductions to the six books

In book I the first proposition is on finding the side of an inscribed pentagon and other chords in a given circle (the first theorem in *Almagest* I 10), "Data circuli dyametro latera decagoni pentagoni ... ". The preceding cosmological section of the *Almagest* (I 3–8) is summarized in the introduction, which in MS Paris BN lat. 16657 runs:[20]

[O]mnium recte phylosophantium non solum verisimilibus et credibilibus argumentis sed et firmissimis rationibus deprehensum est formam celi spericam esse motumque ipsius orbicularem circa terram undique secus globosam in medio imoque defixam. Que quidem etsi omnium cadentium tam gravitate corporis quam quantitate ponderis sit maxima ideoque immobilis, ipsius tamen crassitudo comparatione infinitatis applani respectuque distantie fixorum luminum insensibilis et vicem centri obtinere physica indagatione comperta est. Ad hec duos principales et sibimet invicem contrarios motus superiorum sane animadverti etiam fides oculata comprobavit. Quorum alter semper ab oriente in occidentem pari et eadem concitatione per circulos et inter se et ad eum qui omnium spaciossimus equinoctialem parallelos totum mundane machine corpus movet et agitat, cuius

---

MS Maz. has "figura" and "cata". The term *al-shakl al-qaṭṭāʿ* occurs in the title and opening words of the treatise, MS Paris, BN ar. 2457, f. 164r.

[18]See e.g. [THEON], p. 558, line 2. Examples of his use of the term to specify the form of the theorem used in applications to spherical astronomy occur on pp. 573, 589, 621, etc.

[19]MS Paris BN lat. 16657, f. 87r.

[20]F. 82v. In this and the following extracts I have punctuated the text. Where two manuscripts are collated, no notice is taken of orthographic variants.

circumvolutio circa celestis spere polos indefesse consistit. Alter econtrario solem et lunam et quinque erraticas circa alios diversosque polos circumducit et torquet. Hiis firme adeo fides conciliata est, ut si quis etiam iuste columpnians obviet, aut cavillator verum scienter inficians aut mente captus non indigne estimetur. Que cum ita sint, superest ut propositum aggrediamur.

As may be seen by comparison with other specimens reproduced here, the Latinity of this passage is more classical than the workaday, "medieval" language of the rest of the treatise — though the style is variable and is often more tightly constructed in the enunciations than in the proofs. Two phrases perhaps deserve some comment. First, the opening words "Omnium recte philosophantium" come directly from the opening sentence of PTOLEMY's own preface, οἱ γνησίως φιλοσοφήσαντες.[21] The phraseology suggests a Greek rather than an Arabic source. Secondly, *mundana machina* need not be directly connected with the Lucretian *machina mundi* in *De rerum natura* (V 95–96), a work that was little known in the Middle Ages. For the phrase is found in BOETHIUS' *De consolatione philosophiae*, in poems by HRABANUS MAURUS (d. 856)[22] and PETER OF BLOIS[23] (probably late 1150s), and in treatises on the sphere by the thirteenth-century writers SACROBOSCO, who quotes PSEUDO-DIONYSIUS THE AREOPAGITE, and ROBERT GROSSETESTE.[24]

There are other lists of the same astronomical principles, such as the beginning of PSEUDO-THĀBIT, *De recta imaginatione spere*,[25] but there seems to be no direct connection with the *Almagestum parvum*.

The introduction to book II consists entirely of definitions:[26]

Orizon declivis est cui polus elevatur.

Spera declivis est vel obliqua hiis qui orizonte declivi utuntur.

---

[21]I am grateful to Prof. P. KUNITZSCH (Munich) for this identification and also for pointing out that the true philosophers here mentioned are the peripatetics (see [F. BOLL], p. 153). As he observes, they are cited for matters other than the cosmological principles listed at the beginning of the *Almagestum parvum*.

[22][BOETHIUS], III 12, 14; HRABANUS MAURUS in [HRABANUS MAURUS], carmen 39, line 92. I am grateful to DR. G. BERNT (Munich) for these references. He points out that the phrase is to be found in a number of other writers.

[23]PETER OF BLOIS, "A globo veteri ...". The beginning is printed in DRONKE's introduction in [BERNARDUS SILVESTRIS], p. 9.

[24][SACROBOSCO], pp. 12; SACROBOSCO's text pp. 78, 117, 12; in the commentaries pp. 148, 263, 359, 415. The last words of the *Tractatus de spera* are "Propter hoc legitur Dionysius Ariopagita in eadem passione dixisse: Aut deus nature patitur, aut machina mundi [*in some manuscripts* mundana machina] dissolvetur". Other references are to be found in BARONCINI.

[25][THĀBIT B. QURRA], p. 141. For the author see pp. 118–119.

[26]MS Paris BN lat. 16657, ff. 87v–88r.

Cenit capitum est punctum summitatis capitum et est polus orizontis.

Latitudo regionis est distancia cenit capitum ab equinoctiali, et est arcus meridiani inter cenit capitum et circulum equinoctialem interceptus.

Longitudo regionis est distancia eius ab orientis vel occidentis principio, et est arcus paralelli ad equinoctialem inter cenit capitum et eum circulum qui super amphytritis circuitum in celo est dispositus.

Locus notus dicitur cuius longitudo et latitudo nota.

Speralis angulus dicitur angulus ex duobus arcubus in superficie spere proveniens.

Speralis angulus rectus dicitur cui sub duobus arcubus maiorum orbium contento quarta circuli, supra cuius polum ipse angulus consistit, subtenditur.

In the introduction to book III there are only astronomical principles. The introduction to book IV starts with principles and ends with definitions:[27]

[Book III] Comunia quedam premitenda sunt, quia hic modus demonstrationi est aptior.

Perpetuum motum orbicularem esse[1].

Omnem motum celestis corporis simplicem et verum equalem[2] esse, hoc est, super equos angulos in centro motus consistentes et in equales arcus cadentes equalibus fieri temporibus.

Motum solis vel alterius planete in circulo signorum diversum apparere.

Motum stelle medium esse, cum tota et integra eius revolutio secundum equalia tempora per equales motus fuerit distributa.

Hiis premissis ...

[Book IV] Terram ad lune distanciam sensibilem quantitatem habere, ideoque ad speram lune vicem centri non optinere.

Lunam ab orbe signorum, et ad meridiem et ad septentrionem, declinare et ad orbem signorum reverti.

Circuitiones lune in longum tempore diversas esse.

Circuitiones lune in latum tempore diversas esse.

[27]MS Paris BN lat. 16657, ff. 96r–v and 103r–104r. The emendation is made with the support of MS Berlin lat. Q. 510.

Lunam in omni parte circuli signorum triplicem secundum visum motum habere, modo velociorem, modo[3] mediocrem, modo tardiorem.

Umbram terre semper a solis opposito soli similiter et equaliter moveri.

Lunam a sole menstruum lumen habere.

Faciem lune soli obversam semper a sole illuminari.

Umbram terre causam lunaris eclipsis esse.

Lunam soli et aspectui interpositam defectus solaris causam esse.

Equalis lunatio dicitur reditus lune ad solem secundum utriusque motum medium.

Mensis est equalis lunationis tempus.

Locus verus lune in celo est punctum celi cui linea a centro terre per centrum lune educta in celum occurrit.

Locus lune verus in circulo signorum est communis sectio duorum orbium, quorum unus est ipse orbis signorum et alius magnus orbis per polos circuli signorum et locum verum lune in celo transiens.

Et latitudo lune est arcus istius circuli inter verum locum lune in circulo signorum et verum locum lune in circulo deprehensus.

Motus longitudinis est loci lune in celo, vel in circulo signorum, progressio.

Motus latitudinis est a sectione communi circuli signorum et circuli declinantis lune elongatio.

Motus diversitatis est lune in epyciclo sive lune in ecentrico, cum propter alterutrum istorum modorum diversum motum habeat, ambulatio.

Nodi sunt sectiones circuli signorum et circuli declinantis lune.

Caput est nodus ille per quem transit luna a meridie in septentrione; cauda est nodus oppositus.

1. Celestia corpora perpetuo motu et ideo orbiculari esse mobilia *add. BN lat. 16657 marg., MSS Erfurt Ampl. F 383 Toledo 98–22 in textu*; 2. equalem] *corr. ex* equabilem; 3. qualiter moveri *add. BN lat. 16657.*

The introduction to book V, "Locus lune secundum longitudinem ...", and the greater part of the introduction to book VI, "Superlatio [elongation] lune ad datum tempus ..." consist of definitions. Evidently, it was the style of the treatise to express principles with accusative and infinitive and definitions in the indicative. Thus the last part of the introduction to book VI reads:[28]

> Peticiones due sunt.
>
> Tenebras in circulo lunari ad eam partem orizontis declinare in quam vergit arcus circuli magni transeuntis per duo centra lune et umbre.
>
> Tenebras in solari circulo ad eam partem orizontis declinare in quam cadit arcus circuli magni transeuntis per centrum solis et visum locum lune.
>
> Atque hee declinaciones flexus tenebrarum in utraque eclipsi dicuntur.

4 The manuscripts

There are three main groups of manuscripts. Of these the first two are distinguished by the treatment of the preface. In the first group, which we will call A, and which has the preface essentially as reproduced above, are the following manuscripts:[29]

> Paris, Bibliothèque Nationale lat. 16657, ff. 82v–132r, before 1250. ("minor almagesti", 82v). Expl.: "Explicit hic sextus liber et sexti glosa textus".
>
> Vatican, Reg. lat. 1261, ff. 1r–49r, perhaps late 14c. "incipit liber primus almagesti"; ("liber almagesti demonstratus", list of contents, f. 0v). Very close to MS Paris BN lat. 16657.
>
> Cracow, Jag. Library 1924, pp. 9–163, 13c, (first half?).[30]
>
> Florence, Biblioteca Laurenziana, conv. soppr. 414, ff. 1r–45r, ?14c. ("Incipit liber albategni qui dicitur almagesti parvum", f. 1r). Expl.: ("Explicit hic sextus liber et sexti glosa textus").
>
> Prague, Public and University Library V. A. 11 (802), ff. 1r–59v, 15c. Expl.: "explicit liber almagesti sextus".[31]
>
> Memmingen, Stadtbibl. F 33, ff. 152r–198v, 15c. The text breaks off on f. 152v and begins again on f. 153r. Expl.: "Explicit liber sextus minoris almagesti".

---

[28]MS Paris BN lat. 16657, ff. 122r–v.

[29]The following list contains transcriptions of attributions and titles that I have found in the manuscripts. The names of authors and titles written in a hand different to that of the text appear in brackets.

[30]This manuscript has no "non solum" in the first line and has "inuste" for "etiam iuste" near the end of the preface.

[31]According to [BJÖRNBO, VOGL], p. 129, note 3, the treatise is called *Paruum Almagesti Ptholomei demonstratum per Campanum de primis 6 libris*, but I cannot find this in the film available to me.

Nuremberg, Stadtbibl. Cent. VI 12, ff. 1r–66v, 15c. Written by REGIOMONTANUS.[32]

Leipzig, Universitätsbibliothek 1475, ff. 2r–51v, 15c. Expl.: "Explicit liber sextus minoris almagesti".

Utrecht, Univ. Libr. 725, ff. 4v–8r, 15c. Only a fragment, many propositions without proof. Numerous references to the *Epitome* of PEURBACH and REGIOMONTANUS. Inc.: "Incipit Almagesti abbreviatus ..."

The second group, B, has a slightly different wording in this passage:[33]

> [O]mnium recte philosophancium verisimilibus coniecturis credibilibus argumentis sed et firmissimis rationibus deprehensum est formam celi esse spericam motumque ipsius orbicularem circa terram undique secus globosam in medio imoque defixam. Que quidem etsi omnium cadentium tam quantitate corporis quam gravitate ponderis maxima[1] sit ideoque immobilis, ipsius tamen crassitudo comparatione infinitatis applani respectuque distancie fixorum luminum insensibilis et vicem centri optinere fisica indagatione comperta est. Ad hec duos principales et sibimet invicem diversos superiorum motus sane animadverti etiam occulta fides comprobavit. Quorum alter semper ab oriente in occidentem pari atque eadem concitatione per circulos et inter se et ad illum qui omnium spaciossimus est equinoctialem parallelos totum mundane corpus machine movet atque exagitat, cuius revolutio circa celestis spere polos indefesse consistit. Alter vero solem et lunam et alios quinque erraticos circa alios diversosque polos circumducit et torquet. His ferme fides adeo consiliata est, ut si quis iniuste columpnias obviet aut potius deviet aut cavillator verum scienter inficians aut in huiusmodi disciplina parum excertitatus non indigne estimetur. Que cum ita constent, superest ut propositum aggrediamur.
>
> 1. Seq. ras.

Manuscripts in this group are

Berlin, Staatsbibl. lat. Q 510, f. 114r–175v, 13c. ("Almagesti", list of contents, flyleaf, f. -2v; "Item almagesti tholomeus cum aliis ... Almagesti libri 6 ... ", f. 0v; "Incipit Liber tercius Almagesti", f. 175v).

Erfurt, Ampl. F 383, ff. 1r–51v, 14c. ("Liber magni astrologi in astronomicis", f. 1r; "liber quidam astronomicus ... cuius autor est Geber magnus astrologus".)

---

[32] This manuscript has "iniuste" for "etiam iuste".

[33] Berlin, Staatsbibliothek lat. Q 510, f. 114r.

Paris, Bibliothèque Nationale lat. 7399, ff. 15v–93v, 14c.

Erfurt, Ampl. Q 356, ff. 1r–101v, 15c. Stops in the course of VI 8.

Vienna, Österr. Nationalbibl. 5273, ff. 35v–95v, 16c. "in huius ... exercitatus" is underlined and "aut mente captus" is written over it.

Of the characteristic variants the most striking come towards the end: B has "iniuste" for A's "etiam iuste"; "aut potius deviet" is in B and not in A; and B has "in huiusmodi disciplina parum exercitatus" for A's "mente captus". The phrase "aut potius deviet" seems to be an explanation of the preceding word "obviet" and it may be supposed that it stood as a note in a manuscript and was worked into the text in the process of copying. Further, of the two readings "mente captus" and "in ... exercitatus" the first appears to be the *lectio difficilior*. On both counts B seems to be inferior. As for A's "etiam iuste" for "iniuste", which is preferable, it is possible that a misreading occurred early in the A-transmission, for one of the oldest manuscripts in this group, Cracow 1924, has "iniuste".

Manuscripts of a third group, C, have a characteristic extra passage, just before I 9, on the construction of a table of chords:[34]

> Quia tamen earum numeris et quantitas facilius ex oculo in subiecta figura deprehenditur ... et ad hunc quidem modum tabule ordinentur.

There is also an alternative passage a little before this. The preface is more like A than B. The manuscripts are:

London, Brit. Library, Harley 625, ff. 84r–123r, 132r–135v, 14c. ("almagesti abbreviatum", f. 0r; "hec conclusiones sunt de libro 5to almagesti abbreviato", f. 132r).

Vatican, Reg. lat. 1012, ff. 1–75r, but several folios are missing between ff. 73 and 74. Expl.: "explicit liber sextus" ("amen dico amen ... totus liber continet 146 conclusiones").

Vienna, Österr. Nationalbibl. 5292, ff. 1r–53v, and misplaced folios or commentary up to f. 65v, 15c. ("Epitome Alberti Magni in almagesti ptolomei" on flyleaf).

Dresden, Sächs. Landesbibl. Db 87, ff. 104r–161v, finishing in the middle of the last proposition, 15c. (Same hand? "Incipit almagesti demonstratum de sex primis libris ptolomei", f. 104r; "... almagesti demonstrati ...", at end of book I, f. 108r).

Munich, Bayerische Staatsbibliothek CLM 56, ff. 3r–150r, 15c. Expl.: "Explicit Almagesti minor finitus anno Chrisiti 1434°".

[34] VI 5292, ff. 2v.

("Primo almagesti abbreviatum per magistrum Thomam de Aquino et continet libros sex incipit folio 3°").

Vienna, Österr. Nationalbibliothek 5266, ff. 176r–228v, 15c. Expl.: "Explicit Almagesti minor ... 1434".

In the last two manuscripts the preface is a mixture of A and B traditions.

A fifteenth-century fragment that contains only the passages on instruments is

Vienna, Österr. Nationalbibliothek 5258, ff. 75r–77r, 15c.

5 Irregular texts, commentaries, influence

In several manuscripts the text begins "Omnium recte philosophantium ..." and continues with the usual enunciations, but with different proofs or with proofs that are sometimes different. In other manuscripts the proofs are always different and the enunciations are equivalent to those of the *Almagestum parvum* as described above, but not always in the same words. Thus there is a gradation from slightly irregular text to commentary or completely different text. The manuscripts of these kinds are:

Toledo, Cathedral Library 98–22, ff. 67ra–80vb, 13c, possibly first half.[35] "B" preface.

Basel, Universitätsbibliothek F II 33, ff. 221r–244r, 14c. The folios are out of order. ("Parvum Almagestum", list on f. 0v). Expl.: "Explicit liber Jeber per manus Engelberti. Deo gratias", but "Jeber" is partly erased and there is a note in a later hand "falsissimi scriptoris quia non est verbum correctum nisi sumit malum exemplar". Mixed "A" and "B" preface.

Florence, Ricc. 885, ff. 109r–123v, late 14c. - early 15c. ("almagesti", f. 123r, top margin). Only first five words of preface, then "etc".

Erfurt, Ampl. F 375, ff. 85r–88v, 93r–112v, 113r–126v, 127r–129v, 14c.

Erfurt, Ampl. F 393, ff. 63r–80v.

In the Toledo manuscript there is great divergence from the standard text (if MS Paris BN lat. 16657 may be so called) in book I. Spot checks have not revealed differences other than scribal variants in the other books — though of course a more detailed study may do so.[36] The divergences in book I are not uniform. Thus in Menelaus' theorem for spherical arcs (I 13,

[35][MILLÁS VALLICROSA], p. 202, gives the date as fourteenth century, but this does not seem very likely.

[36]The same goes, of course, for similar remarks made in this preliminary survey, especially in this section.

14) Toledo 98–22 follows the standard text until the penultimate sentence; then in the first case (the second is similar) instead of "Hac igitur linea protracta restat ex kata disiuncta ..." (etc. — see above), there follows "Qua protracta sic argumentare. A puncto A ...". In both cases an elaborate proof follows. In other theorems the treatment appears quite different from the standard — sometimes with different lettering of the diagram (e.g. G and D are interchanged in Prop. 3, and H and T in Prop. 6). An example of more detailed treatment, but with traces of the standard wording is the proof of I 9:[37]

> Verbi gratia in circulo ABG AB et BG arcus continuantur, a quorum communi termino diameter BD procedit, cui perpendicularis GH linea medietas corde arcus duplicantis arcum GB; itemque sit AZ perpendicularis super eamdem diametrum, et sit sinus arcus AB. Quare fient trianguli GEH et AEZ similes. Ergo proportio AZ ad HG est tanquam proporcio AE ad EG per 4 6i; ergo BD sic secat AG in puncto E, ut predictum est. Est autem AG linea duos terminos non continuos continuans. Quare propositum habetur.

Much the same could be said of I 7, in which the standard text's "... proportio GD ad EH, inter quas ZD linea statuatur" (implied is $GD : EH = GD : ZD \cdot ZD : EH$) is also found in the Toledo manuscript, followed by the explanation:[38]

> quoniam proportio GD ad EH constat ex proportione GD[1] ad ZD et ZD ad EH per epistolam Hameti de proportionalitate et per librum Walteri flandrensis[2] de proportionibus ...
>
> 1. GD] GZ MS; 2. Walteri flandrensis (?)] *corr. ex* Walterem flandrensem.

The two names are of some interest: the *Epistola de proportione et proportionalitate* by AḤMAD B. YŪSUF was translated from Arabic into Latin by GERARD OF CREMONA,[39] and there is a copy in Toledo 98–22 just after the *Almagestum parvum*; but the second name is not known among the authors of books on proportions. Suffice it to say here that in the proof of I 13 (f. 68va) the transformation (in modern terms) of $a : b = c : d \cdot e : f$ into $c : d = a : b \cdot f : e$ is proved "ex libro Walteri". Accordingly, WALTER must have written a book on similar lines to THĀBIT's *De figura sectore* or the third part of the AḤMAD B. YŪSUF treatise just mentioned.

In the Basel manuscript the divergences also appear to be mostly at the beginning of the text. Some diagrams are differently lettered, the letters

[37]Toledo, Cathedral Library 98–22, f. 68r.

[38]Toledo, Cathedral Library 98–22, f. 68r.

[39]For the critical edition with commentary, see [AḤMAD B. YŪSUF].

sometimes reflecting the order of the Latin alphabet. An example of a different proof is that supplied for I 9. The beginning runs:[40]

> Sint arcus *AB BG* (?), *AG* sit corda conterminans terminorum illorum, *BD* sit dyameter illius circuli *AG*, ad ad *BD* protrahatur perpendiculariter ab *A* ad *G* alia perpendiculariter ad *BD*, et sit *A*, et *E* similis sectio *AG* et *BD* ...

Clearly, the text has suffered at the hands of the scribes.[41]

MS Ricc. 885 appears to represent a more radical revision, though it stops short in the middle of II 35. There are numerous references to THEODOSIUS as well as to EUCLID.[42] As a specimen I 9 is again reproduced (f. 111v):

> Verbi gratia sit circulus *ABG*, cuius duo arcus continui sint *AB BG*, quorum uterque minor sit semicirculo — quod hic et in sequentibus intelligendum, quare detur memorie. Et sit *D* centrum circuli et linea continuans terminos predictorum arcuum *AG*, quam secet diameter *DB* in puncto *E*. Per duodecimam ergo primi ducatur a puncto *A* [linea] ortogonaliter super *DB*, que sit *AZ*, et a puncto *G* alia, que sit *GH*. Eritque *AZ* corde medietas arcus *AB* duplicati, quod ostendetur per secundam partem tertie tertii et 4 primi et 27^{am} 3^{i}, propter quod dicitur corda mediata. Dicitur etiam ... *ZB* vero dicitur sagitta sive sinus versus ... Probatio: trianguli *AEZ HEG* sunt equianguli ...

With the Erfurt manuscripts we find ourselves further from the standard text. There is no preface, the proofs are different and there is sometimes little left even of the skeleton of the enunciations; none of these texts gets past the second book. In MS F 383, ff. 85r–88 and 93r–112v, *Almagestum parvum* enunciations, sometimes with proofs, appear in the margin of what appears tp be the *Almagest* in GERARD's translation from Arabic into Latin (incipit: "Summa vero principiorum que oportuit nos incipere ... "[43]). There is a table of chords on f. 88v. Folios 113r–126v contain another text, ending in the course of book II, and folios 127r–129r have yet another text. In the last the enunciations are shortened, as in I 1:

> Data circuli dyametro, etc. Et dividatur ...

or substituted by others of equivalent meaning.

[40]MS Basel, F II 33, f. 222r.

[41]The punctuation has been added, but it is hard to see how to make sense of the extract in any form.

[42]E.g. "... adiuvante sexta primi et 6ª secundi Theodosii de speris" in II 8, f. 117v.

[43]Cf. the *Almagest* manuscript Vatican, lat. 2057, f. 8v.

Some of the *Almagestum parvum* enunciations are recognizable in a very late manuscript, Vatican, Pal. lat. 1380, ff. 110r–134r (explicit: "Et sic finunt demonstrationes dictionis prime Almagesti in dei nomine Amen") and on the flyleaf of MS Sinclair 224 of the State Library of Victoria, Australia.[44]

We come now to a series of named texts, with known authors, that use the *Almagestum parvum* enunciations for chord theorems, beginning with the characteristic "Data circuli diametro". First, JOHANNES VON GMUNDEN, the founder of the school of mathematics in Vienna, used the enunciations to compile an elementary treatise on sines and cosines in 1437.[45] The tenuous connection with the *Almagestum parvum* is pointed up by the way the author gives his source:[46]

> ... ad praesens praemittam 6 propositiones quae etiam praemittuntur in principio primi libri Almagesti minoris. Et habuntur finaliter etiam in Almagesti maiori locis et capitulis per allegatum.

Parts of the *Tractatus de sinibus, chordis et arcubus* were taken over word for word by GEORG PEURBACH for his *Tractatus super propositiones Ptolemaei de Sinubus et chordis*,[47] printed in Nuremberg in 1541; and other parts were taken over and altered.[48]

Secondly, in Brussels, Royal Library 1022–1047, ff. 184v–197v, there is a treatise entitled:[49]

> Expositio practice tabule tabularum et propositionum Ptolomei pro compositione tabule sinuum et chordarum necessariarum facta simplicibus per me paulum de gherisheym.

The six propositions begin on f. 189v; on f. 196r the text ends with "Sequitur nunc ipsa tabula rectificata anno domini 1443"; and the table is on ff. 196v–197v.

More influential than any of these was the *Epytoma in Almagestum* by PEURBACH and REGIOMONTANUS. Since PEURBACH was supposed to have compiled the first six books, it is not surprising that, with the exception of the introductory material, it shows great affinity with the *Almagestum parvum*. It is written in the form of proposition and proof, and many of

[44]The last two items I have gleaned from manuscript additions to [CARMODY], apparently in the hand of the author or of F. BENJAMIN. Also mentioned is BE lat. Q 536, 14c. 25b–(47), but Berlin, Staatsbibl. lat. Q 536, ff. 25v–46v is another text.

[45]Edited in [JOHANNES VON GMUNDEN].

[46][JOHANNES VON GMUNDEN], p. 95.

[47][JOHANNES VON GMUNDEN], p. 75.

[48]Copy used: British Library, shelfmark C 78 c 14.

[49]F. 184v. The entry in the list of contents written on the flyleaf is "12. Expositio tabule tabularum M. pauli gerissem". MS Brussels, Royal Library 2962–2978, ff. 211r–212r is a small fragment of this treatise.

the enunciations are similar to those of the *Almagestum parvum*. There is similar material, too, after the theorem on shadows in book II (*Almagest* II 5). The *Epitome* "sometimes follows the *Almagestum minor* nearly word for word, including all its supplements to Ptolemy".[50]

Of the more regular type of citation[51] of the *Almagestum parvum* many are to be found in the *Albion* by RICHARD OF WALLINGFORD.[52] It was no doubt the rich material on eclipses that attracted his contemporary JOHANNES DE JANUA to use it in the compilation of his *Canones eclipsium* (1332). For the *explicit* reads:[53]

> Expliciunt canones eclipsis, quos Magister Iohannes de Ianua compilavit, extrahendo eos partim a canonibus communibus, partim ab Albategni, partim a minori Almagesti, partim a magistro Iohanne de Cecilia in scripto suo super tabulas Ptholetanas et specialiter quantum ad puncta eclipsis minuta casus ac etiam minuta more, AD 1332 in completo 22. die Ianuarii.

6 Comparison with Latin translations of the *Almagest*

It has hitherto been assumed that the *Almagestum parvum* was based upon the translation of the *Almagest* by GERARD OF CREMONA. Thus NALLINO, BIRKENMAJER and HASKINS took 1175 (the date of the translation according to one manuscript) as the *terminus post quem* for its compilation.[54] But comparisons with GERARD's translation scarcely reveal close dependence. The bulk of the preface goes back, ultimately, to the first book of the *Almagest*: from the chapter headings, from extracts from the various chapters or from the list of cosmological principles at the end of *Almagest* I 2 supplemented with a statement about the two principle celestial motions. Of these the last looks the most promising, the supplement being taken from the body of chapter I 8. Since GERARD himself revised his translation, the extracts are taken from MS Vatican, lat. 2057 (13c.) as representive of the revised text with variants from MS Paris, Bibliothèque Nationale 16200 as representative of the unrevised text.[55]

[50] [HELMAN & SWERDLOW], p. 477a.

[51] No attempt has been made to collect these, which must be numerous.

[52] [WALLINGFORD], I 208ff. and II 140.

[53] MS Oxford Digby 97, f. 128v. See [AL-BATTĀNĪ], I xxxvi. NALLINO also mentions (p. xxvii) a *Canones eclipsium* by one JOHANNES DE CAPUA. Whether these are really different I do not know.

[54] See [AL-BATTĀNĪ], p. xxvii; [BIRKENMAJER], p. 144; [HASKINS], p. 104. MS Florence, Bibl. Med. Laur. 89 sup. 45 (13c., beginning) has the date 1175 in the colophon (see [KUNITZSCH 1990], p. 16).

[55] See [KUNITZSCH 1990], p. 4. For a description of the manuscripts see pp. 17 and 12. The passages are MS Vat., ff. 5v35–38 and 8r2–5, 6–8 and MS Paris, ff. 2v, 4v.

> ... quod celum est[1] spericum et motus eius spericus, et quod etiam figura terre cum omnibus suis partibus secundum sensum est sperica, et locus eius in medio totius celi similis est centro, [2]et quod ipsa secundum magnitudinem et spatium est quasi punctum ad orbem stellarum fixarum, et quod ipsa non habet motum localem[2].
>
> ... primi motus celi duo, Quorum unus est qui movet totum semper ab oriente ad[3] occidentem uno modo et revolutionibus equalis velocitatis super orbes adinvicem equidistantes quos duo poli spere celestis revolvunt, quod[4] totum secundum equalitatem volvunt[5], horum autem orbium maior[6] nominatur orbis equationis diei propterea[7] quod ... Alter vero motus est qui movet orbem stellarum currentium contra motum primum videlicet ab occidente ad[3] orientem supra duos alios polos, et non supra polos eius ...
>
> [variants from BN 16200] 1. est] sit; 2–2. et quod ipsa secundum ... localem] et quod ipsa non habet motum localem, et quod ipsa secundum magnitudinem et spatium est quasi centrum quantum ad orbem stellarum fixarum; 3. ad] in; 4. quod] qui; 5. volvunt] revolvunt; 6. maior] maximus; 7. propterea] propter.

For comparison, here are the equivalent passages from the translation made in Sicily from the Greek:[56]

> Quoniam celum et spericum est et sperice fertur. Et quoniam terra figura quidem et ipsa sperica est ad sensum ut secundum omnes partes sumpta. Positione vero media totius celi iacet centro similiter magnitudine aut distantia et puncti rationem optinet ad fixarum speram, ipsa nullum motum faciens transitivum.
>
> ... quoniam due differentie primorum motuum sunt in celo, una quidem sub qua ab oriente in occidentem feruntur universa, semper eodem modo et equa celeritate circumductione faciente secundum equidistantes sibi invicem circulos, huius equaliter universa circumagentis manifestum quoniam scriptos spere polis, quorum maximus circulus equinoctialis appellatur ... altera vero secundum quam stellarum spere secundum contraria predicte lationis circa polos alteros et non eosdem eis qui prime circumductionis quasdam faciunt transmotiones.

A third text, in MS Dresden, Sächs. Landesbibl. Db 87, ff. 1r–71r (c. 1300), supposedly a translation from Greek[57] reads (ff. 2r, 5v):

---

[56]For this text see [Haskins], pp. 157ff. The extracts are from MS Vatican, lat. 2056 (13c.), ff. 2r, 3v–4r.

[57]See [Haskins], p. 109–110.

> quia celum est sperale et suus motus est speralis motus, et terra est speralis est in medio celi et super suum centrum, et mensura sue magnitudinis ad speram fixarum stellarum quasi punctus, nec habet motum secundum locum.
>
> ... quia primi motus qui sunt in celo sunt duo. Unus est motus quo movetur celum ab oriente in occidentem super duos polos, super quos volvuntur omnes parallelici circuli, qui sunt lineati super illos duos polos, equali revolucione. Eorum vero maior circulorum nominatur rectitudo diei ... Alter vero motus est motus stellarum contrarius parti quam diximus esse, ab occidente in orientem super duos polos alios qui non sunt duo poli quos prediximus.

In no case is there any particular similarity to the *Almagestum parvum*. Another possible source for the preface and the introductions to the other books is the treatise on astronomy by ALHAZEN (IBN AL-HAYTHAM) edited by MILLÁS[58] or the corresponding passages in the *Astronomia* of HERMANN OF CARINTHIA.[59] But the wording fits no better than the *Almagest* versions given above. The same goes for the astronomy of JĀBIR B. AFLAḤ in the translation of GERARD OF CREMONA.[60]

We may also compare proposition I 9, as an example of a mathematical theorem:[61]

> [Vat. lat. 2057] [D]escribam etiam circulum supra quem sint *A B G* super[1] centrum *D*, et separabo[2] ex circulo duos arcus *AB BG*, et ponam unumquemque eorum semicirculo minorem — et similiter [3]omnis arcus quem separabo[3] in sequentibus, hanc ergo exceptionem[4] memorie·commendemus —, et protraham duas lineas *AG DEB*. Dico ergo quod proportio *AE* ad *EG* est sicut proportio corde dupli arcus *AB* ad cordam dupli arcus *BG*. Quod sic probatur: protraham enim duas perpendiculares[5] a duobus punctis *A G* ad lineam *DB*, que sint *AZ GH*. Et quia *AZ GH* equidistant et cadit super eas linea *AEG*, erit proportio *AZ* ad *GH* sicut proportio *AE* ad *EG*. Proportio vero *AZ* ad *GH* est sicut proportio corde dupli arcus *AB* ad cordam dupli arcus *BG*, quoniam unaqueque est medietas corde sui dupli.

[58][MILLÁS VALLICROSA], second appendix, pp. 285–312.

[59]The *Astronomia* was the subject of a talk by Prof. R. LEMAY at Hamburg, 1989. I should like to thank him for giving me a complete copy of his edition of this work, which includes a comparison with the ALHAZEN treatise.

[60]There are also passages equivalent to most of the other extracts reproduced here, but they are again no nearer than the *Almagest* quotations. It should be noticed that JĀBIR replaced Menelaus' theorem in its entirety, so that he is an unlikely source.

[61]MSS Vat. lat. 2057, ff. 15r–v; Paris BN lat. 16200, ff. 12vb–13ra; Vat. lat. 2056, f. 8r; Dresden Db 87, f. 12r. The last three words from MS Vat. lat. 2056 are not certain, being very abbreviated.

Ergo proportio AE ad EG est sicut proportio corde dupli arcus AB ad cordam dupli arcus BG. Et illud[6] est quod demonstrare voluimus.

[variants from BN 16200] 1. super] supra; 2. separabo] dividam; 3. omnis arcus quem separabo] dividentur omnes arcus; 4. ergo exceptionem] exceptionem ergo; 5. perpendiculares] lineas perpendiculares; 6. illud] hoc.

[Vat. lat 2056] Rursum esto circulus ABG, cuius centrum D; summanturque in ipsius circumferentia quelibet tria puncta A B G ita ut utraque periferiarum AB et BG semicirculo sit minor ... et ea ergo que recte AE ad EG proportio eadem est proportio eius quod sub dupla perifie AB ad eam que sub dupla perifie BG. Quod oportebat demonstrare.

[Dresden Db 87] 12r. Rursus lineabimus circulum ABC, cuius centrum sit D; et assignabimus super circularem iii puncta A B C, sitque quisque duorum arcuum AB BC minor medietate circuli; et copulabimus duas lineas AC DEB. Dico ...

The last two extracts did not need to be given in full, since the terminology of the first and the diagram-lettering of the second eliminate them. In fact the Dresden manuscript regularly has the diagram letters in the order of the Latin alphabet — a fact that suggests reworking in Latin. MS Vat. lat. 2056 often, perhaps always and certainly in this case, has N for the Greek **H**. Little similarity between the GERARD texts and the *Almagestum parvum* is to be seen: even the similarity of the triangles, which is about all the *Almagestum parvum* gives as proof, is not explicitly mentioned by GERARD. We may notice in passing that MS Florence, Riccardian 885, alone of all the versions of the *Almagestum parvum* that we have seen, has the remark (albeit altered) that the arcs in future will be less than a semicircle. It is likely that the writer, who appears to have elaborated the "standard" *Almagestum parvum*, had access to the *Almagest* as well.

As an example of spherical astronomy, let us compare the *Almagestum parvum* II 3, given above, with the equivalent in GERARD's versions of the *Almagest*:[62]

[Vat. lat. 2057] *Qualiter sciatur altitudo poli per*[1] *hos arcus*[2] *cum ponuntur arcus propter quantitatem diei longioris et econtrario*[3].

Sit illud ergo[4] datum et investigemus inventionem altitudinis poli que sit arcus BZ orbis meridiei. Erit ergo in hac forma proportio corde dupli arcus ET ad cordam dupli arcus TA agregata est ex duabus proportionibus ex proportione corde dupli arcus EH ad cordam dupli arcus HB et ex proportione corde dupli

[62] MSS Vatican, lat. 2057, f. 18v, and Paris, Bibliothèque Nationale lat. 16200, ff. 15vb–16ra.

arcus BZ ad cordam dupli arcus ZA. Duplum vero arcus ET est 37 partes et 30 minuta et eius corda est ...

[variants from BN 16200] 1. per] propter; 2. arcus] *om.*; 3. econtrario] *om.*; 4. illud ergo] ergo illud.

After the values of the various arcs and their chords have been set down, the ratio Cd $2EH$: Cd $2HB$ (where "Cd" means "chord") is "thrown out from" (effectively "divided into") the ratio Cd $2ET$: Cd $2TA$, "Cum ergo proiecerimus ex proportione [Cd $2ET$: Cd $2TA$] proportionem [Cd $2EH$: Cd $2HB$]" to yield the ratio Cd $2BZ$: Cd $2ZA$.[63] Now since ZA is known, since it is a quadrant, BZ, the altitude of the pole, is also known. Now in the *Almagestum parvum* there is no "throwing out" of ratios to eliminate the compound ratio so that a simple proportion is left, but only the multiplication and division of numbers. This proposition is very different indeed in the two texts.

Before leaving mathematical terminology, let us compare the expressions in the *Almagestum parvum* and the *Almagest* versions for what in modern terms would be $x^2 = 2y^2$ or $x^2 = 3y^2$ (or "x when squared is twice or thrice y when squared"):[64]

[*Almagestum parvum*] ... latus tetragoni, existens 90 partium corda, quadratum medie dyametros duplat potentialiter; latus item trigonale, ens 120 graduum corda, medie dyametros quadratum potentialiter triplat ...

[*Almagest*] ... latus quadrati, quod subtenditur 90 partibus, est in potentia duplum medietatis diametri; et latus trianguli, quod subtenditur centum et viginti, est in potentia triplum medietatis diametri ...

The expressions may be represented as

[*Almagestum parvum*] [x] quadratum [y] potentialiter duplat/ triplat

[*Almagest*] [x] est in potentia duplum/triplum [y]

In each case x is nominative and y genitive. The equivalent expressions in Greek and Arabic are:[65]

[x] δυνάμει διπλασία/τριπλασίων [y]

[x] *huwa fī 'l-quwwa mithlā* [y] / *thalāthat amthāl* [y]

---

[63] The Arabic is *fa-idhā alqaynā min nisbat ... nisbat ...* (MS Leiden, ar. 680, f. 14r). For this expression see [WKAS], p. 1191a30ff, under root *lqy*, 4th form.

[64] The *Almagestum parvum* is reported from MS Paris BN lat. 16657, f. 84r; the *Almagest* from MS Vatican, lat. 2057, ff. 9r lines 3–4 and Paris BN lat. 16200, ff. 5v lines 23ff.

[65] [EUCLID, Greek], p. 35, lines 6–11; MS Leiden, ar. 680, f. 5v, lines 5–6.

Not unexpectedly, the translation from Greek in MS Vatican lat. 2056 has "[x] potentia dupla/tripla est [y]". The Dresden text spells out the meaning, as it were, by starting with "quadratum [x]".[66] In these *Almagest* versions, therefore, there is no model for the construction in the *Almagestum parvum.*

The Ptolemaic instruments are fully described in the *Almagestum parvum*. For the armillary instrument in book V, for instance, the description begins:[67]

> Queruntur primum due armille convenientis[1] mensure orbiculares ambe similes et equales per omnia, et sit utriusque tam interior, que centrum respicit, quam exterior superficies politissima et equalis per totum latitudinis et utraque armilla eiusdem ubique spissitudinis; et sic inseratur altera alteri ut sese secent ortogonaliter et per equalia.
>
> 1. convenientis] convenientes MS

In GERARD's *Almagest* this passage reads:[68]

> [Vatican, lat. 2057] Accipiemus[1] duas armillas mensurate magnitudinis sapienter et bene rasas, quarum superficies quadrate equales similes omni modo, et componemus unam earum in[2] alteram in duobus locis oppositis orthogonaliter super superficies suas.
>
> [Variants in Paris BN 16200] 1. Accipiemus] Accipias; 2. in] infra.

These texts hardly seem related here. Nor does the translation from Greek in the Vatican, "Duos circulos sumentes diligenter tornatos ..."[69] provide any more similar readings.

Finally, let us compare the descriptions of the second instrument in book I for finding the obliquity of the ecliptic. We notice, incidentally, that the description of the two instruments is displaced in the *Almagestum parvum*, coming after the *kata* propositions instead of immediately after the calculation of chords, as in the *Almagest*. In the *Almagestum parvum*:[70]

> [*Almagestum parvum*] Paratur et aliud commodius et facilius instrumentum, laterem scilicet ligneum vel lapideum vel eneum, quadratum, quere, cubitalis latitudinis et apte altitudinis, ut super latus sine tortuositate et inclinatione erigi possit; sitque

[66]MSS Vatican, lat. 2056, f. 4v, line 17ff; Dresden Db 87, f. 7r, line 1ff.

[67]MS Paris BN lat. 16657, f. 111r. The emendation is made with the support of MS Berlin, Staatsbibl. lat. Q 510, f. 145v.

[68]MS Vatican, lat. 2057, f. 68r; Paris BN lat., f. 64r.

[69]MS Vatican, lat. 2056, f. 41r. The Dresden text stops with book IV.

[70]*Almagestum parvum* from MS Paris BN lat. 16657, f. 86v; the emendations are made with the support of MS Berlin, Staatsbibl. lat. Q 510. The *Almagest* extract is from MSS Vatican, lat. 2057, ff. 14v–15r, and MS Paris BN lat. 16200, ff. 12rb–va.

una superficierum levissima et equalis. Positoque centro in uno angulorum super ipsum quartam circuli describe[1] et ab eo centro duas lineas rectas angulum rectum continentes et quartam circuli includentes protrahe, et quartam circuli in xc partes et unamquamque partium in minuta quot poteris partire. Deinde duas pinnulas tornatiles pyramydales equales longitudine et grossicie[2] quere, et unam in centro orthogonaliter infige et alteram extremitati linee a centro descendentis. Quo completo erige instrumentum super latus suum duabus pinnis ad orientem conversis et ea que in centro est superiori et alia deorsum inferiori; sitque superficies in qua fixe sunt observa orienti. Tunc perpendiculo a superiori pinna in inferiorem demisso ad meridiani superficiem et orizontis equidistantiam adapta, umbramque pinnule in centro existentis quorsum in meridie cadat diligenter attende. Et per hoc sicut superius distantiam tropicorum et remotionem summitatis capitum ab equinoctiali contemplare.

1. describe] describere MS; 2. et grossicie] grossiore MS.

[*Almagest*, Vat. lat. 2057] Assumam[1] etiam loco duarum armillarum aliud considerandi instrumentum, quod velocius fiat et levius et vicinius assumatur, faciam enim laterem lapideum aut ligneum quadratum mensurate latitudinis et altitudinis, ut supra superficiem sue basis absque tortuositate et declinatione erigatur[2]. Sitque una superficierum eius vehementer plana et levis et equalis; ponam autem apud unum angulorum huius superficiei punctum quod constituam centrum et describam super[3] ipsum quartam circuli et protraham[4] ab ipso duas lineas rectas ad duas extremitates quarte descripte continentes angulum rectum, qui est sub quarta. Et dividam arcum[5] quarte in nonaginta partes, et dividam partes in partes suas, deinde post hoc[6] faciam duos paxillos parvos rotundos tornatiles[7] cum turno equales in quantitate et crossitie[8], et figam eos in duabus extremitatibus unius duarum rectarum linearum[9] erecte supra superficiem orizontis, et sit eius locus a latere ad partem[10] meridiei, Et ponam medium[11] extremitatis unius duorum paxillorum[12] super[3] medium puncti centri[13] quarte, et ponam medium extremitatis[14] paxilli alterius supra medium puncti quod est in extremitate altera inferiori linee. Deinde erigam hanc superficierum lateris, in qua est hec linea, super[3] lineam descriptam in terra equidistantem linee meridiei ut sit superficies equidistans linee meridiei. Et ponam lineam que est inter duos paxillos libratam cum perpendiculo erectam supra superficiem orizontis ortogonaliter, nobis eam rectificantibus per ea que ipsam[15] firment donec supra ipsam cadat filum quod descendit cum perpendiculo a paxillo superiori ad paxillum inferiorem. Postea considerabimus in medietatibus dierum umbram paxilli superioris qui est in centro, et ponemus sub quarta descripta aliquid ut sit locus umbre vehementer manifestus. Et considerabimus medium umbre super quas partes quarte cadat, et per hoc indicabitur transitus solis in linea orbis meridiei in latitudine[16].

[variants from BN 16200, ff. 12rb13–12va8] 1. Assumam] Accipiam; 2. et declinatione erigatur] erigatur et declinatone; 3. super] supra; 4. protraham] protrehaam (?); 5. arcum] angulum; 6. post hoc] *om.*; 7. tornatiles] piramidales; 8. crossitie] grossitie; 9. rectarum linearum] linearum rectarum; 10. ad partem] ex parte; 11. medium] locum; 12. duorum paxillorum] paxillorum duorum; 13. centri] quod est in centro; 14. medium extremitatis] medietatem extremi; 15. ipsam] eam; 16. latitudine] terra.

Here at last we find some striking words common to GERARD and the *Almagestum parvum*: "sine tortuositate" and "tornatiles". But this is a poor harvest from such a long passage.

In general, if one of the GERARD texts is the basis of the *Almagestum parvum*, the compiler must have been at some pains to change the terminology as much as possible. The alternative is another source.

7 Who was *quidam vir* ?

The attributions in the manuscripts of the *Almagestum parvum*, being so various, can at best be used for confirmation. It is unlikely that ALBATEGNI, CAMPANUS, GEBER, ALBERTUS MAGNUS and ST. THOMAS AQUINAS all had a hand in its composition.

On the question of authorship the best discussion is still [BIRKENMAJER]. After dismissing WEISSENBORN's hypothesis that the *Almagestum parvum* was the lost translation of ROBERT OF CHESTER on the grounds that AL-BATTĀNĪ is often cited, BIRKENMAJER argues against AVERROËS as author because of his known aversion to epicycles and eccentrics. Then he looks for a Christian author writing between 1175 and 1277: Christian because of the passage in III 16 on choosing the era for tables:[71]

> ... aut potissimum annos Christi, qui est rex regum et dominus dominantium,

1175 as the date of GERARD's translation of the *Almagest*, and 1277 as the supposed date of the *Speculum astronomiae* (which he assumes was written by ROGER BACON). His first candidate for authorship is CAMPANUS (d. 1296), because of the inscription in Prague V. A. 11 (see above, n. 31) and a note in MS Oxford, Ashmole 424:

[71]MS Paris BN lat. 16657, f. 101v. The proposition is labelled "17" in the margin. This and the next quotation are fuller in [BIRKENMAJER].

> ... [Wilhelmus de Morbeka] socius fuit magistri Campani qui fecit Parvum Almagesti et commentavit Geometriam Euclidis ...

But against this hypothesis is the date of MS Paris BN lat. 16657, which is included in the *Biblionomia* of RICHARD DE FOURNIVAL, and so must be earlier than about 1250. BIRKENMAJER thought that CAMPANUS began his career as writer in the 1260's. Finally, the *Biblionomia* attributes the work to one "Galterus de Insullis", but none of the Walters of Lille are suitable. We may note that the attributions in the *Biblionomia* are not always reliable — e.g. "Theodotii libri de speris, ex commentario Boetii",[72] a most unlikely text. Is it possible that the writer has confused the *Almagestum parvum* with the treatise on proportions by WALTER OF FLANDERS? Two unknown mathematicians called WALTER at this early date does seem unlikely.

The AVERROËS text is perhaps more likely than BIRKENMAJER supposed, since the second part of the astronomical treatise attributed to him in the Hebrew translation of JACOB ANATOLI is a summary of the *Almagest*.[73] But the two texts are too dissimilar for a simple transfer. For instance, the *kata* theorems are differently treated — though it must be admitted that AVERROËS's treatment bears some similarity to that of the Toledo manuscript of the *Almagestum parvum* (and also of the "translation" of the *Almagest* in Dresden in MS Db 87). Again, several theorems by THEODOSIUS and MENELAUS are inserted in book I, presumably following the example of JĀBIR B. AFLAḤ. Another difference — not to move from book I — lies in the laboured manipulation of proportions in AVERROËS' treatment of declination (equivalent to *Almagestum parvum* I 16).[74]

Let us now consider what new information has been gathered here. First there is the puzzling relationship with GERARD's translation of the *Almagest*. There is a little common terminology, though in our examples not much; and the lettering of the more complicated diagrams would indicate a translation from the Arabic that used GERARD's system of translating diagram letters. We are left with the conclusion that the compiler either had some other access to the *Almagest* in addition to GERARD or deliberately and radically altered the wording, perhaps with the intention of simplifying or modernizing it. There is plenty of evidence of reworking, if only in the substitution of sines for chords and the addition of material from AL-BATTĀNĪ and others; and it is quite possible that changes in terminology

---

72 [BIRKENMAJER], p. 165.

73 The first part, on cosmological and astronomical principles, has been edited and studied by Mme. JULIANE LAY for her doctorate.

74 MS Paris, BN heb. 696, ff. 86r–v for *kata* theorems, 86v–88r for THEODOSIUS and MENELAUS, and 88r for the theorem on declination. I should like to thank Dr. H. ISAACS (Cambridge) and Dr. M. SEGRE (Munich) for their help at various times with translating passages from this text.

and style were made at the same time. As for the possibility of a second translation of the *Almagest*, neither of the other two versions considered here answers the purpose. The two occurrences of the letter H for E provides evidence, albeit weak evidence, that the corresponding propositions were seen in Arabic, either by the compiler or by his lost source.

The name *Almagest* was known in the West long before 1175: we are told, for instance, that GERARD went to Toledo "amore almagesti", though it should be remembered that this was written after his death, maybe forty years after he made the journey.[75] Again, HERMANN OF CARINTHIA cites it by name several times in his *De essentiis*, written in 1143. But some, if not all, these citations must refer to another work.[76] The Sicilian translation of the *Almagest*, the one quoted in this article from MS Vatican, lat. 2056, was apparently little known in the Middle Ages, but R. LEMAY has put forward arguments to show that HERMANN was the translator. Among them is an early dating of a note to a copy of the work that ascribes the version to HERMANN in the reign of King ROGER (d. 1154). Another argument concerns interpolated matter in MS Vatican, lat. 2056: not only was HERMANN wont to interrrupt a translation with passages of his own — properly marked as such, as in this case —, but this interpolation refers to AL-BATTĀNĪ ("... ut ait Albategni ..."). In two works, the *Astronomia* and *Regulae in canonem astronomiae*, identified and edited by LEMAY, HERMANN "modifie les tables ptoléméenne apparemment sous l'influence d'al-Battānī";[77] and al-Battānī is also quoted in the *De essentiis*. If indeed HERMANN did translate the *Almagest*, then the work was known in the West, at least in HERMANN's circle, about the middle of the twelfth century.

Secondly, there is a sprinkling of Greek terms in the *Almagestum parvum*. Curiously, this is not at variance with the occurrence of the Arabic term *kata*. For if the supposition adumbrated above, that the term came from the "Intellexi quod dixisiti ..." translation of THĀBIT's treatise, is correct, then our text is again connected with Greek vocabulary. For this translation from Arabic has many Greek terms (e.g. "apodixis") and even copies the constructions of translations from Greek, e.g.:[78]

> Dico quoniam proportio eius qui sub dupla periferia AE ad eam que sub dupla periferia AZ ...,

[75]For early knowledge of the name *Almagest* see [KUNITZSCH 1974], p. 115ff.

[76][HERMANN], pp. 120, 134, 136, 140. The first of these is on nesting the celestial spheres, not in the *Almagest*; the third says that the body of the Sun is 170 times that of the Earth (the Ptolemaic figure, not the Albategnian), but continues with the bodies of the planets, not in the *Almagest*. For arguments that HERMANN did know the *Almagest*, see [HERMANN], pp. 287 (under 67rB) and 334 (under 77rD).

[77][LEMAY], pp. 464, 466-467. Before presenting the positive evidence he demolishes HASKIN's arguments against the ascription.

[78][THĀBIT B. QURRA], p. 162, para. 40; p. 160, para. 40. For "apodixis" see p. 161, para 1, and p. 162, para. 40 (last word).

Compare this with the enunciation of the first form of Menelaus' theorem for spherical arcs in the Sicilian translation of the *Almagest*:[79]

> dico ergo quoniam eius sub que dupla periferie GE ad eam que sub dupla periferie EA proportio ...

In this passage from THĀBIT GERARD has

> Dico ergo quod proportio corde dupli arcus AE ad cordam dupli arcus AZ ...,

which is a word-for-word translation of the Arabic.[80] To transform the simple Arabic wording in this way, the translator or redactor must have had a copy of such a text as the Sicilian *Almagest* in front of him, if indeed he was not the translator of it himself. We note that the word *al-jayb* in its single occurrence is transliterated, "elgeib"; GERARD has "sinus".[81]

Thirdly, we have noted the short form of some of the proofs, which is similar to the style of *Version II* of the *Elements*. Fourthly, there are texts of the *Almagestum parvum* with the same or similar enunciations, but different proofs. A comparison with *Version II* is suggested. For the definitions and enunciations of this compilation were collected from several sources, including the Boethian tradition and translations from Arabic; and, later, proofs — of the sketchy kind noted — were supplied. Also, there are versions of the *Elements* — of which the best known is by CAMPANUS — with the same definitions and enunciations, but with different proofs.[82] True, the phenomenon of the skeleton text with different flesh is known elsewhere,[83] but the similarity remains.

It is tempting to link at least the preface with the name of BOETHIUS himself, since according to CASSIODORUS he made PTOLEMY known to the Latins. In a letter to BOETHIUS THEODERICH writes:[84]

> ... Translationibus enim tuis Pythagoras musicus, Ptolemaeus astronomus leguntur Itali ...

But the earliest manuscript of the *Almagestum parvum* is of the thirteenth century; there appears to be no other manuscript attestation of the "Ptolemaeus" translation; and another work by PTOLEMY might be meant. Besides, the Latin style is not quite right for a text written in antiquity and

---

[79]MS Vatican, lat. 2056, f. 8r, lines 33–34.

[80]Paris, BN ar. 2457, f. 166r, lines 10–11.

[81][THĀBIT B. QURRA], p. 162, para. 4; p. 160, para. 4.

[82][EUCLID Version II], Introduction, pp. 11–31.

[83][KNORR], part III, chapters 8–10.

[84][CASSIODORUS], I 45, 4. I am grateful to Prof. M. FOLKERTS for this reference. It is a pleasure to acknowledge his help in various parts of this paper and in typing it into the computer.

would only just be appropriate for BOETHIUS, since the words "orbicularis" and, to a lesser extent, "concitatio" are non-classical.[85] Therefore the preface is more likely to be a product of the later Middle Ages.

In conclusion, it is tentatively suggested here that the preface, most of the earlier part of the book (where the proofs are short), the enunciations and perhaps the introductions to books II–VI were the work of a scholar in the HERMANN–ROBERT circle in the mid-twelfth century and that the treatise was filled out later on the basis of a form of GERARD's translation of the *Almagest*. The whole was finished by about 1200.

References

[AḤMAD B. YŪSUF] SCHRADER, SR. M.: *The* Epistola de proportione et proportionalitate *of Ametus filius Iosephi*, PhD thesis, University of Wisconsin 1961.

[ALBERTUS MAGNUS] ALBERTO MAGNO: *Speculum astronomiae*. Edizione a cura di STEFANO CAROTI, MICHELA PEREIRA, STEFANO ZAMPONI, sotto la direzione di PAOLA ZAMBELLI. Domus Galilæana, Quaderni di storia e critica della scienza, Nuova serie 10, 1977.

[BARONCINI] BARONCINI, GABRIELE: *Note sulla formazione del lessico della metafore "Machina mundi"*, Nuncius, Annali di storia della scienza, 4 (1989), fasc. 2, 3–30.

[AL-BATTĀNĪ] *Al-Battānī sive Albatenii Opus Astronomicum*, ed. C. A. NALLINO, Milan 1899–1907.

[BERNARDUS SILVESTRIS] BERNARDUS SILVESTRIS: *Cosmographia*, edited with introduction and notes by PETER DRONKE, Leiden 1978.

[BIRKENMAJER] BIRKENMAJER, ALEKSANDER: *La Bibliothèque de Richard de Fournival*, Studia Copernicana, 1 (1970), 118–210. Originally published in Polish in 1919.

[BJÖRNBO, VOGL] *Alkindi, Tideus und Pseudo-Euklid. Drei optische Werke*. Herausgegeben und erklärt von AXEL ANTHON BJÖRNBO und SEB. VOGL,

[85]This sentence is based upon the judgement of Dr. G. BERNT. I am most grateful to him for his help.

Leipzig–Berlin 1912 (*Abhandlungen zur Geschichte der mathematischen Wissenschaften*, Heft 26.3).

[BOETHIUS] *Anicii Manilii Severini Boethii Philosophiae Consolatio*, ed. L. BIELER, Turnhout 1957.

[F. BOLL] BOLL, FRANZ: *Das Epigramm des Claudius Ptolemaeus*, in *Kleine Schriften zur Sternkunde des Altertums*, Leipzig 1950. Originally written in 1921.

[CARMODY] CARMODY, FRANCIS J.: *Arabic Astronomical and Astrological Sciences in Latin Translation. A Critical Bibliography*, Berkeley and Los Angeles 1956.

[CASSIODORUS] CASSIODORUS, *Variae*, ed. TH. MOMMSEN, *Monumenta Germaniae Historiae*, Auctores antiquissimi, XII, Berlin 1894.

[EUCLID, ADELARD] BUSARD, H. L. L.: *The First Latin Translation of Euclid's* Elements *Commonly Ascribed to Adelard of Bath*, Toronto 1983.

[EUCLID, Greek] *Claudii Ptolemaei opera quae exstant omnia*, Vol. 1, *Syntaxis mathematica*, ed. J. L. HEIBERG, Leipzig 1898–1903.

[EUCLID Version II] *Robert of Chester's (?) Redaction of Euclid's* Elements, *the so-called Adelard II Version*. Edited by HUBERT L. L. BUSARD and MENSO FOLKERTS, Basel–Boston–Berlin 1992.

[GEBER] *Gebri filii Affl'a Hispalensts, de astronomia libri IX*, Nuremberg 1534.

[HASKINS] HASKINS, CHARLES HOMER: *Studies in the History of Medieval Science*, New York 1927.

[HELMAN & SWERDLOW] HELMAN, DORIS and SWERDLOW, NOEL: *Peurbach (or Peuerbach), George*, Dictionary of Scientific Biography, 15, 437b–479a, New York 1978.

[HERMANN] HERMANN OF CARINTHIA: *De essentiis*, a critical edition with translation and commentary by CHARLES BURNETT, Leiden 1982.

[Hrabanus Maurus] *Poetae Latini Aevi Carolini*, II, *Monumenta Germaniae Historiae*, Berlin 1964.

[Johannes von Gmunden] Busard, H. H. L.: *Der Traktat de sinibus, chordis et arcubus von Johannes von Gmunden*, Österreichische Akademie der Wissenschaften, Math.-Naturwiss. Kl., Denkschriften, 116. Band, 3. Abhandlung, Vienna 1971.

[Knorr] Knorr, Wilbur: *Textual Studies in Ancient and Medieval Geometry*, Boston–Basel–Berlin 1989.

[Kunitzsch 1974] Kunitzsch, Paul: *Der Almagest. Die Syntaxis Mathematica des Claudius Ptolemäus in arabisch-lateinischer Überlieferung*, Wiesbaden 1974.

[Kunitzsch 1990] Claudius Ptolemäus: *Der Sternkatalog des Almagest. Die arabisch-mittelalterliche Tradition. II. Die lateinische Übersetzung Gerhards von Cremona*, herausg. und bearbeitet von Paul Kunitzsch, Wiesbaden 1990.

[Kunitzsch 1991] Kunitzsch, Paul: *Letters in Geometrical Diagrams, Greek – Arabic – Latin*, Zeitschrift für Geschichte der Arabisch-Islamischen Wissenschaften, 7 (1991), 1–20.

[Lemay] Lemay, Richard: *De la scolastique à l'histoire par le truchement de la philologie: itinéraire d'un médiéviste entre Europe et Islam*, La diffusione delle scienze islamiche nel medio evo europeo (Convegno Internazionale, Roma 1–4 Ottobre), Accademia Nazionale dei Lincei, Rome 1987.

[Millás Vallicrosa] Millás Vallicrosa, José: *Las traducciones orientales en los manuscritos de la Biblioteca Catedral de Toledo*, Madrid 1942.

[Thābit b. Qurra] *The Astronomical Works of Thābit b. Qurra*, ed. Francis J. Carmody, Berkeley and Los Angeles 1960.

[Theon] *Commentaires de Pappus et de Théon d'Alexandrie sur l'Almageste*. Texte établi et annoté par

A. Rome. Tome II: Théon d'Alexandrie, *Commentaire sur les livres 1 et 2 de l'Almageste*, Città del Vaticano 1936. (*Studi e Testi* 72.)

[Sacrobosco] Thorndike, Lynn: *The* Sphere *of Sacrobosco and Its Commentators*, Chicago 1949.

[Wallingford] North, J. D.: *Richard of Wallingford. An edition of his writings with introductions, English translation and commentary*, 3 vol., Oxford 1976.

[WKAS] *Wörterbuch der klassischen arabischen Sprache*, Band II Teil 2, Wiesbaden 1991.

VI

The Astronomy of Jābir ibn Aflaḥ

1. Introduction

The astronomer "Geber" was cited as an authority in astronomy from the later middle ages until well into the sixteenth century[1]. His name was associated with trigonometrical methods not found in the *Almagest* and with criticism of some of Ptolemy's astronomy, particularly his placing Venus and Mercury below the Sun. Copernicus called him an "egregious calumniator of Ptolemy"[2]. This article is intended to give a general idea of "Geber"'s work; detailed papers on his trigonometry and other topics are being prepared.

"Geber" is Jābir ibn Aflaḥ, who wrote a lengthy exposition and critique of Ptolemy's *Almagest*. Almost nothing is known of his life. The only reliable reference seems to be by Maimonides in his *Guide for the Perplexed*, II 9. Discussing the order of the planetary spheres, he says[3]

> More recently some Andalusian scholars concluded, from certain principles laid down by Ptolemy, that Venus and Mercury were above the Sun. Ibn Aflaḥ of Seville, with whose son I was acquainted, has written a famous book on the subject; also the excellent philosopher Abu-Bekr ibn-Alẓaig, one of whose pupils was my fellow-student, has treated this subject and offered certain proofs – which we have copied – of the improbability of Venus and Mercury being above the sun.

The term "of Seville" (al-Ishbīlī) is confirmed by MSS of his treatise on the *Almagest* and by al-Biṭrūjī's citation[4], where he is called "Abū Muḥammad Jābir ibn Aflaḥ al-Ishbīlī". From Maimonides' acquaintance with his son it is reasonable to conclude that Jābir lived in the twelfth century or the last decades of the eleventh. It is probable that Abū Bakr ibn al-Ṣā'igh, or Ibn Bājja (Latin Avempace), discussed the question of Venus and Mercury in reply to Jābir, who was criticising the standard view, and that Jābir was therefore active before 1138, the year of Ibn Bājja's death (possibly by poison)[5].

In a *Compendium of the Almagest*, attributed to Ibn Rushd (1126–98), there is a remark that Ibn Aflaḥ lived in the same century[6]. If we believe that Ibn Rushd wrote the *Compendium* despite his known dislike of epicyclic astronomy[7]—and it should be remembered that at least one other opponent of epicycles, Henry of Hesse, also wrote "secundum ecentricos et epiciclos"[8]—, or if it was written by a contemporary, then Jābir must have flourished after 1106 (500 H.). Taking the evidence of the Maimonides quotation into account, we may therefore suppose that Jābir lived in the first half of the twelfth century.

Jābir ibn Aflaḥ should be distinguished from other writers of a similar name. For instance, the poet Ibn Aflaḥ (d. ?141)[9], that is Abū-l-Qāsim ʿAlī ibn Aflaḥ al-Absī of Baghdad, seems to have no connection with Jābir. According to Ibn Khallikān,[10] the poet Ibn Aflaḥ wrote satirical verses about the physician Ibn al-Tilmīdh, complaining about the starvation he inflicted. It is therefore likely that the Ibn Aflaḥ reported by Ibn al-Qifṭī[11] as writing a satirical poem on the Jewish physician Abū-l-Barakāt is the poet. This is especially likely since the two physicians are easily confused: they share the name Hibat Allāh, and both died about 1165 at the age of nearly a hundred[12]. Identification of the two Ibn Aflaḥs led Sanchez Perez to claim[13], on the authority of Ibn al-Qifṭī, that Jābir was a contemporary of Abū-l-Barakāt. The only other Aflaḥ likely to be mistaken for the astronomer is Abū Aflaḥ ha-Saraqosṭī, the author of *The Book of the Palm*, extant in Hebrew as *Sepher ha-Tamar*. He quotes Ibn Rushd, and according to Solomon Gandz[14] probably lived in the thirteenth or fourteenth century.

Other Jābirs must also be distinguished from the astronomer. He has occasionally been identified with al-Battānī, that is Muḥammad ibn Jābir al-Battānī[15]. Far more common is the wrong identification with the alchemist Jābir ibn Ḥaiyān (c. 721–815), also known in Latin as "Geber" and probably the author of some of the alchemical works that exist under this name. The two Gebers have seldom been distinguished until recently. Nicolas Antonio[16] attributes the *Summa Perfectionis* etc. to "Geber, alias Mahomed Geber filius Afflah" immediately after discussing the genuine *Expositio in Ptolemaei Novem Libros*. It was to Jābir ibn Ḥaiyān or a composite Geber that Dr. Johnson referred in his definition of "Gibberish" as "... anciently written *gebrish*, ... probably from the chymical cant, and originally implied the jargon of *Geber* and his tribe".[17]

It is appropriate to dismiss here the derivation of the word "algebra" from "Geber". In fact the word comes from "*al-jebr*, the redintegration or reunion of broken parts, from *jabara* to reunite, . . ."[18]. Again, the *Tabule Jahen*, which appear in a work entitled *De Eris* translated by Gerard of Cremona, are tables for Jaen[19] in Southern Spain, but there is no reason to connect them with Jābir.

A legend credits Jābir with building the tower now belonging to the cathedral in Seville—and sometimes the Torre del Oro as well. Antonio writes[20]

> My Sevillian friends are persuaded that he constructed in this city an exceedingly high and beautiful tower, which is certainly still standing, and attached to it a mosque, which – with some rebuilding by the Christians – has now been converted into the largest church in Spain, sacred to the Virgin Mother of God; and that he constructed another tower on the bank of the river Guadalquivir. To this the citizens gave the name "Golden" because of its beauty or because the Indian treasure was once preserved there. The same man is said to have erected two other towers similar to the one in Seville—one, I think, in Morocco and the other in Rabata [Amman ?] in Arabia Petraea.

In *Conquista de la Betica*, Juan de la Cueva recounts the presentation of Jābir's tower, abandoned by the Moors to the Christians. He describes the tower as guarding the treasure of kings past and present and hence known as "golden"[21]. There are several glowing references to "Geber" in the prefacing "Descripcion de Sevilla". The evidence of these and similar references[22] has no real weight since they usually credit Jābir with the invention of algebra as well as building towers. Besides, according to modern opinion, the Torre del Oro was built in the early thirteenth century[23]; and the cathedral tower, once a minaret for the mosque, was completed in 593H. (1196–7 A.D.) during the reign of the Muwaḥḥidī ruler Abū Yūsuf Ya'qūb al-Mansūr[24]. A further legend that Jābir used the minaret for observation seems to go back no further than Ashbach, writing in 1837[25]. The legends do, however, testify to a strong tradition connecting Jābir with Seville.

2. Texts

(a) Arabic

Jābir's fame rested on one work, a critique of the *Almagest* in nine books. The Arabic text[26] is extant in three MSS: Berlin 5653, written at Damascus in 1229 A.D. and entitled "Iṣlaḥ Almajisṭi" or "Correction of the Almagest"; Escurial 910 and Escurial 930. The Escurial MSS, with no title other than "Book of Astronomy", are written in a fourteenth century Maghribine hand, but are otherwise undated. There is a fragment of book V in Hebrew script in the Bibliothèque Nationale, heb. 1102, ff 12r–16r. Steinschneider suggests that it is part of an incomplete Arabic MS in Hebrew script that in 1871 he saw in the possession of the traveller Saphir[27]. For want of a better-attested title, the work will be referred to here as the *Iṣlaḥ Almajisṭi* or *Iṣlaḥ*.

None of the MSS is perfect. In Escurial 910, for instance, theorem 11 in book I and the description of Jābir's instrument in book V are displaced. Of more importance is the omission in all the MSS of the passage in book VI corresponding to *Almagest* VII 4 (middle)—VIII 4, which includes the chapters on the Milky Way and on the solid sphere.

The Escurial MSS are very close. But an examination of the important trigonometrical passage in book I reveals the following differences from the Berlin MS. For brevity the Berlin MS and the common reading of the Escurial MSS are called B and E respectively.

(i) E has an extra theorem in plane trigonometry. It shows how to solve a triangle in which two angles and one side are shown..
(ii) The corollaries of proposition 12, the fundamental theorem of Jābir's spherical trigonometry, are different in E and B.
(iii) In prop. 13, the Sine-theorem for spherical triangles, the special case of a right triangle is proved first. B makes three cases of this, in which the triangle has three right angles, two right angles, and one right angle; but in E the three cases are taken together.
(iv) The idea of the proof of prop. 13 is the same in E and B, but the details are quite different. For instance, in B two similar results are proved simultaneously, but E proves one and lets the other follow by similarity. In general B is more prolix.
(v) For prop. 14 E proves a different theorem and gives as a corollary the theorem found in B.

Such differences can only have come about by deliberate revision. And since E gives a more complete but less prolix exposition, one must assume that it represents the revised text. It is possible that this revision was done by no less a figure than Maimonides, together with his favourite pupil Joseph ben Yehudah ben Shamʿūn[28]. For in his biographical note on Maimonides, Ibn al-Qifṭī (1172/3–1248) says (in al-Zawzānī's abridgement)[29]

> ... And he polished [the style of] the treatise *Istikmal* of Ibn Aflaḥ of Seville on astronomy—and he did it very well—for it was in origin disorganised ...

and in his section on Joseph ben Jehudah[30].

> ...And when the Jews and Christians in that country were obliged to become Muslims or emigrate, he was cunning and hid his religion. When he was able to move and travel to the state of Egypt, he gathered all his money and possessions and went to Egypt. He sought out Moses ben Maimon of Cordova, the leader of the Jews in Egypt, and studied some time under his supervision—staying some time at his house. He asked him to correct the astronomy of Ibn Aflaḥ of Andalus, which he had brought with him from Sabta; and they were occupied together—he and Moses—in correcting this work and giving an accurate edition of it. Having left Egypt, ...

The story is repeated by Barhebraeus in his *History of the Dynasties*, and squares well enough with the dedicatory introduction to the *Guide*[31]

> My dear pupil, ever since you resolved to come to me, from a distant country, and to study under my direction, I thought highly of your thirst for knowledge, ... But when you had gone with me through a course of astronomy, after having completed the [other] elementary studies which are indispensable for the understanding of that science, I was still more gratified by the acuteness and the quickness of your apprehension. Observing your great fondness for mathematics, I let you study them more deeply, for I felt sure of your ultimate success ...

Besides, Ibn al-Qifṭī was a personal friend of Joseph—he says so later in Joseph's biography—and his source of information may well be personal knowledge. Nonetheless there are some problems. It seems[32] that Joseph left Cairo for Aleppo about 1187, for Jehuda al-Ḥarizi says in a letter, written when he left Aleppo for Jerusalem in 1217, that Joseph had been in Aleppo for thirty years. His astronomical studies with Maimonides therefore took place before the *Guide* was written, i.e. 1190 or 1191[33], and this is confirmed by the end of the introduction[34],

> When, by the will of God, we parted, and you went your way, our discussions aroused in me a resolution which had long been dormant. Your absence has prompted me to compose this treatise for you and for those who are like you, however few they may be. I have divided it into chapters, each of which shall be sent to you as soon as it is completed. Farewell!

But when Maimonides mentions Ibn Aflaḥ's "famous book" on the position of Venus and Mercury, he gives no indication that he and Joseph had studied it together, let alone revised it. The passage gives the impression that Maimonides knew the work only by hearsay, if only because it implies that Ibn Aflaḥ's book is entirely devoted to the question of Venus and Mercury. Furthermore, in part II chapter 24, which concerns the reality of excentrics and epicycles, we find[35]

> You know of astronomical matters what you have read under my guidance and understood from the contents of the "Almagest". But there was not enough time to begin another speculative study with you. ...

Now we could explain this passage by taking "the *Almagest*" (literally "the book of the *Almagest*"[36] as referring to Jābir's *Iṣlaḥ Almajisṭī*. The chapter on the order of the spheres may have been written before Maimonides met Joseph, and incorporated into the *Guide* from old notes. Another possibility is that ʿAlī al-Zawzānī abridged Ibn al-Qifṭī's biographies carelessly (the original is not extant) and that Maimonides and Joseph revised the *Iṣlaḥ* at a later date. Of course, their revision may anyway not be responsible for the differences of the Escurial and Berlin MSS noted above.

(b) Latin

The *Iṣlaḥ* was translated into Latin by Gerard of Cremona (1114–87). At the end of the Madrid MS, Bib. Nac. 10006, f. 152r, the scribe has written, after "finitur",

> Hunc librum transtulit in Toledo magister G. Cremonensis de Arabico in Latinum.

Much the same appears in other MSS. "Geber's astronomy" also appears on a list of Gerard's translations appended to the eulogy of him beginning[37]

> Sicut lucerna relucens in abscondito non est ponenda, ...

There are some seventy works in the list, some of them substantial. It is thought that he could translate so many because he worked with interpreters; for the Almagest, for instance, he had the help of the Mozarab Galippus[38]. According to the eulogy, Gerard was drawn to Toledo by his enthusiasm for the *Almagest*.

> ... amore tamen *Almagesti* quem apud Latinos minime reperiit, Tolectum perexit. ...

His translation was completed in 1175[39]. Since Jābir's many quotations from the *Almagest* do not repeat this translation, it could be argued—but with no certainty—that Gerard translated the Jābir work first. The quotations seem to have been translated afresh since the opening passage of Jābir's book IV does not tally with the so-called second translation of the *Almagest* from the Arabic given in the margin of the Wolfenbüttel MS Gud. Lat. 147[40].

There is a curiosity in the fifth book. The instrument described in the Latin version is quite different from the one described in the Arabic MSS[41], but the diagrams have been taken over without change. The passage in question must have existed in Arabic since it also appears in the Hebrew translations. Of the two instruments the one described in the Arabic MSS is the original since it corresponds to the description in book I.

There are some twenty MSS of "Geber" in Latin, two of them in Regiomontanus' hand, and five fragments[42]. The number of MSS itself shows a considerable interest in the text, and in 1534 it was printed, together with Peter Apian's *Instrumentum Primi Mobilis*, at Nuremberg.

The Latin text begins

> Scientia species habet, quarum melior, post scientiam fidei, est cuius scita fixa sunt, ...

thus omitting the religious preface beginning

> Praise be to God, the First [but] without beginning and the Last [but] without end, ...

which is found in the Arabic MSS and is quoted by Ḥājjī Khalīfa[43]:—

> Vir quidam recentiorum alium commentarium composuit qui ita incipit: Laus Deo primo sine initio etc., ...

The style of the translation is literal. Where a title is given it is usually nondescript, e.g. "Astronomia Gebri" or "Liber Geber super Almagesti".

Quite distinct, but equally vague as to title, is a reworking of the first six books of the *Almagest*, called by Haskins the *Almagestum Parvum*.[44] Its author, frequently cited by Richard of Wallingford as "commentator almagesti", is called simply "quidam vir" in the *Speculum Astronomie*, which describes the work as being derived from the Almagest and Al-Battani's astronomy:[45]

> Ex his quoque duobus libris collegit quidam vir librum secundum Euclidis stilum, cuius commentarium continet sentenciam utriusque, Ptolemaei scilicet atque Albategni, qui sic incipit, "Omnium recte philosophantium" ...

The *Almagestum Parvum* was probably composed in Latin. It is true that the reference to Christ in the third book as "rex regum et dominus dominantium" could have been inserted, and the style polished to remove the usual traces of translation from Arabic, but on the other hand no Arabic MS nor any reference to it in Arabic has been found. Only two MSS ascribe the work to Jābir. In one the scribe's explicit reads[46]

> Explicit liber Jeber per manus Engelberti. Deo Gratias.

but "Jeber" is partly erased, and a note in a somewhat later hand, possibly of the mid-fifteenth century, accuses the scribe of gross error, the only excuse being that he had a bad MS to copy from

> falsissimi scriptoris quia non est verbum correctum nisi sumit malum exemplar.

The other ascription to Jābir occurs in a list of contents of a codex[47], not in the text and written in a different hand. Other MSS attribute the *Almagestum Parvum* to Al-Battānī, Campanus and St. Thomas Aquinas[48]; and the rest are anonymous. It is therefore safe to assume that Jābir was only thought to have written it by confusion with the genuine work translated by Gerard of Cremona. Birkenmajer[49] suggests two candidates for authorship, Campanus of Novara and a Walter of Lille. It is possible that Walter composed the text and Campanus re-edited it, for there are several variants to account for.

(c) Hebrew

The *Iṣlaḥ* was translated into Hebrew twice, by Moses ibn Tibbon in 1274 and by his nephew Jacob ben Maḥir (1236–1304). The unique MS of Moses ibn Tibbon's translation may be an autograph since the colophon reads[50]

> ... And I have translated it from the Arabic language into Hebrew language, I, Moses ben Rabbi Samuel ibn Tibbon the Spaniard.

No doubt this sentence could have been copied with the rest, but it does use the first person in an emphatic manner and the writing is of the right age. Most of the passage in book VI missing in the Arabic is also missing here. The beginning of the MS is lost, but this probably amounts to only one folio. A spot check on a couple of passages shows that the translation is not always literal. The instrument in book V is the same as in the Latin version, but there are no diagrams at all. From an examination of the trigonometrical section it is clear that it was translated—again like the Latin—from the unrevised text. Moses' translation cannot have been widely disseminated, for Samuel ben Jehuda of Marseille, who revised Jacob ben Maḥir's translation, knew of it only by hearsay[51].

Samuel gives an account of the circumstances of this revision at the end of the translation[52]. He and his brother David went to Trinquetailles, a suburb of Arles, and procured an Arabic text of the *Iṣlaḥ*. They managed to copy an eighth of this in two days, working in a school-room and living on "the bread of adversity and water of affliction". They were in a hurry to return the manuscript and to attend to affairs at Tarascon, where they had set up house, and in their haste they got the pages out of order. After a great deal of travelling Samuel found an autograph of Jacob ben Maḥir's translation, which he found had many faults, including the omission of the passage on the Milky Way and the Solid Sphere. These faults he corrected by comparison with the Arabic MS he had used before. Samuel finished the task at Aix in 1335. Altogether, he found the revision of Jacob's translation more trouble than preparing a completely fresh one.

The translation, at any rate in Samuel's revision, follows the Escurial MSS in the section on trigonometry. Theorem 11 of book I is placed after theorem 15—the translation follows the order of Escurial 910 in this respect. Samuel complains so bitterly about Jacob ben Maḥir's inaccuracy that it is tempting to speculate that Jacob worked from an unrevised text. It is even possible that Jacob never translated the *Iṣlaḥ* and that Samuel revised Moses ibn Tibbon's translation. But this is unlikely—although translations should not be multiplied beyond necessity—because the extant MS, be it autograph or no, is clearly labelled as his work. In book V Samuel gives both instruments—in BN 1024 the page is divided into two columns with an explanatory note in the margin.

Extant in Hebrew translation, but otherwise unknown, are short commentaries, apparently by Jābir ibn Aflaḥ, on Thābit ibn Qurra's tract on Menelaus' theorem[53], beginning

> Know that Thābit ... in connection with the eighteen modes to which he refers ...

and on Menelaus' *Spherics*[54], beginning

> Know that of the triangles which are on a sphere, every arc being less than a semicircle ...

There are also extracts of a commentary on Euclid attributed to Jābir[55].

3. *The Iṣlaḥ Almajisṭī*

Jābir redivided the substance of the *Almagest* into nine books. The first contains an introduction and theorems in trigonometry. The second contains general philosophical considerations and spherical astronomy, and corresponds to *Almagest* I 2–8, 12. 14–16 and book II. The third is on the Sun (*Almagest* III) and the fourth on the Moon (*Alm.* IV–V 10). In the fifth book Jābir describes the instrument that will "replace all instruments discussed in this book" and continues with parallaxes, distances of Sun and eclipses (*Alm.* V 11 – VI). Book VI is on the fixed stars (*Alm.* VII and VIII). Book VII contains Jābir's famous re-ordering of the planetary spheres and in general covers the ground of *Almagest* IX–XI; book VIII is on stations and retrogradations (*Alm.* XII); and the last book, IX, corresponds to *Almagest* XIII—on planetary latitudes and heliacal risings and settings.

Despite the observing-instrument, which Jābir claims replaces Ptolemy's four instruments, the *Iṣlaḥ* is entirely theoretical. There are no tables in it. Where Ptolemy works through an astronomical calculation with numerical data, Jābir's practice is to show how the calculation is done, and, where appropriate, to give the answer at the end: "scientia" is disentangled from "operatio". Again, mathematics and astronomy are separated as far as possible, the mathematical theorems being put into book I. The astronomical argument can thus be seen more clearly, and each mathematical result is proved once and not every time it is required. By giving a few propositions for solving plane triangles in book I, Jābir removes the clutter of perpendiculars from the diagrams of subsequent books—to say nothing of obviating the need to imagine a circle drawn round every right triangle to be solved. But his most influential improvement is to spherical trigonometry. He replaced Menelaus' theorem, with all its awkwardness and prolixities, by simpler rules involving only four quantities[56]:

> ... ipse [Ptolemy] utitur in plurimo suarum probationum figura sectore [i.e. Menelaus' theorem], que est figura difficilis et partitur in ramos plurimos, et diversificatur in ea compositio proportionis varietate extranea; quapropter sit difficilis aspicienti in ipso rememoratio eius et ipsius comprehensio et concludere ea que concluduntur ex ea. ...
>
> Acciderunt ergo nobis per gratiam Dei et bonitatem auxilii eius propositiones faciles et breves, quibus excusamur a libro Milei et a figura sectore et a plurimo libri Theodosii, et quibus extrahitur ignotum ex noto per quatuor numeros proportionales, non per sex numeros sicut preparantur in figura sectore.
>
> ... in most of his proofs he used the sector-figure, which is subdivided into numerous ramifications and involves a proportion that can be composed in a variety of diverse ways. It is therefore difficult for the student to remember it, to understand it, and to deduce those things that are deduced from it. ...
>
> By the grace of God and by the excellence of His help we have been granted easy and short propositions, with which we can avoid using Menelaus' book and the sector-figure and most of Theodosius' book. With these propositions the unknowm is extracted from the known by means of four proportional numbers, not six as are provided by the sector-figure.

After describing the difficulties of the *Almagest* and the improvements outlined above, the introduction continues with a list of Ptolemy's "errors". To justify this procedure, Jābir compares himself with Aristotle, who said that Plato and Truth were both friends but Truth was the greater friend[57]. But in the first of the listed "errors", Ptolemy's choice of eclipses for calculating the basic periods, Jābir complains[58]

> ... ipse nuper ostenderat quod valde longinquum est ab electione et quod est una trium positionum a quibus cavere monet in inquisitione horum spaciorum, et precipit abstinendum ab eis. Iam ergo elegit et non percepit quod cavere ab eo monuerat et preceperat ab eo abstinendum.
>
> Sermo vero eius: "Hec est via qua incessit ille qui fuit in inventione harum rerum, et possibile est tibi ut scias quod hec via non est facilis incessus neque proxime acceptionis; imo est necessaria in ea consideratio vehemens et comprehensio exquisita eius". Quod narro et quod continuatur cum hoc est sermo non comprehensus; et illud est quoniam non oportet ut huiusmodi dicat sermonem, ut si ipse iam veniat cum via alia faciliori hac et non sit necessarium in ea illud quod in hac via necessarium est de pertransitione, et cum hoc non indigeat via qua venerunt antiqui. Sed non fuit ei possibile de illo, ...
>
> ... he had just shewn that this type of interval was far from acceptable and that it

> was one of the three positions which he warned us to beware of when finding these periods and which he admonished us to avoid. He chose [the intervals] and did not realise he had warned us to beware of them and admonished us to avoid them.
>
> Again, he says: "This is the way our predecessor worked in finding these things, and you can see that this method is neither easy nor simple to grasp – indeed, it demands intense concentration and a fine understanding". What I have quoted and what continues here is incomprehensible. This is because he should not make such a remark as if he had come with another easier method that obviates steps necessary in this method, and as if he did not need the method the ancient astronomers used. But he could not do it, . . .

The tone of criticism is scarcely that of a friend. Before this, in the section on the sphericity of the heavens, Jābir describes Ptolemy's reasoning as[59]

> sermo in ultimo aggregationis et abbreviationis et est cum hoc vanitas sermonis; et hoc est primum quod apparet de vanitate eius in arte geometrie.
>
> extremely compact and abbreviated – hence the worthlessness of his reasoning. This is the first time that his worthlessness in the art of geometry appears.

What Ptolemy said was, that of different figures of equal perimeter, those with more angles have greater capacity, and so the circle is the greatest of the plane figures—and similarly the sphere is the greatest of solid figures. Jābir reasonably complains that this implies that circles and spheres are full of angles, and that it would be far easier to compare the sphere with a polyhedron of equal surface directly (as he does himself in book I). But in its context in the *Almagest* the point is a small one: Ptolemy's slight clumsiness does not merit such a violent reaction.

On the other hand, there are many places where Jābir fills out Ptolemy's exposition without any vituperation. He also excuses some of the obscurities of the *Almagest*, which he says has suffered in translation. Words have been shuffled about and ideas separated that should be together, so that the reader's attention flags[60]:

> . . . propter permutationem interpretum ipsius de lingua ad linguam accidit in eo antecessio et postpositio verborum et separatio inter intentiones eius, quod facit ambiguum legentem ipsum et facit eum hesitare et facit ipsum stare, cum non possit componere intentiones quesitas propter separationem earum; quare illud removet eum quandoque ab assiduatione in aspiciendo.

Jābir evidently went to a great deal of trouble to find out what Ptolemy had really said. On a point in lunar theory, in which the text was "errata in ultimo erroris"[61]

> ... possibile ergo est quod translator libri non intellexit illud quod voluit Ptolemeus; ergo alteravit narrationem propter illud: ergo alterata est intentio quam ipse voluit. Quamvis ego iam perquisivi ab hoc in libris pluribus translationibus Hunani et translationis Alhahazeg, et non inveni in eis diversitatem nisi parvam in dictionibus. In intentione vero non inveni in duabus translationibus diversitatem omnino, ...

> It therefore seems possible that the translator did not understand what Ptolemy meant, and therefore altered the text, so that the meaning is altered. But however much I pursued the matter in numerous copies of Ḥunain's and al-Ḥajjāj's translations, I could only find small differences in wording and no difference in meaning at all.

The most famous of Jābir's criticisms concerns the position of Venus and Mercury. Ptolemy placed them between the Moon and the Sun, which thus divided the planets of limited elongation from those which could have any elongation. They have never eclipsed the Sun because they have never been in line with it—just as most conjunctions of Sun and Moon produce no eclipse. The question cannot be settled, Ptolemy says, because Venus and Mercury show no parallax, from which their distance from Earth could be measured, and placing them nearer than the Sun would not entail any parallax even at their perigees. Jābir begins[62]

> ... ego quidem miror omni admiratione de esse huius viri; et hesito hesitatione magna propter illud quod apparet de contradictione eius et ipsius inquietudine, et ipse non percepit illud. Et res huiusmodi quidem elongatur ultima longitudine ab eo qui studet—sicut ipse studuit—ex sermone in istis rebus notabilis quantitatis. Et ipse non percepit contradictionem suam, ...

> ... I am very puzzled by what kind of man this is. I am quite at a loss to account for this inconsistency and confusion of his—which he did not notice. Such a thing must be very alien to anyone who makes any considerable study of these things, as he did. He did not see his inconsistency, ...

He then points out that Ptolemy himself had reckoned a maximum solar parallax at 2′ 51″, and that if Venus and Mercury were within the Sun's sphere, Venus would have a parallax of about $\frac{1}{3}°$ (the figure is almost exactly 19′, actually) at its perigee and Mercury 7′. Jābir reasons that Venus' apogee must at best be the same distance from the Earth as the Sun's perigee, and scales up the parallax accordingly—a procedure that is accurate enough for such small angles. He is evidently considering Mercury as fitting beneath the Sun separately, for if he had calculated for both planets' being beneath the Sun (when Mercury's apogee is the same distance from the Earth as Venus' perigee), then Mercury's maximum parallax becomes just over 47′. Jābir considers it possible that the parallaxes he has calculated have not been observed, since to show them the planets would have to be in conjunction with the Sun. So he calculates the parallaxes for maximum elongation, 6′ for Venus and 4′ for Mercury (actually, 7′ and $4\frac{1}{2}$′ are better figures). The argument is brought to a triumphant climax[63]:

> Cum ergo non inveniatur eis utriusque diversitas aspectus cui sit quantitas de qua sit curandum, secundum quod ipse dixit, et soli sit diversitas aspectus sensibilis cui sit quantitas de qua est curandum, quomodo sunt sub sole?

> Since, therefore, no parallax worth bothering about (according to Ptolemy) is to be found in either of them, and the Sun does have a sensible parallax worth bothering about, how can they be below the Sun?

Delambre[64] points out that Ptolemy's value for solar parallax is deduced, not observed, and that parallax of this order could not be measured with the instruments of the time; but he is wrong in saying that Jābir forgot the impossibility of observing Venus in inferior conjunction. By a lengthy demonstration Jābir now disposes of Ptolemy's statement that Venus and Mercury are never actually in line with the Sun. As for the "natural" order of the planets, he says that the two luminaries should be together, and the five planets—in many ways so alike—should also be together. Besides, if it were natural, as Ptolemy claims, that the Sun should separate the planets of limited elongation from the others, then the Moon should be with the outer planets.

Although the order of the planets adopted by Jābir was the order given by some of Ptolemy's predecessors, the astronomy of the *Iṣlaḥ* seems to be original where it differs from the *Almagest*. However, Jābir is not original in his spherical trigonometry and his insistence on separating theory from calculation. In both these matters he was anticipated by Abū-l-Wafā',[65] who flourished at Baghdad in the tenth century. Direct copying of Abu-l-Wafā''s work may be ruled out since his approach is more sophisticated than Jābir's, but the two authors could have been using a common source. For similar reasons, Jābir could not have been copying al-Bīrūnī's *Canon Masudicus*. It was through Jābir, however, that the Latin West learnt the spherical trigonometry developed by the Arabs.

Duhem[66] suggested that the *Iṣlaḥ* was a wholesale plagiarism from a lost Greek original—on the grounds that no Arabic author is quoted in it and that the order of the letters used in the theorems follows the Greek alphabet. The latter argument can be dismissed since the order of the letters is equally that of the "abjad" (or numerical) order of the alphabet, which is normally used for such purposes[67]. It is as difficult to disprove the plagiarism as it is to prove it, but until evidence of the Greek original—in the form of a MS, quotation, or reliable report—turns up, the presumption is that the *Iṣlaḥ* was originally written in Arabic by Jābir ibn Aflaḥ.

4. *Influence of the Iṣlaḥ Almajisṭī*

(a) In Arabic

An epitome of the *Iṣlaḥ Almajisṭī* was made by the Persian astronomer and physician Quṭb al-dīn al-Shīrāzī (1236–1311). An early MS is Thurst 3 in the Bodleian, ff 75r–93r, apparently of the thirteenth century[68]. It is hard to read—for one thing there is often an imprint of the opposite page confusing the text, as though the volume had been shut before the ink had dried. But an examination of the trigonometrical section shows that al-Shīrāzī followed the unrevised text and achieved great compression without omitting anything essential. His teacher, Nāṣir al-dīn al-Ṭūsī, did not, however, mention Jābir in his celebrated *Kitāb Shakl al-qaṭṭāʿ* on Menelaus' theorem and spherical trigonometry.

Jābir is mentioned in a *Compendium of the Almagest* attributed to Ibn Rushd. The attribution has some foundation, for, apart from ascriptions in the MSS themselves, the *Compendium* appears on a list of Ibn Rushd's works in MS Escurial 879[69]. It is now unknown in the original, but is extant in at least eleven MSS in a Hebrew translation (1231) by Jacob Anatoli. Samuel ben Judah of Marseille says at the end of Jābir's *Iṣlaḥ* that he was unable to find the *Compendium* in Arabic and had to make do with an inaccurate Hebrew translation, which he ascribed to one Nathan of Cento (ha-Meati). He says further[70]

> that the good found in [Ibn Rushd's book] was gleaned from this book, more precious than gold, yea, than fine gold [cf Ps. 19.11 and 119.12]; most of the wonderful good things found in this book were rejected from that book; and almost that there is no good in all the book except what is gleaned from this.

"This book" appears to be Jābir's *Iṣlaḥ* as Steinschneider thought, rather than the *Almagest* itself as Renan thought. Samuel is liberal with his insults—witness the scurrilous remarks he made about his contemporary

Levi ben Gerson[71]—and the *Compendium* may be much better than his remark suggests. Steinschneider gives[72] a synopsis of the work deduced from the headings in the MSS. It is divided into two parts: the first is on the spheres—their number, positions, and causes,— and the second is on the motion of the spheres and the measure of their separation. In the first part the author cites Jābir for his hypothesis about Venus and Mercury, and castigates him for failing to understand Ptolemy properly.

The *Compendium* was virtually unknown in the Latin West, though Pico della Mirandola, who read it in Hebrew, referred to it in his *Disputationes adversus Astrologos*.[73] One of his remarks was copied by Copernicus into his *De Revolutionibus*, I 10. There is a curious annotation to a copy of the first (1496) edition of the Peurbach-Regiomontanus *Epitome of the Almagest* in the British Museum. It occurs at the end of book I and applies to one of the first theorems proved by Menelaus' theorem, or the Rule of Six Quantities:—

> Nota quod iste modus proportionum sex quantitatum hic datus reductus est leviori via ad quattuor quantitates ab Averroe Commentatore in suo Almagesto, et in hoc libro in 33 at 34 huius; ideo facilius erit per quatuor quantitates, agendo per regulam trium rerum per 15 sexti.

Thus "Averroes the Commentator in his Almagest" reduces the six quantities to four by an easier method than that used at the end of Peurbach's first book. However, the author of the note is probably referring to the *Almagestum Parvum*, or some similar compilation, and not the *Compendium* at all. Al-Biṭrūjī (fl. late 12c.), who tried to make homocentric astronomy work quantitatively, was also well acquainted with the *Iṣlaḥ*. In the introduction to the *Principles of Astronomy*, Jābir is cited as one of two astronomers who opposed Ptolemy (Azarquiel was the other). Jābir's criticisms concerned[74]

> the order of the spheres of the sun, Venus, and Mercury. Moreover, Jābir corrected certain matters in (Ptolemy's) book according to the principles which Ptolemy had presented.

(b) In Hebrew

Jābir became known to Jewish astronomers both directly and indirectly through the Hebrew translations of Ibn Rushd and al-Biṭrūjī. His theories are discussed in the astronomical section of the encyclopaedic work *Mi-*

drash ha-Ḥokmah[75] (1247) by Judah ben Solomon ha-Kohen, where the criticisms of a certain David (perhaps David ibn Naḥmias) are reported, and in Levi ben Gerson's *Milḥamot Adonai*[76] (finished 1329). Again, the *Iṣlaḥ* was sometimes used as an elementary textbook. Samuel b. Judah studied it in his youth under Abba Marie, also called Sen Astruc[77]; and Kaleb Afendopolo (who in 1482 copied Paris 1024, the best MS of Samuel's version of the *Iṣlaḥ*) studied it under Mordicai Comtino and wrote a commentary on it, which is unfortunately lost except for an abstract in MS Paris 1024[78].

In the introduction to a set of astronomical tables for Perpignan (1361), the author, Jacob ben David ben Yomtob, lists Jābir's book V among his sources—others are Maimonides and Levi ben Gerson[79]. None of these names, however, seems to have been passed to the anonymous Latin translation, "Tabulae Jacobi filii David Bonaediei", at any rate in the Bodleian copy[80]. Other references to Jābir by Jewish scholars include a list of passages in the *Almagest* corrected by Jābir, which was compiled by Profiaṭ Duran about 1400[81], anonymous annotations to a Hebrew *Almagest*[82] and extensive quotations of the *Iṣlaḥ* by Serach ben Nathan, the seventeenth century Keraite writer[83].

(c) In Latin

In the Latin West the most interesting references belong to the fourteenth century or later. But in a tract starting[84]

> Incipit canones Magistri Gerardi Cremonensis in motibus celestium ...
>
> Quoniam cuiusque actionis quantitatem temporis ...

which is obviously connected with Azarquiel's *Canons*, Jābir's methods for finding certain arcs on the celestial sphere are given directly after Ptolemy's. Unfortunately there is no guarantee that the canons are Gerard's.

The reference in the *Speculum Astronomie* attributed to Albertus Magnus is disappointing. In a paragraph on improvements to the *Almagest*, before a description of al-Biṭrūjī's *Principles of Astronomy* we find[85]

> Corrigitur eciam apud Thebit motus octave spere seu stellarum fixarum in libro qui sic incipit "Ymaginabor speram ..." etc., et apud Iohannem Gebri Hispalensem motus Veneris et Mercurii in libro quem nominavit *Flores* suos, et apud alium ...

The motion of the eighth sphere—of the fixed stars—is also corrected by Thābit in a

> book which begins, "I shall imagine a sphere ..." etc.; and the motion of Venus and Mercury is corrected by Johannes Gebri of Seville in a book which he called his *Flowers*; and another ...

The author does not give an incipit for Jābir's treatise, as he does for nearly all the others (e.g. Thābit in the above quotation), and gratuitously calls Jābir "Johannes" and his book "Flores". He is evidently confusing Jābir with John of Spain, who translated Abū Maʿshar's *Flores Astrologiae*, and probably took the information about Venus and Mercury from the first page of al-Biṭrūjī's *Principles of Astronomy*, which had been translated by Michael Scot in 1217[86]:

> Et non diuersificati sunt aliqui ab eo preter ... Azarkel ... et Abu Mahomet Jeber Aven Aflah ispalensis in ordinatione celi Solis et Ueneris et Mercurii, et in locis particularibus sui libri in quibus accidit Tholomeo error; et sanavit ea Jeber et compleuit secundum radices quas posuit Tholomeus.

In other references Albertus Magnus seems to know him only as an Authority. On the question of the spherical figure of the Heavens, about which Jābir did have something to say, "Geber" appears with three other names to support a metaphysical statement with which he was not concerned[87]; and, together with Ptolemy and Thābit, he is cited in connection with gems[88].

Of uncertain date and provenance is the pseudo-Thābit's *De Quantitatibus Stellarum et Planetarum et Proportio Terre.* In this short tract Jābir is quoted in Gerard's translation for his statements of the relative diameters and volumes of Sun, Moon, and Earth.[89] Of course the real Thābit could not have quoted Jābir, but it is possible that his text was interfered with after translation into Latin.

Again of uncertain date is an anonymous commentary on the *Iṣlaḥ* in the Bibliothèque Nationale, written in a fourteenth century hand[90]. It begins with the correlation of the first six theorems in book I with the appropriate theorems of Theodosius' *Spherics*:—

> Geber in libro 30 figurarum ad probacionem sui propositi commodius habendam quedam premittit: prima autem eius proposicio eadem est primi Theodosii in primo; secunda autem proponit sensum quinque proposicionum primi libri Theodosii, scilicet 7e 8e 9e 10e 11e. Sue autem tercie ...

Jābir's theorems are re-stated in a more compact form, which resembles Al-Shīrāzī's treatment, but the style—particularly in the passages of commentary—is not unusually Arabic in character for an astronomical work. The commentary is therefore probably not a translation. Unfortunately, the unique MS ends abruptly in the middle of the fourth book, without an "explicit".

The planetary theory in the *Iṣlaḥ* did not pass unnoticed. In the eighth book of his *Tractatus de Intelligentiis et Motoribus Celorum*, Thierry of Freiberg (d. about 1311) says it is remarkable that Ptolemy's order for the planets is generally accepted when Jābir's is more reasonable[91]:—

> ... ordo iste communiter tenetur et acceptus est a Tholomeo de Almagesti, quod mirum videtur, cum alius ordo plus invitatur rationem quem ponit Geber approbans ipsum in sua astrologia.

He then repeats the argument about parallax and finishes with high praise for Jābir and his insight:[91]—

> Ex hoc ergo rationabiliter arguitur Venerem et Mercurium esse sursum Solem secundum Geber, qui fuit mirabilis considerator in hac facultate.

Jābir's trigonometry is discussed and developed in a few paragraphs beginning

> Nota in figura tercia decima primi libri Geber quod arcus AG subtensus recto potest esse quarta circuli et minor quarta et maior quarta.

and sometimes appended to Latin MSS of the *Iṣlaḥ*, and also in various collections of notes on astronomy and spherics.[92] Its further progress in the fourteenth century may be seen in the works of Richard of Wallingford[93] and Simon Bredon[94]. Even after the systematization of trigonometry by Regiomontanus in the 1460's (without giving Jābir any credit) and the publication of his *De Triangulis* in 1533, "Geber" was still referred to for his trigonometrical results by such mathematicians as Nuñez[95] and Savile[96].

George of Trebizond, the translator of the *Almagest* whom his contemporary Regiomontanus once called a "shameless and ill-tempered babbler"[97] was not at all sympathetic to Jābir's programme of correcting the *Almagest*[98]:—

> Geber vero unum librum premisit ut (sic ipse ait) excusaret se a figura sectoris; adde quod multa eorum quae sequuntur hac demonstratione probantur, quare hac figura reiecta multa labuntur. Unde Geber etiam ausus fuit statim incipiendo longissima uti oratione in qua saepius atque saepius errasse Ptolomeum repetit.

> Jābir set forth a book to obviate (so he said) the Sector Figure. Add that many of the subsequent results which are proved by this theorem will go unproved without this figure! Jābir also had the effrontery to begin straight away with a long diatribe in which he again and again asserts that Ptolemy was mistaken.

Copernicus, too, thought Jābir's criticisms of Ptolemy were ill-judged, but it is almost certain that his spherical trigonometry came ultimately from Gerard's translation of the *Iṣlaḥ*—whether or not through Regiomontanus' *De Triangulis* as intermediary.

NOTES

1. E.g. Henry Bates, *Magistralis compositio astrolabii* (1274), Venice 1485, ff c8r and d4r-v (see E. Littré, *Histoire Littéraire*, 26 (1873) 558–62);
   Roger Bacon, *Communia Mathematica*, pars 1 dist. 3 cap. 5, *Opera Hactenus Inedita* XII (1950) p. 49, ed. Steele.
   Henry of Hessen, *Tractatus contra Astrologos*, pars II cap. 5, H. Pruckner's *Studien zu den astrologischen Schriften des Heinrich von Langenstein* (1933) p. 181.
   Regiomontanus, annotation to *Almagestum Parvum* in MS Nuremberg Cent VI 12 f 59r, and elsewhere.
   Maurolycus, preface to *Sphaerica* (1558), f 45v.
2. On the title-page of his copy of Jābir's astronomy (pr. 1534) now in the Uppsala University library, shelfmark W 16; also reported on the title-page of a copy in the Bodleian (Savile X 3).
3. M. Friedländer's translation, 1904 (Dover reprint 1956), p. 164.
4. See notes 74 and 86.
5. See, e.g., M. Steinschneider, *Die hebraeischen Übersetzungen des Mittelalters und die Juden als Dolmetscher*, 1893 (repr. 1956), p. 356; and P. Duhem, *Le Système du Monde*, 1913–17 (repr. 1965), p. 131.
6. S. Munk, footnote to his edition of the *Guide des Égarés*, 1856, pp. 81–2.
7. F. J. Carmody, "The Planetary System of Ibn Rushd", *Osiris* 10 (1952), 556–86. See Section 4 (a).
8. *De Motibus Planetarum*. See *De Reprobatione Eccentricorum et Epiciclorum*, MS Prag 1272, f. 45vb.
9. Ibn Khallikān, *Biographical Dictionary*, tr. De Slane, 1843–71, II, 324–6.
10. *Ibid*, II, 599–600.
11. *Ta'rīkh al-ḥukamā*, ed. Lippert, 1903, p. 343.
12. *Encyclopaedia of Islam*, I (1960) 111 and III (1971) 956–7.
13. "Cheber Benaflah de Sevilla", *Revista de la Sociedad Matemática* Española, Nov. 1911, pp. 113–120. p. 115; *Biografias de Matemáticos Árabes que florecieron en España*, 1921, pp. 69–73, p. 70.
14. Reported by G. Sarton, *Introduction to the History of Science*, III 1521. See also M. Steinschneider, *Zur Pseudepigraphischen Literatur*, 1862, pp. 14–7; and Scholem's edition and German translation, 1926 and 1927.

15. E.g. by Ahlwardt, the cataloguer of the Berlin Arabic MSS.
16. *Bibliotheca Hispana Vetus*, 1788, II 404.
17. *A Dictionary of the English Language*, 1755.
18. *O.E.D.*, 1933.
19. Heinrich Hermelink, "Tabulae Jahen", *Arch. for the Hist. of the Exact Sciences*, (1964), 108–12.
20. *Loc. cit.* (note 16).
21. f 463r in the first edition (1603).
22. E.g. Rodrigo Caro, *Antiguedades ... de Sevilla*, 1634, cap. 11.
23. M. Watt, *A History of Islamic Spain*, 1965, p. 80; *Enciclopedia Universal Illustrada Europeo-Americana*, LX 878.
24. A. Sayili, *The Observatory in Islam and its Place in General History*, 1960, Publications of the Turkish Historical Society no. 38, pp. 182–3.
25. *Ibid.*
26. An edition is being undertaken by M. Hugonnard-Roche in Paris.
27. *HU* (see note 5) p. 543; "Vite di matematici arabi tratte da un'opera ined. di B. Baldi con note", Boncompagni's *Bullettino* (1877), p. 526.
28. Not identical with Ibn ʿAqnīn. See Encyclopaedia Judaica, 1971, II 757 (art. by S. T. Rabinowitz); S. Munk, "Notice sur Joseph b. Jehouda", Journal Asiatique, 3rd series, XIV (1842), pp. 5–70, 8 foll.
29. Lippert (see note 11), p. 319.
30. *Ibid.* 392–3.
31. Friedländer (see note 3), p. 1.
32. Munk, "Notice" (see note 28), p. 20 foll.
33. *Ibid.*
34. Friedländer (see note 3) p. 1.
35. Translation by S. Pines (1963, p. 322), which is more literal than Friedländer's (p. 196).
36. Munk gives "kitāb al-majisṭī" and "le livre de l'Almageste" in "Notice" (see note 28) p. 34.
37. B. Boncompagni, *Delle vita e delle Opere di Gherardo Cremonese ...*, 1851; K. Sudhoff, "Die kurze "Vita" und das Verzeichnis der Arbeiten Gerhards von Cremona", *Arch. für Gesch. der Medizin* 8 (1914) 73–82.
38. Daniel of Morley, *Philosophia*, near the end. See V. Rose, *Hermes* VIII pp. 332 and 343.
39. C. H. Haskins, *Studies in the History of Medieval Science*, 1924 (repr. 1960), p. 104 note 139.
40. *Ibid.* pp. 107–8.
41. Including the fragment BN heb 1102.
42. See the entries in F. J. Carmody, *Arabic Astronomical and Astrological Sciences in Latin Translation*, 1956, and Thorndike and Kibre, *Catalogue of Incipits of Medieval Scientific Writings in Latin*, 1963. Two other MSS are Seitenstetten 2° 53 in Regiomontanus' hand, and Milan Ambros. Heb. 97 ff 20 r (56v) – 31r (45v) of about 1500 A.D.
43. *Lexicon Bibliographicum*, ed. with Latin tr. by G. Fluegel, 1850, vol. V pp. 387–8.
44. *Studies* (see note 39) p. 104. This text is now being edited.
45. *Opera Omnia*, 1891, X 631–2.
46. Basel F II 33, f 244r.

47. Erfurt, Amplon. F 383. See Schum's catalogue.
48. MSS Florence, Laur. Conv. Sopp. 414; Prag 802 or V.A. 11 (according to Birkenmajer – see note 49 – p. 145); and Munich CLM 56 respectively.
49. Aleksander Birkenmajer, "La bibliothèque de Richard de Fournival", *Studia Copernicana*, I 117–210, tr. from his 1922 art. in Polish, pp. 145–7.
50. Bodleian, Opp. Add. fol. 17, ff. 1–100r.
51. In his remarks at the end of the *Iṣlaḥ*. See note 52.
52. B. N. heb. 1024, ff 107r–108r. There is a transcription (though not from this MS) by Neubauer and Renan on pp. 562–3 of "Les Écrivains Juifs Français", *Histoire Litéraire* XXXI (1893) pp. 351–789. They give a synopsis in French on pp. 560–1. Steinschneider gives a German synopsis in *Mathematik bei den Juden*, 1964 reprint, pp. 134–5. The MSS of Samuel's revision are in the Bibliothèque Nationale, nos. 1024, 1014, 1025, 1036.
53. Bodleian MSS Hunt. 96, ff 40v–42r, and Heb. d 4, ff 165r–169r, Neubauer's cat. nos. 2008 and 2773 respectively; Berlin Q 747 (Steinschneider's cat., no. 204), ff 156–8.
54. Following the Thābit tract in Hunt. 96, ff 42r–v, and Heb. d 4, ff 169r–177r.
55. Berlin Q 747, ff 133v–134v. Mentioned by A. G. Kapp, "Arabische Übersetzer und Kommentatoren Euklids", *Isis* 23 (1935) 54–100, p. 71.
56. Printed edition, pp. 1 and 2; MS Madrid 10006, ff 1r and 1v. The quotations in this section are in Gerard's translation – according to the printed text unless the Madrid MS offers a more plausible reading. Thus "figura" and "compositos" are supplied from the MS.
57. Referring to *Nicom. Eth.* 16, 1096[a] 11–17.
58–63. Printed edition, pp. 48, 21, 1, 59–60, 103, 104; MS Madrid 10006, ff 39r, 14v, 1r, 51r, 101r, 101v, respectively. "legentem ipsum et facit eum hesitare et facit ipsum stare" is a Madrid 10006 reading (f 1r).
64. *Histoire de l'Astronomie du Moyen Age*, 1819 (repr. 1965), p. 184.
65. See Carra de Vaux, "L'almageste d'Abū-l-Wéfa" *Journal Asiatique*, 8th series, XIX (1892), pp. 408–71.
66. *Système du Monde*, II 172–9, esp. 177–9.
67. See F. J. Carmody in his introduction (p. 27) to his edition of al-Biṭrūjī's *De Motibus Celorum*, 1952.
68. Other MSS are Marsh 713 in the Bodleian and Florence Pal. 315 (in Persian). See Brockelmann, *Gesch. der Arabischen Litteratur*, Supplement I, p. 858.
69. E. Renan, *Averroës*, prem. partie, chap. I, ix – pp. 64–79 of fourth edition.
70. See note 52. In the transcription, "bō" is missing in the last line of this passage, but occurs in MS BN 1024.
71. Neubauer & Renan (See note 52) p. 567.
72. *HU* (see note 5) pp. 546–9.
73. Book X, chap. iv.
74. B. R. Goldstein's translation, 1971, Vol. I, p. 53. See also pp. 98, 113, 122, 124, 125.
75. Steinschneider *HU* (see note 5) pp. 1–4, 524, 544; *Math. Jud.* (see note 52) p. 100.
76. Duhem, *Système du Monde*, V 206–7.
77. See note 52.
78. Steinschneider, *HU* (see note 5) p. 519; *Math. Jud.* (see note 52) p. 205.
79. Steinschneider *Math. Jud.*, (see note 52) p. 153.

80. Can. misc. 27 ff 1–122r (15c).
81. Steinschneider, *HU* (see note 5) p. 544. The MS contains other items by Profiat Duran. See Steinschneider, *Op. Cit.* p. 547.
82. Steinschneider, *Baldi* (see note 27) p. 526. MS is Munich 70.
83. Steinschneider, *HU* (see note 5) p. 544.
84. Bodleian, Can. misc. 51 ff. 73r–85v.
85. MS Erfurt Amplon. Q 223 ff 106r–v. Erfurt Amplon. Q 349 differs only trivially from it. The 1891 edition is corrupt ("Ulgembum" for "Gebri").
86. Carmody, *Al-Biṭrūjī* (see note 67) p. 71.
87. *Comm. in II Sent.*, art VI, *Op. Om.* (1894) vol. 27.
88. *Mineralium* II iii 3 (*Op. Om.* V 189). See Thorndike, *Hist. Magic and Exp. Sci.* II 557.
89. See F. J. Carmody, *The Astronomical works of Thābit ben Qurra*, 1960, pp. 145–6 and 119; F. S. Benjamin, "The *De Quantitatibus Stellarum* of Thabit ben Kourrah", *Essays in medieval life and thought presented in honour of A. P. Evans*, pp. 91–8.
90. Paris, BN 7406 ff 114ra–135rb.
91. MS Vatican Apost. Lat. 2183, ff 191ra and b: Chapter 8 is "De ordine sperarum celestium et ratione quam ponit Geber super uno modo ordinis quem ponit" (f 188rb). For further reference see Krebs, "Meister Dietrich (Theod. Teut. de Vriberg), sein Leben, seine Werke, seine Wissenschaft", *Beiträge z. Gesch. der Phil. des Mittelalters*, 1906, Band V, Heft 5–6; also Duhem, *Système du Monde*, III 386 foll.
92. E.g. Madrid 10010 ff 49r–52r, Bodleian Digby 168 F 126v. Also as marginal notes to Brit. Mus. Harley 625.
93. Principally in the *De Sectore*, a variant of the *Quadripartitum*, and the *Tractatus Albionis* (book I theorems 12, 13, 16, 18, 21). I am indebted to Dr. J. D. North for these references.
94. In his *Commentary on the Almagest*, of which the best MS is Bodleian Digby 168 ff 21r–39r, 14c.
95. *De Arte atque Ratione Navigandi*, 1573, chapters 21–23.
96. In his lecture-notes: e.g. Bodleian MS Savile 27, ff 87v, 88r, 114r, 116r, etc.
97. Ernst Zinner, *Leben und Wirken des Joh. Müller von Königsberg, genannt Regiomontanus*, 2nd edn., 1968, p. 156.
98. *Commentum in C Ptolemaei Magnam Constructionem*, Bodleian MS Laud Latin 111 f 20r.

VII

Appendix 1 to item VI

The Manuscripts of Jābir b. Aflaḥ's Treatise

1. Arabic

To the manuscripts given on p. 88 may be added[1]:

London, British Library, or. 10725, ff. 92v–175v
Cambidge, University Library, Add. 1191, ff. 19r–131r, 1354 AD

They are both in Hebrew script. In the London manuscript the pages are out of order and some parts seem to be missing. The Cambridge manuscript is in the same group as Berlin 5653 (= Landberg 132) in the trigonometric passage in book I.

2. Latin

The following is a list of all manuscripts of Gerard's translation known to me:

Cambridge, University Library Mm. II. 18 (2313), ff. 1ra–49vb, 14c.
Cesena, Biblioteca Malatestiana, Plut. sin. XXVII 2
Cracow, Biblioteka Jagiellońska 569, pp. 132–235, 14c.
Cracow, Biblioteka Jagiellońska 1921, pp. 8–242 (ff. 5v–122v), 14c.
Darmstadt, Hessische Landes- und Hochschulbibliothek 1987, 46ff., 15c.
Dresden, Sächsische Landesbibliothek Db 87, ff. 162r–268r, ca. 1300
Erfurt, Wiss. Bibliothek, Amplon. F 382, ff. 3r–110r, 15c.
Florence, Biblioteca Laurenziana, San Marco 215, ff. 1r–124r, ca 1400
Jena, Universitätsbibliothek, El. fol. 67, ff. 93r–139r
London, British Library, Harley 625, ff. 8v–84v, 14c.
London, Royal College of Physicians 383, ff. 162r–178v, 13c.

---

[1] I am grateful to Professor B. Goldstein for telling me about these manuscripts before publishing the information in Goldstein (1979), p. 32 (see bibliogrpahy to Appendix 2).

Madrid, Acad. de la Historia, Bibl. Cortes 2787 (formerly 676)
Madrid, Biblioteca Nacional 10006, 150ff., 14c.
Milan, Biblioteca Ambrosiana, G 70 sup., ff. 65ra–121ra, 14c.
Nuremberg, Stadtbibliothek Cent. III 25, 96r–130r, 1460. Written by Regiomontanus.
Oxford, Bodleian Library, Ashmole 357, ff. 97r–178v (end missing), 14c.
Oxford, Corpus Christi College 233, ff. 32r–64r
Paris, Bibliothèque de l'Arsenal 1035, ff. 1r–79v, 14c.
Paris, Bibliothèque Nationale, lat. 16198, ff. 74r–122v, 14c.
Philadelphia, University of Pennsylvania, lat. 3, 13c.
Seitenstetten, cod. Folio 53, 110 ff., 15c. Written by Regiomontanus.
Vatican, lat. 2059, ff. 1r–192v, 14c.
Vatican, lat. 3096, ff. 14v–140r, 14c.
Vatican, Ottob. lat. 2234, ff. 1ra–40vb, 41ra–42ra
Venice, BN, Marc. lat. XI 60 (=VIII 14) , ff. 1–47, 14c.
Vienna, Nationalbibliothek 5392, ff. 1r–218r, 15c.
Vienna, Nationalbibliothek 10905, ff. 1r–142v, 16c. Written by Johannes Vögelin.

Not all the above are complete. Further fragments and excerpts are to be found in:

Erlangen, Universitätsbibliothek 836 (formerly 786), ff. 98v–109r. Written by Praetorius.
Milan, Biblioteca Ambrosiana, D 114 Inf 59r(56v)–69v(46r) 16c.
Oxford, Bodleian, Digby 178, ff. 107r–108r, 15c.
Paris, Bibliothèque Nationale, lat. 7267, ff. 24ra–26ra, 14c.

VIII

Appendix 2 to item VI

Jābir ibn Aflaḥ and the Establishment of Trigonometry in the West

1. Introduction and Summary

It was largely through Gerard's translation of Jābir's *Iṣlāḥ al-majasṭī* that the Latin West learnt of the systematic trigonometry developed by mathematicians in the Arabic-speaking world. In spherical trigonometry the principal alternative was Menelaus' theorem (or the "sector figure"), which had been used by Ptolemy. A compound proportion involving six quantities and referring to a complete spherical quadrilateral, this was less easy to apply than Jābir's simple proportions referring to a spherical triangle. It is very unlikely that Jābir discovered or re-discovered any of these results, but his *Iṣlāḥ* was the vehicle by which they were transmitted to the West. In plane trigonometry Jābir's procedures were esssentially Ptolemaic, but he demonstrated them explicitly as mathematical procedures and did not just apply them when necessary, as in the *Almagest*. Jābir's spherical trigonometry was expressed in sines and his plane trigonometry in chords. He did not use the shadow functions (tangent and cotangent) and in these respects his mathematics may be considered more primitive than that of many of his predecessors.

There are some twenty manuscripts of the astronomy of "Geber" in Latin; and in 1534 it was printed by Peter Apian. Both the spherical and the plane trigonometry were the subject of marginal annotations and commentaries. In some cases the writers extended the results to triangles with obtuse angles. Apart from isolated notes in "Geber" manuscripts two series of such extensions are considered here: in a disparate collection of notes and in an incomplete commentary. For plane trigonometry there is also the anonymous work *De tribus notis*. The end of the medieval phase of Western trigonometry came in the 1460s, when Regiomontanus (Johann Müller of Königsberg) wrote his *De triangulis omnimodis libri quinque*, which systematized the trigonometry available to him — a substantial part being that transmitted by the *Iṣlāḥ* and commentaries on it. Printed in 1533, long after Regiomontanus' death in 1576, the *De triangulis* became the standard work on the subject in the Latin world. Copernicus presented what is essentially Jābir's trigonometry in the first book of his *De revolutionibus orbium coelestium*, printed 1543, and also in his *De lateribus et angulis triangulorum*, an extract (with trivial variations) of the trigonometrical portion of the *De revolutionibus* printed a year earlier. Although Copernicus' direct dependence on Regiomontanus has been questioned[1], his trigonometry is clearly based upon the tradition of which the *De triangulis* was a notable part.

2. Spherical Trigonometry

Jābir reduces all problems in spherical astronomy to solving triangles, usually by dividing them into two right-angled traingles. His spherical trigonometry comprises three main results (theorems 13–15 below) together with the so-called Rule of Four Quantities (th. 12), from which they are derived, and a set of rules (th. 11) for deciding whether the arcs determined from the theorems are less than, equal to or greater than a quadrant of a great circle and whether the angles determined are acute, right or obtuse.

In the summary below the following conventions are used. All circles are great circles on a sphere. The pole of a circle is one of the two points on the surface of the sphere equidistant from points on the circle; the pole of the circle of which AB is an arc is here denoted by pole(AB). The arc of an angle between two great circles intersecting at A (say) is the arc cut off by these great circles from the circle of which A is the pole; it is denoted by $\hat{A}$; rt. (of angles) means right. $\frac{1}{4}\odot$ means a quadrant of a great circle on the sphere.

Two or three diagram letters together (say AB or ABG) mean the arc (AB or ABG), except the diameter AB and the sides of triangles LGK and DMT in th. 12, which are straight lines. The complement of an arc (AB say) is denoted by overlining ($\overline{AB}$). Sin AB means $R \sin AB$ in modern terms (with R = the radius of the sphere); Cos AB = Sin $\overline{AB}$ = $R \cos AB$.

Only the essential mathematical argument is reproduced. In theorems 12, 14 and 15 formal enunciations are omitted and the results stated only by

[1] See e.g. Zeller, pp. 53–56.

reference to the diagrams. Square brackets, [ ], are used for things implied by the text, but not explicitly stated, or for summaries and obtrusions by the compiler. Subdivisions of theorems, indicated by numbers in brackets, have been added for clarity.

In the main the printed text of 1533 has been followed, though Arabic and Latin manuscripts have also been consulted. The numbering of the propositions is that of the printed edition; it is also found in many Latin manuscripts.

Jābir's propositions on spherical trigonometry

11 In the solution of triangles — and other figures are resolved into triangles and these are divided into right triangles — the unknown arcs [and angles] are found only as sines. A sine is common to two arcs, one less than $\frac{1}{4}\odot$ and one greater. Therefore rules for determining whether an arc found in a right triangle is greater or less than a $\frac{1}{4}\odot$ are given here.

[1] In a right triangle each side containing the right angle imitates the angle opposite it: if the angle $>$, $=$ or $<$ rt., the side $>$, $=$ or $<$ $\frac{1}{4}\odot$ and *vice versa*. [The three case are proved individually with similar proofs. E.g.]

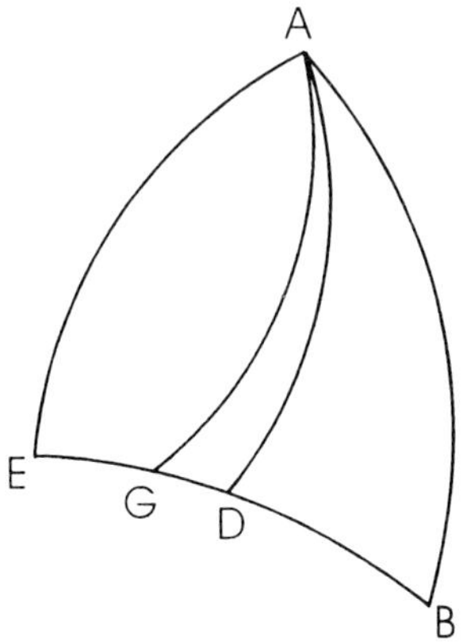

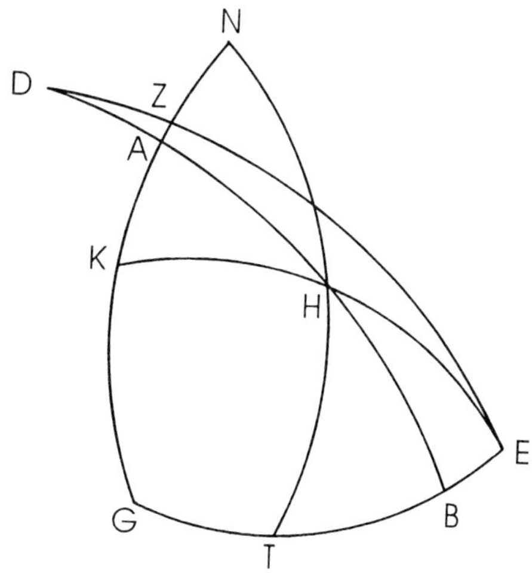

Let $\hat{B}$ be rt. [See first diagram]
Let $\hat{A} > \frac{1}{4}\odot$. *I say*: $BG > \frac{1}{4}\odot$
Proof. Let $B\hat{A}D$ be rt.
$\therefore D$ is pole (AB) as above [i.e. $\because \hat{B}$ is rt., pole (AB) lies on $\odot DB$; and $\because \hat{A}$ is rt., pole (AB) lies on $\odot AD$]
$\therefore BD = \frac{1}{4}\odot$
$\therefore BG > \frac{1}{4}\odot$

[2] In a right $\triangle$ ABG, $\hat{B}$ rt., [i] if one of AB, $BG = \frac{1}{4}\odot$, AG also $= \frac{1}{4}\odot$; [ii] if both AB, BG [a] $<$ or [b] $> \frac{1}{4}\odot$, $AG < \frac{1}{4}\odot$; [iii] if one of AB, $BG >$ and the other $< \frac{1}{4}\odot$, $AG > \frac{1}{4}\odot$. [Again the proofs are similar. E.g. ii, a]

AB, $BG < \frac{1}{4}\odot$ [See second diagram]
Let $BD = GE = \frac{1}{4}\odot$: join DE and let DE cut $\odot AG$ at Z
$\because \hat{B}$ is rt. and $BD = \frac{1}{4}\odot$, $D =$ pole (BG)
$\therefore D\hat{E}G$ is rt.
$\because D\hat{E}G$ is rt. and $EG = \frac{1}{4}\odot$, G is pole (DZE)
$\therefore GAZ = \frac{1}{4}\odot$
$\therefore AG < \frac{1}{4}\odot$

Similar conditions apply to the angles because of their imitation of the sides opposite them.

The converses — i.e. if $AG = \frac{1}{4}\odot$, one of $AB, BG = \frac{1}{4}\odot$, etc. — are also true [and are proved by *reductio ad absurdum*].

12 [The "Rule of Four Quantities"[2]] If $AGDB$ and $AEZB$ are two great circles, neither passing through the poles of the other [and intersecting at A, B] and if G, D are points on one of the circles (or one on each circle) and GE, DZ are perpendiculars to the other circle, then

$$\text{Sin } AG : \text{Sin } GE = \text{Sin } AD : \text{Sin } DZ^{3}$$

[1] First let G, D be both on $\odot AGDB$; and let EG and ZD be arcs through the poles of $\odot AE$ [and $\therefore$] cutting $\odot AEZ$ at right angles.

Proof. Let GK, $DT \perp$ plane of $\odot AEB$

[2] It has yet to be established that this term applied exclusively to this result and not to this and other results involving four quantities. It is used here for convenience.
[3] Subsequently this result will be indicated by the notation $A{:}GE,DZ$.

Let GL, $DM \perp$ diam. AB of $\odot AGD$
Draw KL, TM
$\because GK \parallel DT$ and $GL \parallel DM$, $\therefore L\hat{G}K = M\hat{D}T$
But $\hat{K} = \hat{T} =$ rt.
$\therefore \triangle LGK \sim \triangle DMT$
$\therefore GL : GK = DM : DT$
But $GL = \text{Sin}\, AG$, $GK = \text{Sin}\, GE$, $DM = \text{Sin}\, AD$, $DT = \text{Sin}\, DZ$
$\therefore \text{Sin}\, AG : \text{Sin}\, GE = \text{Sin}\, AD : \text{Sin}\, DZ$ Q.E.D.

[Corollaries] $\because \text{Sin}\, AG = \text{Sin}\, GB$ and $\text{Sin}\, AD = \text{Sin}\, DB$
$\therefore \text{Sin}\, BG : \text{Sin}\, GE = \text{Sin}\, BD$[4] $: \text{Sin}\, DZ$
and $\text{Sin}\, AG : \text{Sin}\, GE = \text{Sin}\, BD : \text{Sin}\, DZ$

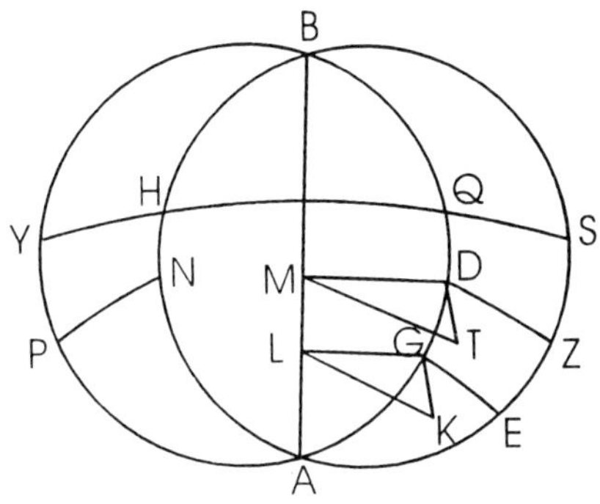

[2] Let G be a point on $\odot AGD$ and N a point on $\odot AEZ$. Arc NP is drawn at right angles to $\odot AGD$

I say: $\text{Sin}\, AG : \text{Sin}\, GE = \text{Sin}\, AN : \text{Sin}\, NP$

Proof. Draw great circle $YHQS$ through the poles of circles AGD, AEZ: it bisects their semicircular arcs at right angles
$\therefore AQ = AS = BQ = BS = AH = AY = BH = BY = \frac{1}{4}\odot$
And $\because$ great circles bisect each other [so that $YQ = SH$], $YH = SQ$
$\therefore$ Sin (any of the 8 arcs [AQ etc]) : Sin (YH or SQ) is the same ratio
$\because N$, H are on $\odot ANH$, $YH \perp [\odot APY]$,
$\therefore \text{Sin}\, AN : \text{Sin}\, NP = \text{Sin}\, AH : \text{Sin}\, HY$ [A:NP,HY]
Similarly $\text{Sin}\, AG : \text{Sin}\, GE = \text{Sin}\, AQ : \text{Sin}\, QS$ [A:GE,QS]

[4]MSS Berlin 5653, Madrid, BN 10006 and British Library, Harley 625 have *AD* here.

But $\text{Sin}\,AH : \text{Sin}\,HY = \text{Sin}\,AQ : \text{Sin}\,QS$
$\therefore \text{Sin}\,AG : \text{Sin}\,GE = \text{Sin}\,AN : \text{Sin}\,NP$

13 [The sine rule for spherical triangles]

Sin (each side of a triangle) : Sin (angle opposite it) is one ratio.

The triangle may have 3, 2, 1 or no right angles.

[1] If there are three right angles, their arcs must be each $\frac{1}{4}\odot$ and the sides must be each $\frac{1}{4}\odot$. So the above ratio is that of equality in each case.

[2] If there are two right angles, the point of the remaining angle is the pole of the side subtending it, which is therefore the arc of this angle. So the above ratio in this case is that of equality. This is also true of the other two cases, $\because$ each arc $= \frac{1}{4}\odot$ and each subtended angle is right.

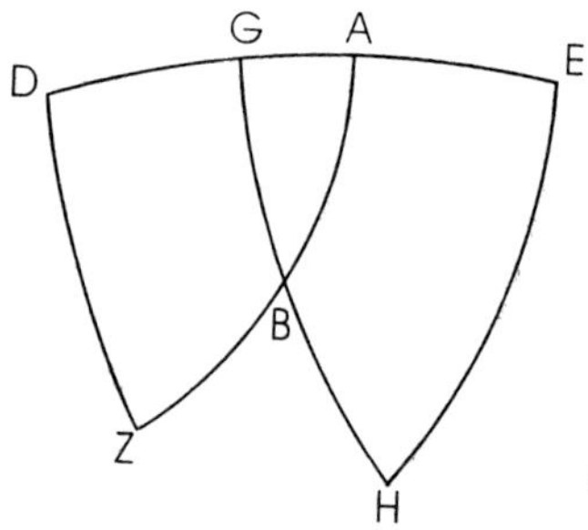

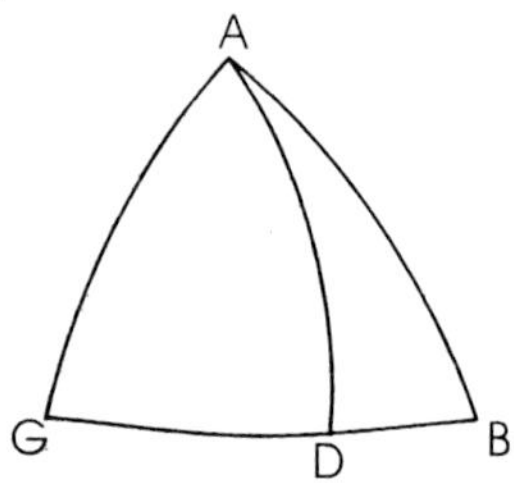

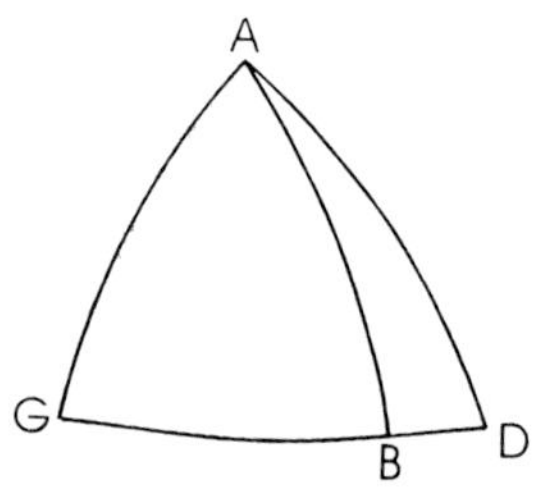

[3] If there is one right angle, let it be $\hat{B}$ in $\triangle ABG$ [See first diagram]

I say: $\text{Sin}\,AB : \text{Sin}\,\hat{G} = \text{Sin}\,BG : \text{Sin}\,\hat{A} = \text{Sin}\,AG : \text{Sin}\,\hat{B}$
Proof. Let $AD = GE = \frac{1}{4}\odot$ and $AZ = GH = \frac{1}{4}\odot$; draw EH, ZD.
$\therefore$ G is pole(EH) and A is pole(DZ)
$\therefore$ G:AB,EH
$\therefore \text{Sin}\,AG : \text{Sin}\,GE = \text{Sin}\,AB : \text{Sin}\,HE$
Similarly $\text{Sin}\,AG : \text{Sin}\,AD = \text{Sin}\,GB : \text{Sin}\,DZ$ [A:GB,DZ]
But $GE = AD = \frac{1}{4}\odot$ and $\therefore$ each is arc of $\hat{B}$ (rt.)
And EH is arc of $\hat{G}$ and DZ is arc of $\hat{A}$
$\therefore \text{Sin}\,AG : \text{Sin}\,\hat{B} = \text{Sin}\,GB : \text{Sin}\,A$

Similarly Sin AG : Sin $\hat{B}$ = Sin AB : Sin $\hat{G}$
$\therefore$ Sin BG : Sin $\hat{A}$ = Sin AB : Sin $\hat{G}$[5] Q.E.D.

[4] Let there be no right angles in $\triangle ABG$

Draw AD from A [to BG] through pole(BG)
[i] [D] falls between B and G [See second diagram]
In $\triangle ADG$ $\hat{D}$ is rt.
$\therefore$ Sin AG : Sin AD = Sin $\hat{D}$(rt.) : Sin $\hat{G}$
And in $\triangle ADB$ $\hat{D}$ is rt.
$\therefore$ Sin AD : Sin AB = Sin $\hat{B}$: Sin $\hat{D}$
$\therefore$, *ex aequali perturb.*[6], Sin AG : Sin AB = Sin $\hat{B}$: Sin $\hat{G}$
$\therefore$, *alternando*, Sin AG : Sin $\hat{B}$ = Sin AB : Sin $\hat{G}$
Similarly, if a $\perp^{r}$ is drawn from B to AG,
Sin AB : Sin $\hat{G}$ = Sin BG : Sin $\hat{A}$[7]
[ii] If $\perp^{r}$ AD falls outside the triangle, the same argument applies [see third diagram]:
$\therefore$ Sin AG : Sin AB = Sin $A\hat{B}D$: Sin (arc $\hat{G}$)
But Sin $A\hat{B}D$ = Sin $A\hat{B}G$, $\because$ the angles are supplementary
$\therefore$ Sin AG : Sin AB = Sin $A\hat{B}G$: Sin $\hat{G}$ Q.E.D.

In theorems 14 and 15 $\hat{B}$ of $\triangle ABG$ is right.

---

[5] For the last three lines in MSS Berlin 5653 and Cambridge add. 1191 there is the equivalent of the simpler "Sin AG : Sin $\hat{B}$ = Sin AB : Sin $\hat{G}$ = Sin BG : Sin $\hat{A}$".

[6] The printed text has: "ergo in proportione æqualitatis secundum proportionem muthrariba". In Sir Henry Savile's copy of the printed text, Bodleian Library, Savile X3, "aequalitatis" is changed to "aequa" and there is a note in his hand (as is confirmed by comparison with his lecture notes in the Bodleian manuscript Savile 27) against the last word, which is evidently a corruption of the Arabic *muḍṭariba* (there are other corruptions in the manuscripts): "perturbatam τετραγμένην". See Heath's translation of Euclid, II 115,136. A similar Arabic term confused Suter in his translation of the same part of Abū Naṣr's proof of the sine rule (see Suter (1909), p. 159, n.2). The complete phrase in Berlin 5653, f.10v, l. 14, is *fa-fī nisbat al-musāwāt ʿalā al-nisba al-muḍṭariba* (the reading fa-fī is not clear in the Berlin MS and has been taken over from MS Cambridge, add. 1191). The Paris commentary has in this place "secundum proportionalitatem mutharyba (id est indirectam)" and "per indirectam proportionalitatem" for a similar passage later. Regiomontanus has "per aequam proportionalitatem indirectam" and "per aequam indirectam".

[7] This line, "Similarly ... Sin $\hat{A}$", seems misplaced, but Berlin 5653 also puts it here.

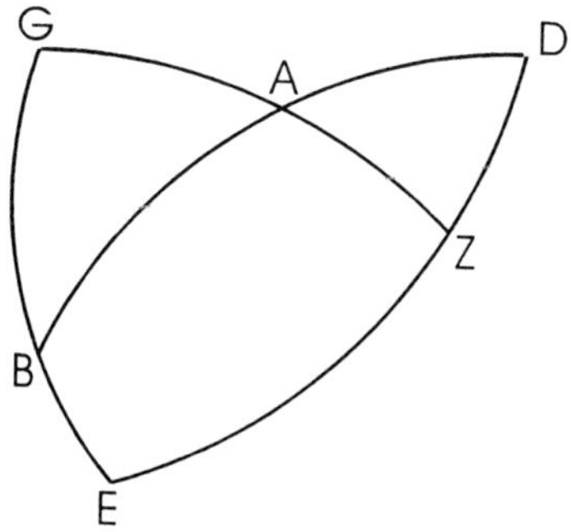

14 $\text{Sin}\,\hat{A} : \text{Sin}\,\hat{B}\,(\text{rt.}) = \text{Cos}\,\hat{G} : \text{Cos}\,AB$[8]

Proof. Let $DB = \frac{1}{4}\odot$; let $DZ \perp AG$ and let it meet BG at E
$A{:}GB,DZ$
$\therefore \text{Sin}\,BG : \text{Sin}\,AG = \text{Sin}\,DZ : \text{Sin}\,AD$
But $\text{Sin}\,BG : \text{Sin}\,AG = \text{Sin}\,\hat{A} : \text{Sin}\,\hat{B}\,(\text{rt.})$ [th. 13]
And DZ is the arc of the complement of $\hat{G}$[9]; and $AD = \overline{AB}$
$\therefore \text{Sin}\,\hat{A} : \text{Sin}\,\hat{B}\,(\text{rt.}) = \text{Cos}\,\hat{G} : \text{Cos}\,AB$ Q.E.D.

15 $\text{Cos}\,AG\,(\text{opp. rt.} \angle) : \text{Cos}\,BG = \text{Cos}\,AB : \text{Sin}\,\frac{1}{4}\odot$

Proof. In the same figure $D{:}AZ,BE$
$\therefore \text{Sin}\,AZ : \text{Sin}\,EB = \text{Sin}\,AD : \text{Sin}\,DB$
But $AZ = \overline{AG}$, $EB = \overline{BG}$[9], $AD = \overline{AB}$, $DB = \frac{1}{4}\odot$
$\therefore \text{Cos}\,AG : \text{Cos}\,BG = \text{Cos}\,AB : \text{Sin}\,\frac{1}{4}\odot$ Q.E.D.

As a contrast we may give two forms of Menelaus' theorem (see diagram for Props. 14–15):

$$\text{Sin}\,GA : \text{Sin}\,AZ = (\text{Sin}\,BG : \text{Sin}\,EB) \cdot (\text{Sin}\,DE : ZD)$$
$$\text{Sin}\,GZ : \text{Sin}\,ZA = (\text{Sin}\,EG : \text{Sin}\,BE) \cdot (\text{Sin}\,DB : AD)$$

[8] In the Escorial manuscripts it is proved that $\text{Sin}\,AG : \text{Sin}\,BG = \text{Cos}\,AB : \text{Cos}\,\hat{G}$ in place of this result. In MS Escorial 910 this is proved by omitting the line depending on th. 13.

[9] Since $\hat{G}$ is pole(DZE), for $\hat{Z}$ is right by definition and $\hat{E}$ is right because D is pole(GBE).

In fact the second relation yields Jābir's theorem 15, but it is clear even in this case that it is easier to apply the ready-made result to triangle ABG than to complete the quadrilateral in the right way and to contrive the right form of Menelaus' theorem.

The origin of Jābir's spherical trigonometry

Jābir was not the original discoverer of these results, for about the year 1000 AD a group of mathematicians in the East — notably Abū 'l-Wafā' al-Būzajānī, Abū Naṣr ibn ʿIrāq (al-Bīrūnī's teacher) and Abū Maḥmūd Ḥāmid al-Khujandī — also produced similar series of theorems, starting with essentially the same proof of the Rule of Four Quantities[10]. At the turn of the century Braunmühl suggested that Jābir reinvented his spherical trigonometry and that his evident acquaintance with Thābit's treatise on Menelaus' theorem, containing as it does almost a complete proof of the Rule of Four Quantities, would have been sufficient preparation for this[11]. But Jābir's aims — to simplify spherical astronomy by obviating the obscurities and prolixities of the "sector figure" and to separate theory from calculation — are too similar to those of Abū 'l-Wafā' to be a coincidence. True, Abū 'l-Wafā' himself could hardly be Jābir's source, since his mathematics is more developed — he uses, for instance, the tangent functions and Jābir does not. In fact, no direct dependence on any extant work is suggested here. The absence of Jābir's theorem 14 from some of these series of theorems means little, since it was not unknown in the East, but would not be included by every writer since it had relatively little application. In the whole of Jābir's Book II (equivalent to *Almagest* I 14 to the end of II) it is not used once[12].

A comparison of Jābir's spherical trigonometry with that of Ibn Muʿādh, who worked in Seville a few decades before Jābir, reveals surprisingly little in common between them beyond the solution of triangles as the method of solving spherical problems generally. Ibn Muʿādh took Menelaus' theorem as the basic result and developed his trigonometry from that and not from the Rule of Four Quantities[13].

---

[10]The best account of this mathematical activity is Debarnot (1985), especially pp. 13–20 of the introduction.

[11]Braunmühl (1897) 42, 45-46 and (1900) 81-82.

[12]It was not introduced in vain. It is used twice, for instance, at the end of book VI (printed edition, pp. 102-103) in the determination of a star's distance from the equator and of the co-culminating point of the ecliptic (cf. *Almagest* VIII 5).

[13]For Ibn Muʿādh, see Villuendas.

Regiomontanus' debt to Jābir

Hieronymus Cardanus (1501–1576) claimed that Regiomontanus' *De triangulis* owed more to theft than to ability[14]:

> Fuit et nostra ætate Ioannes Monte regius, qui & ipse quædam de trigonis reliquit: sed ita, ut potius furtum quam ingenium suum ostenderet.

and in a well-known passage charging Regiomontanus with plagiarism in various works he indicates the source[15]:

> Libri de triangulis sphericis inuentio est tota Hebri Hispani.

"Heber Hispanus" is elsewhere classed as the tenth of the *viri subtilitate praestantes*[16]. The extent of Regiomontanus' borrowing from Jābir has been indicated in more sober terms by Braunmühl in 1897 and 1900. Propositions 11–15 of Jābir's book I are all to be found in Regiomontanus' book IV with essentially the same proof and sometimes the same diagram letters[17]: Jābir's Props. 11(1), 11(2) and the converses are reproduced in Regiomontanus' Props. 3, 4, 5–8 respectively; and Jābir's 12, 13(1–3), 13(4), 14, 15 in Regiomontanus' 15, 16, 17, 18, 19. Braunmühl gives additional correspondences between Jābir's Props. 2–5 and 7–10 and various propositions from the third book of the *De triangulis*[18], but the matter is complicated by Jābir's having as source Theodosius, an author also available to Regiomontanus[19].

In his proof of the sine theorem for spherical triangles with one right angle Regiomontanus considers three cases, according to whether the non-right

---

[14]In *Encomium geometriae*. See *Opera omnia* IV 443.

[15]In *De exemplis centum geniturarum*. See *Opera omnia* V 498. The passage is quoted by Zinner (1968) 290, Eng. tr. 190.

[16]The first nine are Archimedes, Aristotle, Euclid, Scotus, the Calculator [Richard Swineshead], Apollonius, Archytas, the author of *Algebra* [al-Khwārizmī] and al-Kindī; the last two are Galen and Vitruvius. *De subtilitate*, near the end of book 16, p. 803.

[17]The cases of identity of lettering may arise through both authors' following the Arabic *abjad* alphabet used for diagrams. As commonly translated into Latin, at any rate by Gerard of Cremona, the order of the letters is $A, B, G, D, E, \ldots$. See Kunitzsch.

[18]Braunmühl (1897) 59, n. 1

[19]In the margins of his copy of the printed edition (Bodleian, Savile X 3), Savile noted the correspondences with Theodosius for the first ten propositions and to the fourth book of *De triangulis* for Propositions 11–15 (e.g. "transtulit in 15am 4i triangulorum suorum Regiomontanus" against Prop. 12).

angles are acute or obtuse; but, although three diagrams are drawn, only one proof (Jābir's) is given. The same procedure is adopted for the equivalent of Jābir's Prop. 15, but in the equivalent of Prop. 14 a proof is given — essentially the same — for each of the three diagrams drawn. As will be seen, the inclusion of diagrams and proofs for triangles with obtuse angles was already practised in the fourteenth century at the latest.

One result in no way dependent on Jābir is the equivalent of the cosine formula for spherical triangles in V 2. It is thought that Regiomontanus derived this result from al-Battānī's astronomy[20]. It has long been realized that Regiomontanus' achievment in the *De triangulis* was not originality of detail, but systematization of the material available to him.

Regiomontanus probably wrote the *De triangulis* in the years 1462–1464. 1462 is the date of completion of the *Epytoma ... in almagestum Ptolomei*, begun by his teacher Peurbach; in the dedication he refers to a book on triangles that he planned to write[21]. In April 1465 he includes in a letter to Jakob Speier what appears to be a reference to his own work on spherics together with those of Jābir and Menelaus[22]:

> Neque id satis erit; primum enim Gebri Hispalensis vidisse oportebit aut tertium Menelai de sphericis, sive librum triangulorum spheralium nove traditionis.

Zinner dates the completion of the part on spherical triangles to before 1564[23]. By the 1460s Regiomontanus was acquainted with Jābir's trigonometry. For he not only refers to it in his correspondence[24], but he actually copied the work twice: MS Seitenstetten, Folio 53, ca. 1460 and MS Nuremberg, Stadtbibliothek Cent. III 25 somewhat earlier[25]. Further, three of the annotations to his own transcription of the *Almagestum parvum*[26], probably made a little later, refer to Jābir's trigonometry[27].

[20]See, e.g., Braunmühl (1900) 130–131.
[21]Zinner (1968) 80, 86; Eng. tr. 52, 55.
[22]Curtze (1902) 304.
[23]Zinner (1968) 103-104; Eng. tr. 65
[24]Curtze (1902) 243; Gerl 87.
[25]Zinner (1968) 76, 310; Eng. tr. 48, 206.
[26]For this work, see Lorch (1992).
[27]Nuremberg, Stadtbibliothek Cent. VI 12, ff. 4v, 12r, 13r. On this codex see Zinner (1968) 75–76, 315; Eng. tr. 48, 210

Medieval generalizations of Jābir's spherical trigonometry

In most manuscripts of Jābir's astronomy the diagrams for props. 13(3) and 14–15 were drawn for triangles with all sides less than a quadrant, i.e. for triangles with one right angle and two acute angles. The mathematical argument of the text was sometimes understood to apply only to such triangles and to need generalizations to accommodate other configurations. Regiomontanus realized that all that was required was more diagrams, though, curiously, in his equivalent of Prop. 14 he repeated the argument for each configuration. The fourteenth-century annotator of MS British Library Harley 625 also realized this, for within a note (see below) to Prop. 13 he says of Jābir's proof (*probatio*), "que licet valeat in casu quo uterque est maior ... ", the *uterque* referring to the sides (*arcus*) other than the one opposite the right angle. The diagrams he drew sufficed for the purpose and were more complicated than those implied by the note that he goes on to repeat. For Props. 14–15 he drew four diagrams and gave an explanation in a marginal note (f. 13r):

> Nota pro 14° et 15° theoreumate quod contingit arcus *AB BG* continentes angulum rectum utrosque esse minores quarta circuli, ut patet in prima figura, et utrosque contingit esse maioris 4ª circuli ut in 2ª figura, et contingit esse *AB* esse minorem 4ª circuli et *BG* maiorem 4ª circuli ut in 3ª figura, item contingit econtra *AB* esse maiorem 4ª circuli et *BG* minorem 4ª ut patet in 4ª figura.

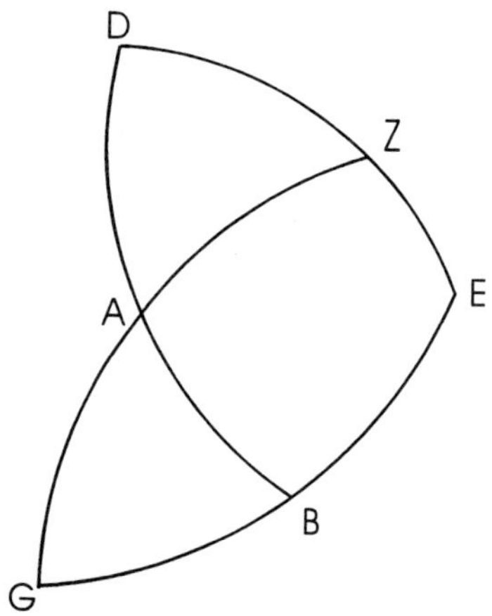

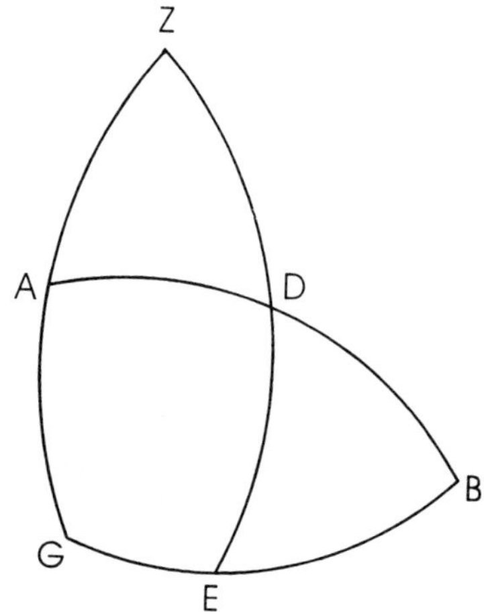

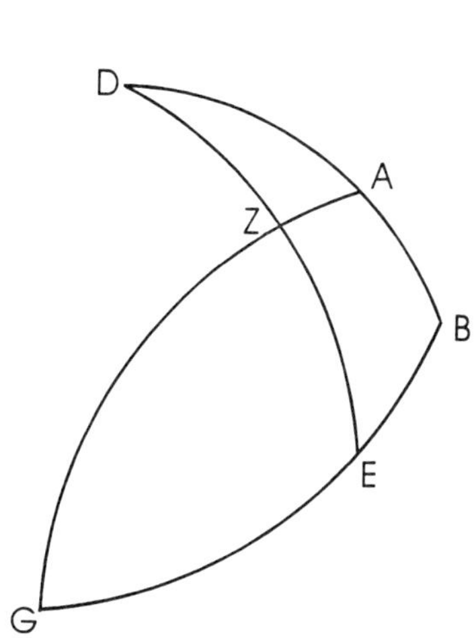

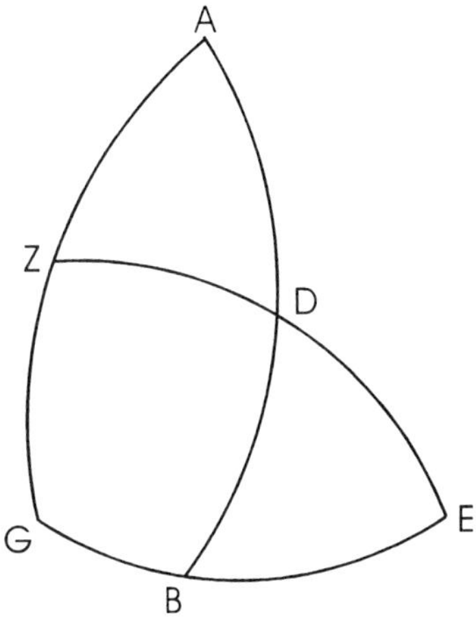

The implication is that the argument of the text applies equally to all four, as indeed it does.

Generalization to triangles in which at least one side is greater than a quadrant can also be achieved in another way. From Prop. 11 it is clear that either no sides or just two sides of a right spherical triangle are greater than a quadrant. In the latter case the two long sides may be extended to semicircles, so that a triangle is formed — from the third side and the supplements of the two long sides — in which each side is less than a quadrant. The result applied to this triangle is easily converted to apply to the original triangle. Proofs of this sort are found in at least two sources: in a series of notes and in a commentary in MS Paris BN lat. 7406. Since the argument of the notes is more straightforward, they will be considered first.

In collected form these notes appear in three manuscripts:

> Madrid, Biblioteca Nacional 10010, ff. 51v–52v, 14c. (siglum: M)
> Vatican, Reg. lat. 1268, ff. 207v–211r (f. 209 should come before f. 208)
> Paris, Bibliothèque Nationale, lat. 7377B, ff. 38r–60v, 15c. (siglum: P)

Tummers has shown that the Vatican manuscript is a copy of the Madrid manuscript[28]. Since in addition the deviations from Madrid are trivial, the derivative manuscript will not be reported further. Common to the three manuscripts are notes on Theodosius' *Sphaerica*, on proportions, on Jābir's spherical trigonometry, on his plane trigonometry, on Prop. 17 of his book I, and on Thābit's work on the sector figure. Paris continues with notes on the surface area of bodies, on Theodosius' *Sphaerica* III 11 (referred to as III 12), on similar sectors, on Menelaus' *Sphaerica* (numerous notes), on Aḥmad b. Yūsuf's work on the sector figure, and on *Almagest* II 6. These multitudinous notes have an unknown origin or origins. Suffice it to say here that the additional letters used in the notes on Jābir, F and C, are close to the beginning of the Latin alphabet, but are not equivalent to any near the beginning of the Arabic alphabet. The second of the Theodosius comments that begin the series is similar in content, but not in form, to a note appended to some manuscripts of Gerard's translation of the *Sphaerica*[29] and the note on the *Sphaerica* III 11 and the following note in the Paris manuscript are in essentially the same formulation as in MS Bodleian, Digby 168, f. 126vb (where they are given without proof)[30].

The notes on Jābir's spherical trigonometry are on Props. 13(3) and 14, the note on the latter comprising five separate comments, here labelled 14(i) ... 14(v). These notes, though only up to the very beginning of 14(iii) are also collected at the end of MSS

> Vatican, lat. 2059, ff. 192v–193v, 14c. (siglum: V)
> Vienna, Nationalbibliothek 5392, ff. 218r–220r, 15c. (siglum: W)
> Vienna, Nationalbibliothek 10905, ff. 143r–144r, 16c.

In the Vienna manuscripts the notes on spherical trigonometry are followed by a note on Jābir's VII 14. The whole is in turn followed (on a new page) by a different series of "Annotationes in Gebrum". On f. 146v of MS 10905 there is a colophon dated January 1527 and signed "Joh. Vögelin Heylpronensis"; and since the manuscript is very close to codex 5392, though with a few additional deviations, it seems that Vögelin copied this manuscript from it. It is not reported in the apparatus to the text. Note 14(ii) also appears in

> British Library, Harley 625, f. 12v, 14c. (siglum: H)

directly following the note cited above.

---

[28]See the introduction to his Anaritius, pp. xxi–xxv.

[29]E.g. Cracow, Biblioteka Jagiellońska 1924, p. 257. See Lorch (1995), towards the end. The note is on *Sphaerica* II 11.

[30]For the general history of the lemma on *Sphaerica* III 11, see Knorr (1986).

Of the five manuscripts MPHVW only M and P present a full text. In W there is one diagram, that of 13(3) in Jābir's text with the addition of point F on line HBG produced and of the arc FA. Otherwise only H has diagrams, for the note on Prop. 13, and of these the first is the same as Jābir's for Prop. 13(3) and the second and third, reproduced here, are of a curious hybrid species. On the one hand they correspond to the cases of the note (the case in which $AG > \frac{1}{4}\odot$, $BG > \frac{1}{4}\odot$ and $AB < \frac{1}{4}\odot$ is neither considered in the note nor represented in the diagram) and point F of the note is dutifully presented. On the other hand arcs HE and DZ are drawn and these form no part of the note, but rather make it superfluous (since $G{:}AB,EH$ and $A{:}GB,ZD$ together prove the proposition for these diagrams as for the first). We may conclude that these diagrams are due to the annotator of manuscript H and that in all probability the note originally had no diagram. In none of the manuscripts is there a diagram for 14(iii)–14(v).

M 51v-52r, P 41r–42v, V 192v–193v, W 218r–219r, H 12v *marg.* for note 13 and 13r *marg.* for note 14(ii) only

[13] Nota in figura tercia[1] decima primi libri geber quod[2] arcus AG subtensus recto potest esse quarta circuli et minor quarta et maior quarta. Sed si fuerit quarta circuli, tunc probatum est quod intendit, quia[3] uterque[4] duorum reliquorum est quarta circuli. Quando autem est minor quarta[5] circuli, tunc[6] uterque duorum[7] reliquorum[8] arcuum[9] vel[10] est[11] minor quarta circuli[12] vel[13] est maior[14]; et ipse posuit[15] probationem secundum[16] quod uterque est minor quarta[17] circuli[18] et nos ponamus probationem secundum[19] quod uterque est maior quarta circuli[20], que sit hec[21]. Quia uterque duorum arcuum[22] AB BG est maior quarta circuli, ergo uterque duorum arcuum qui remanent ad complendos semicirculos est minor quarta circuli; et sit unus eorum[23] GF[24] et alter AF. Ergo[25] secundum quod iam probatum est[26] erit proportio[27] sinus arcus[28] AG[29] ad[30] sinum arcus anguli F[31] recti cui ipse subtenditur sicut proportio sinus arcus GF ad sinum arcus anguli

[1]tercia decima: 13 **V**, tredecima **W** [2]quod: qui **M**; Nota in ... quod: Nota quod in hac figura 13ª **H** [3]quia: quod **P** [4]uterque: tunc alter **H**, alter **V**, utraque **W** [5]quarta: *supra* **P** [6]tunc: vel *supra* **H** [7]duorum: *corr. ex* duo **M** [8]duorum reliquorum: reliquorum duorum **V** [9]arcuum: *om.* W [10]vel: *marg.* **M**, *scr. et del.* **H** [11]est: supra **P** [12]circuli: ut in prima figura *add.* **H** [13]vel: uterque *add.* **H** [14]est maior: maior est **W**; quarta ut in secunda figura *add.* **H** [15]posuit: po^t **H**, ponit **V**, invenit **W** [16]secundum: hoc *add.* **MP** [17]quarta: p *add. et del.* **V** [18]quarta circuli *om.* **H** [19]hoc *add.* **MP** [20]s *add. et del.* **V** [21]et nos ponamus ...hec: que licet valeat in casu quo uterque est maior tamen in illo casu potest aliter sic probari **H** [22]*ABG corr. in AB add. et del.* **V** [23]eorum: *supra* **M**; unus eorum: eorum unus **P** [24]ad sinum arcus anguli *add. et del.* **P** [25]ergo: *om.* **W** [26]est: *om.* **V** [27]proportio: *supra* **V** [28]arcus: *supra* **M** [29]semiarcus *add. et del.* **V** [30]ad: *supra* **V** [31]*F*: *om.* **HW**, *B* **V**; cui *add. et del.* **V**

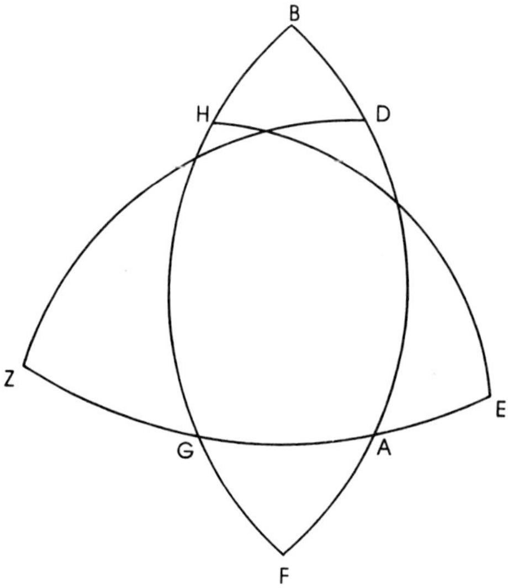

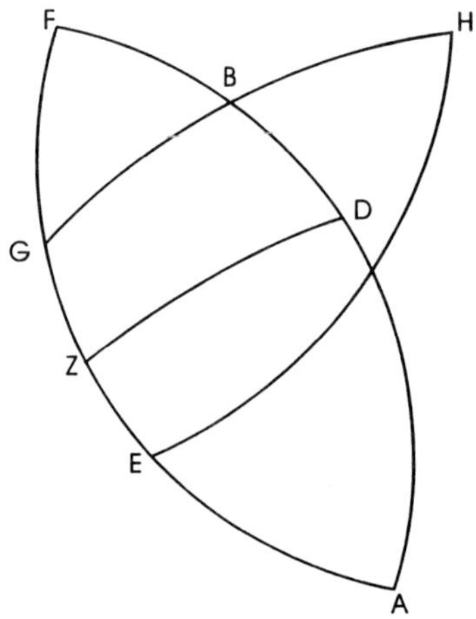

A cui ipse subtenditur et sicut proportio sinus arcus AF ad sinum arcus anguli G cui ipse subtenditur. Sed sinus arcus anguli F est[32] ipse idem qui est sinus[33] arcus anguli B, quia[34] uterque est quarta circuli[35], et sinus arcus GF est ipse idem qui est sinus arcus BG, et sinus arcus anguli FAG est ipse idem qui est sinus arcus[36] anguli[37] BAG, quoniam aggregatio[38] arcuationis[39] eorum amborum est semicirculus, et sinus arcus[40] AF est ipse idem qui est sinus arcus AB, et sinus arcus anguli AGF est ipse idem qui est sinus[41] arcus anguli AGB, quia[42] aggregatio arcuationis[43] eorum[44] est semicirculus. Ergo proportio sinus arcus AG[45] ad sinum arcus anguli B est sicut proportio sinus arcus BG[46] ad sinum arcus anguli A et sicut proportio sinus arcus[47] AB ad sinum arcus anguli G.

Cum autem arcus AG[48] subtensus recto est maior[49] quarta circuli, tunc duorum reliquorum arcuum unus est maior quarta circuli et alter minor[50], et sit arcus AB maior quarta circuli. Et arcus AG similiter[51] est maior quarta circuli[52]. Ergo

---

[32]est: *supra* **M** [33]subtenditur et ...sinus *om.* **V** [34]quia: *corr. ex* qui **M** [35]est quarta circuli: rectus **V** [36]AB et sinus arcus *add. et del.* **P** [37]anguli: *in corr.* **P** [38]aggregatio: *corr. ex* aggregationi **P** [39]aggregatio arcuationis: aggregacionis **HW** [40]arcus: *supra* **P** [41]sinus: *supra* **M** [42]quia: *corr. ex* que **P** [43]arcuationis: aromationis **W** [44]aggregatio arcuationis: amborum **H** [45]AG: *corr. ex* anguli AGB **M**, AGB **P** [46]BG: anguli G **W** [47]anguli *add.* **P** [48]est *add.* **W** [49]est maior: repet. et del. **M** [50]minor: *in corr.* **P** [51]et arcus AG similiter: ut in 3ª figura cum ergo similiter arcus AG **H** [52]et arcus AG ...circuli: *om.* **V**

duo reliqui arcus qui remanent ad complendos semicirculos[53] sunt minores duabus quartis et etiam uterque[54] est minor quarta circuli[55]. Et arcus *BG* est minor quarta circuli, et sic iam probatio redibit[56] ad illam probationem quam diximus et[57] ad probationem libri[58].

Probationes necessarie in theoremate 14° eiusdem libri.

[14(i)] Protractio perpendicularis *DZ* super arcum *AG* est necessaria quando angulus *DAZ* est acutus[59], sed quando est expansus[60] non est necessaria eius protractio[61] Et tum[62] probabitur[63] theorema. Quod sic cadat[64] perpendicularis cum angulus *A*[65] est acutus, sicut dicit liber, ita probatur: quia[66] angulus *DAZ*[67] est acutus ergo[68] angulus *DAG* est expansus; ergo polus circuli *GA* cadit inter circulum *GA* et circulum *DA*. Cum ergo figurabitur circulus transiens per polum et[69] per punctum *D*, transibit sicut transit[70] circulus *DZ*.

[14(ii)] Quod[71] perpendicularis *DZ* sit arcus complementi anguli *G*[72] sic probatur. Quia si protraxero[73] arcum *DG*, erit angulus *DGB* rectus, et complementum anguli *BGA* erit[74] angulus *DGA*, et arcus *DZ*[75] est arcus anguli *DGA* quia[76] uterque duorum arcuum *DG* et *ZG* est quarta circuli.

[14(iii)] In eadem autem figura potest contingere quod uterque duorum arcuum *AG AB* sit maior quarta circuli et quod[77] uterque sit minor[78], sed Geber[79] probat cum uterque duorum arcuum *AB GA* est minor quarta circuli et nos probemus cum uterque[80] est maior. Quia uterque duorum arcuum *AB AG* est maior quarta circuli, ergo uterque duorum arcuum qui remanent ad complendos semicirculos est minor quarta circuli; et sit unus eorum *AC* et alter *GC*. Ergo secundum hoc quod ipse iam probavit[81], erit proportio[82] sinus arcus anguli *BCG* ad sinum arcus anguli *CBG* sicut proportio sinus arcus complementi anguli *BGC*[83] ad sinum arcus complementi arcus *BC* subtensi angulo *BGC*[84]. Sed sinus arcus anguli *BCG* est ipse idem qui est sinus arcus anguli *BAG*, et sinus arcus anguli *CBG* est idem ipse qui est sinus arcus anguli[85] *ABG*, et sinus arcus[86] complementi anguli *BGC* est ipse idem qui est sinus arcus complementi anguli *AGB*, et sinus arcus[87] complementi

---

[53]qui sint *GF FB add.* **H** [54]uterque: *in corr.* **P**, utrumque **W** [55]quarta circuli: circuli quarta **P** [56]probatio redibit: redibit probatio **HW** [57]et: *om.* **P**; alillam ...et *om.* **W** [58]ad illam ...libri: *om.* **V** [59]acutus: a centro **W** [60]expansus: pansus **P**, expens° **W** [61]eius protractio: protractio eius **VW** [62]tum: tunc **V**, tamen **W** [63]probabitur: *corr. ex* probatur **P** [64]cadat: cadit **V** [65]*A*: *N* **W** [66]quia: quod **W** [67]angulus *DAZ*: *DAZ* angulus **P** [68]ergo: *supra* **P** [69]et: *om.* **P** [70]transit: *in corr.* **V** [71]Quod: Et quod **H** [72]*G*: *supra* **H** [73]si protraxero: protraxio **V** [74]erit: est **H** [75]*DZ*: *corr. ex DGZ* **V** [76]quia: et **V** [77]quod: *om.* **V** [78]*seq. ras.* **V**; *desin.* **V** [79]*desin.* **W** [80]est *add. et del. supra* **P** [81]probavit: probababavit **P** [82]sa *add. et del.* **M** [83]est ipse idem quid est sinus arcus complementi *add. et del.* **P** [84]ad sinum ... *BGE marg.* **M** [85]*BAG add. et del.* **M** [86]anguli *ABG* arcus: *supra* **P**; anguli *add. et del. supra* **P** [87]complementi anguli *BGC* ...arcus: *marg.* **P**

lateris BC est idem qui est sinus arcus complementi lateris BA: ergo proportio sinus arcus anguli A ad sinum arcus anguli B recti est sicut proportio sinus arcus complementi anguli G ad sinum arcus complementi lateris AB subtensi angulo G.

[14(iv)] Et hoc quidem verum est si angulus BAG fuerit acutus; sed si fuerit expansus, tunc angulus BAZ erit acutus et redibit probacio ad idem, cum idem sit sinus arcus anguli BAZ et arcus anguli BAG et reliqua istius trianguli, cuius tres anguli sunt angulus BAZ et angulus oppositus angulo BGA[88] et angulus CBA[89], sunt ut[90] ea que sunt trianguli ABG[91].

[14(v)] Quod autem duorum arcuum AB et AG unus sit maior quarta circuli et alter minor est impossibile, quia, si inmaginati fuerimus circulum magnum transeuntem per polos duorum circulorum[92] AB et AG, secabit arcus duorum circulorum[93] ABC AGC super ipsorum medietates; et erit angulus quem continet arcus AB cum circulo eunte per polos ipsius rectus; et angulus BAG est acutus, quapropter erit arcus circuli euntis per polos ei subtensus minor quarta circuli. Si ergo descripserimus circulum transeuntem per polum circuli AB et per punctum B, non secabit arcum AGC nisi in partes inequales cuius minor pars sit versus punctum A, quia si non[94] fuerit ita tunc iam secabunt[95] se duo[96] circuli magni, quorum unus[97] transit per polos duorum circulorum AB AG et alter per polos circuli AB et[98] per punctum B, in partes minores medietatibus ipsorum – quod est impossibile, cum ipsi circuli magni sint. Et completa est eius declaratio.

In the note to Prop. 13 the various cases for a right triangle ABG are determined first by the size of AG, the side opposite the right angle. When $AG < \frac{1}{4}\odot$, both the other sides $< \frac{1}{4}\odot$ or both $> \frac{1}{4}\odot$. For the first case the theorem is proved in the text. In the second, the two long sides AB, BG are extended to complete semicircles BAF, BGF and the text result is applied to $\triangle AGF$ (see the first figure in the text). When $AG > \frac{1}{4}\odot$, one of AB, BG $> \frac{1}{4}\odot$ and the other $< \frac{1}{4}\odot$. If $AB > \frac{1}{4}\odot$ (the other case is similar), the long sides AB, AG are extended to complete the semicircles ABF, AGF and the text result is applied to $\triangle FBG$ (see the second figure in the text).

Note 14(i) is hard to interpret: the possibility of DAZ's being obtuse ("sed quando est expansus") does not arise in any of the four diagrams drawn in MS Harley, which, although intended for another note, we may use as exhausting the possibilities. Further, the deduction that $D\hat{A}G$ is acute because $D\hat{A}Z$ is

[88] BGA: *corr. ex* BAG **M** [89] CBA: EBA **MSS** [90] ut: *supra* **P** [91] et angulus EBA ... ABG *marg.* **M** [92] duorum circulorum: circulorum duorum **P** [93] circulorum: *supra* **M** [94] non: *supra* **M** [95] arcum AGC nisi in partes equales cuius minor pars *add. et del.* **P** [96] duo: *repet.* **P** [97] unus: *corr. ex* unius **P** [98] et: etiam **P**

obtuse is valid only for the first two of these configurations. The purpose of the note appears to be to justify the positioning of line DZ in the text diagram.

In note 14(ii), to justify the assumption of the text that DZ is the complement of $\hat{A}$, it is assumed that $D\hat{G}B$ is right and that G is pole(DZE)[31].

In 14(iii) Prop. 14 is proved for the case of $\hat{A}$ acute – see the beginning of 14(iv) – and AB, $AG > \frac{1}{4}\odot$. This is done by extending AB and AG to meet at C and applying the text result to $\triangle BGC$ (see Fig. 1). The note begins with the statement that AG and AB can be both greater or both less than $\frac{1}{4}\odot$. This follows immediately from Prop. 11, but it is also justified by 14(v). Here, too, the reasoning is not entirely clear. The sides AG, AB (one of them $< \frac{1}{4}\odot$ and the other $> \frac{1}{4}\odot$ are extended to meet at C; a circle through the poles of AGC and ABC will cut these semicircles into equal parts [at X, Y in Fig. 2] and the side BG cuts AGC into unequal parts (the note says unnecessarily that the smaller part is on the side of A). The argument concludes by saying that, if this were not the case, the two circles drawn to intersect the semicircles would not bisect each other, which is impossible. As it stands, this perhaps means that both the circles could not cut the two semicircles into equal parts [at X, Y], since it has already been shown that $XY < \frac{1}{4}\odot$. The resulting situation was no doubt considered impossible because the intersection of the two circles would then be pole(AB), since $\hat{B}$ and $\hat{Y}$ are right, but be less than $\frac{1}{4}\odot$ away from it.

Fig. 1

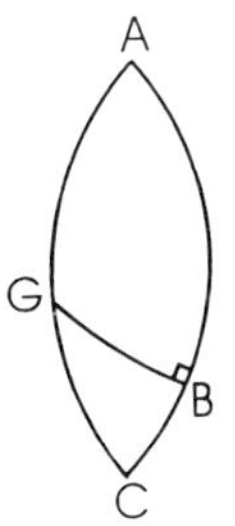

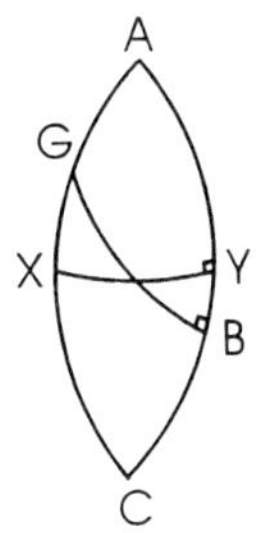

Fig. 2

In 14(iv) the case of $\hat{A}$'s being obtuse is considered. The idea is to extend GA and GB to meet at C to form $\triangle ABC$, of which $\hat{A}$ is acute. No doubt

[31] See the second note to the summary of Jābir's Prop.14 above.

the second of the four configurations drawn in MS Harley, and reproduced above, is intended, for if $AG > \frac{1}{4}\odot$, Z lies between G and A, so that $B\hat{A}G$ and $B\hat{A}Z$ are the same angle.

Altogether these notes at best give the impression of being battered in the course of time by copying, but the principle of applying the text result to a neighbouring triangle and then converting it to the triangle under discussion is clear enough.

In MS Paris, Bibliothèque Nationale, lat. 7406, ff. 114ra–135rb there is a commentary on Jābir's astronomy beginning "Geber in libro 30 figurarum ... ". The manuscript is written in a fourteenth-century hand, but when the commentary was written is not known. It was probably written in Latin and not translated from Arabic. Little can be deduced from the extra diagram letters used (e.g. Q in I 13), but some quotations from Jābir's text show too great a similarity to Gerard's translation to be coincidental. Examples of this are the quotation of the enunciation of I 17 (exact) and of I 3 (almost exact)[32]. In general, however, the commentator rephrases the enunciations of the Latin Jābir. This is particularly noticeable in the propositions in which Jābir's enunciations are specific to a diagram. Ironically, in the two cases in the theorems on spherical trigonometry in which the wording of the enunciation is commented upon, the words are the commentator's[33].

There are two further indicators that the commentary was written in Latin. One is the remark at the end of I 16(f. 117vb): "Sed istud et alia circa hoc invenies demonstrata in libro ysoperimetrorum". This is most probably the treatise *De isoperimetris* translated from Greek[34]. The other is the enunciation of the first of the propositions on the calculation of chords (f. 118vb):

---

[32]Ff. 117va–vb and 114ra respectively. There is a considerable difference in wording and a slight difference in content between Jābir's I 3 and Theodosius' *Sphaerica* I 17 (in Heiberg's numbering), from which it was taken. See MS Paris, Bibliothèque Nationale, lat. 9335, f. 4ra, for the translation by Gerard of Cremona and MS Salamanca, University Library 221, f. 186r for the "long" or Campanus text. This excludes the Latin Theodosius as possible source.

[33]The two cases are: (1) "Inclinatis super se duobus circulis " in Prop. 12, where Jābir has "Cum sint duo circuli ... et non transit unus eorum per polum alterius" and (2) "habente unum angulum rectum" in Prop. 14 (see below), where Jābir has "in quo est unus angulus rectus".

[34]See Busard, esp. p. 62.

> Dato circulo latera decagoni exagoni pentagoni tetragoni trianguli omnium equilaterorum et equiangulorum ab eodem circulo circumscriptorum reperire.

This is strikingly similar to the enunciation of the first proposition in the *Almagestum parvum*[35]:

> Data circuli dyametro latera decagoni exagoni pentagoni tetragoni atque trianguli omnium ab eodem circulo circumscriptorum reperire.

Clearly the phraseology was borrowed in Latin; it seems that the commentator of Jābir took it from the *Almagestum parvum* or its source.

For the theorems on spherical trigonometry the Paris commentary keeps the numbering of the text. Propositions 11 and 12 are essentially as in Jābir. Propositions 13–15 run:

13. In omni triangulo ex arcubus magnorum sinus laterum ad sinus arcuum suorum angulorum sunt in proportione una.

In hac propositione distinguendum est inter latus et arcum: vocat[1] hic Geber arcum anguli illum arcum qui subtenditur angulo contento duabus 4^{is}; latus autem est arcus ubicumque respiciens angulum. Unde omnis arcus latus est, non econtrario.

Volo igitur ostendere quod que est proportio sinus alicuius lateris ad sinum arcus sui anguli eadem est proportio sinus alterius lateris ad sinum arcus sui anguli et eadem est sinus tercii lateris ad sinum arcus sui anguli.

Sit ergo triangulus datus ABG. Qui si omnes angulos habet rectos, planum est tunc quod omnia latera sunt 4^e et ipse sunt arcus rectorum. Item si sint tantum B et G recti, ergo A est polus BG circuli; ergo BG arcus est et latus et arcus. Sic et AB quia G est angulus rectus; sic quoque AG quia B rectus est angulus.

Sed sit tantum B rectus. Oportet ergo ex 4^a parte 11^e latus AG minus aut maius esse 4^a circuli. Sit primo minus: ergo utrumque [115vb] laterum AB et GB aut minus est 4^a aut maius. Sit utrumque minus primo. Protraham ergo GA super[2] E ut sit EG 4^a et iterum AG super D ut sit AD 4^a; item sit GH 4^a et AZ 4^a; deinde

[1]vocat: *in corr.* **M** [2]super: utrumque *add. et del.* **MS**

[35]MS Paris, Bibliothèque Nationale, lat. 16657, 13c., f. 83r. See Lorch (1992), p. 421f. There are several texts with this enunciation as *incipit*.

pingam *EH* et *DZ* arcus magnorum. Ergo *A* polus est circuli *DZ*; ergo angulus *Z* rectus est et *B* rectus; ergo ex proxima que est proportio sinus lateris *AG* ad sinum

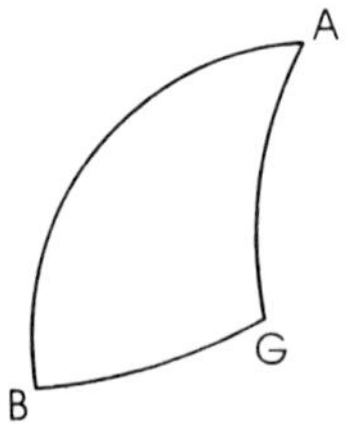

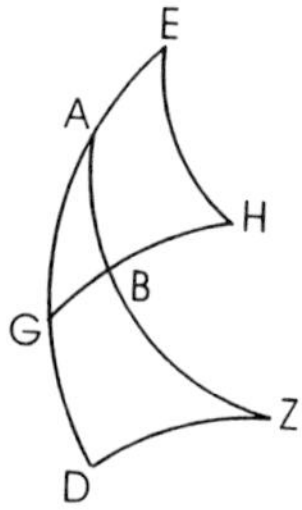

arcus *AD* eadem est sinus lateris *GB* ad sinum arcus *DZ*, qui est arcus anguli *A*. Arcus autem *AD* est arcus *B* recti anguli quia est 4ª. Similiter *G* est polus circuli *EH*; ergo *H* angulus rectus est; ergo ex proxima que est proportio sinus lateris *GA* ad sinum arcus *GE* eadem est sinus lateris *AB* ad sinum arcus *EH* anguli *G*. Quod si vides, vides propositum secundum hanc positionem.

Sit autem tam *AB* quam *GB* maius 4ª. Ducam tum *AG* super *D* ut sit *AD* 4ª, sitque *AZ* 4ª, et ducam arcum magni circuli *DZ*: est igitur *A* polus circuli *DZ*. Et non est angulus *A* rectus, sed *D* et *Z* recti sunt et *B* rectus. Ergo ex premissa que est proportio sinus lateris *AG* ad sinum arcus *AD* eadem est sinus lateris *GB* ad sinum lateris arcus *DZ*. Ex hoc patet propositum secundum hoc positum.

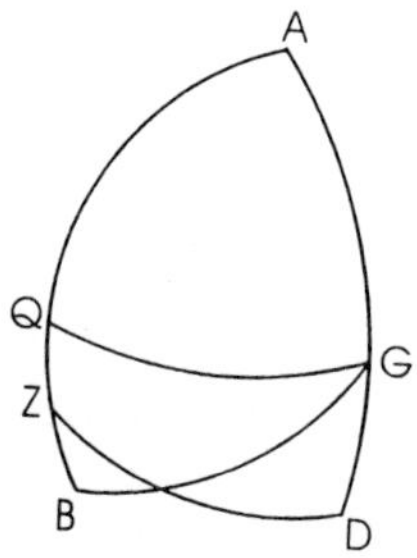

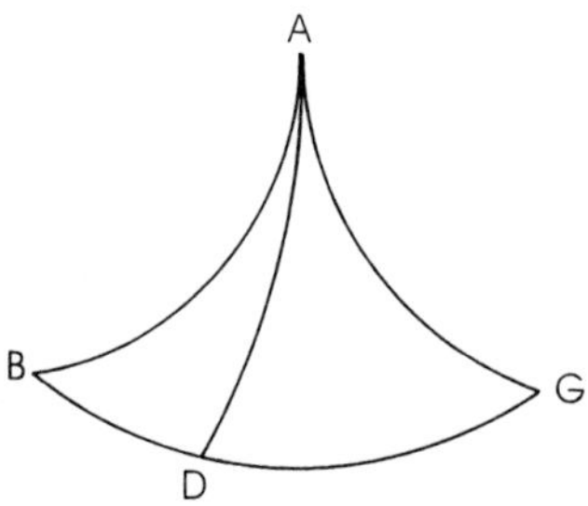

Ponam modo ut latus recto[3] angulo oppositum sit maius quarta et, ne oporteat fieri novam figuram, sit triangulus datus *AZD* habens *Z* solum rectum, cui opponitur latus *AD* maius 4ª. Ergo oportet ut vel maius vel minus 4ª sit *AZ*. Sit minus,

[3]recto: rectum **MS**

et fiat *AB* 4^a et sit *AG* 4^a; et curvabo arcum magni inter *G* et *B*. Ergo tam *G* quam *B* est rectus, quia *A* polus est circuli *GB*. Apparet igitur ex premissa quod que est proportio sinus lateris *AD* ad sinum arcus *AG* eadem est sinus lateris *DZ* ad sinum arcus *GB*.

Si autem etiam *AZ* maius est 4^a, sit *AQ* 4^a. Ergo *Q* angulus rectus est; ergo iterum, cum etiam *Z* sit rectus, proportio sinus lateris *AD* ad sinum arcus *AG* est sicut proportio <sinus> lateris *DZ* ad sinum arcus *GQ*, quod probare volui.

Age nunc, nullum rectum habeat datus triangulus *AGB*. Ducam ergo per polum circuli *GB* et super *A* punctum arcum circuli magni, qui aut cadet inter *G* et *D* aut [116ra] extra. Si inter ea puncta, ergo uterque angulus *D* est rectus. Ergo que est proportio sinus lateris *AB* ad sinum lateris *AD* eadem est proportio sinus arcus anguli *ADB* recti ad sinum arcus anguli *B*; et que est proportio sinus lateris *AD* ad sinum lateris *AG* eadem est sinus arcus anguli *G* ad sinum arcus anguli *ADG* sive anguli *ADB*, uterque enim rectus. Ergo ex 24^a 5^i Euclidis secundum proportionalitatem mutharyba[4] que est <proportio> sinus lateris *AB* ad sinum lateris *AG* eadem est sinus arcus anguli *B* ad sinum arcus anguli *G*. Ergo permutatim et habes propositum.

Similiter duc super polum circuli *AB* et super *G* punctum arcum magni inter *A* et *B* et proba ut modo sinus lateris *AG* ad sinum arcus anguli *B* est in proportione sicut sinus lateris *GB* ad sinum arcus anguli *A*.

Sed contingat, sicut potest, quod arcus ductus super polum et angulum non cadat inter alios angulos sed extra et, ut maneat figura, fuerit datus triangulus *ADG*. Ducam ergo circulum cuius arcus <est> arcus *AB* super polum arcus *GDB* et super *A* angulum; ergo *B* est rectus angulus. Proba igitur ut modo per indirectam[5] proportionalitatem quod sinus lateris *AG* ad sinum lateris *AD* est sicut sinus arcus anguli *ADB* ad sinum arcus anguli *G*. Sed idem est sinus anguli *ADB* qui est sinus anguli *ADG*, quia eorum arcus simul iuncti faciunt circumferentie medietatem, quelibet autem due partes semicircumferentie unius habuit eundem sinum. Ex hoc apparet propositum secundum ultimam omnium positionum.

14. In omni triangulo[6] ex arcubus circulorum magnorum habente unum angulum rectum proportio sinus arcus unius duorum reliquorum angulorum ad sinum arcus anguli recti est sicut proportio sinus complementi[7] arcus anguli reliqui ad sinum complementi lateris eiusdem[8].

---

[4]mutharyba: id est indirectam *supra* **MS** [5]indirectam: indirecta **MS** [6]h *add. et del.* **MS** [7][ ]lementum arcus [ ]id quo exce[ ] idem arcus [ ] l' excedit [ ]m *marg.* **MS** [8]id est subtensi *supra*, subtensi *marg.* **MS**

Verbi gratia: sit datus triangulus *ABG* cuius angulus *B* sit rectus. Producam autem arcum *AB* super *A* usque dum compleam 4^{am} et sit complementum eius *AD* arcus; complementum arcus *AG* sit *AZ*; ducam autem arcum *DZ* circuli – occurrat [116rb] arcus *GB* super punctum *E*. Ergo *D* polus est circuli *GBE*; ergo angulus *E* rectus est et *DZE* quarta est. Arcus autem *ZAG* in triangulo *ZGE* respicit angulum rectum et est 4^a. Ex 4^a parte 11^e huius triangulus *ZGE* habet etiam alium rectum. Sed *G* non est rectus, ergo *Z* est rectus, ergo *G* est polus circuli *DZE*, ergo *GBE* est 4^a, ergo *ZE* est arcus anguli *G* et complementum arcus est arcus *ZD*. Dico ergo quod proportio sinus arcus anguli *A* ad sinum arcus anguli *B* est sicut proportio <sinus> *DZ* arcus, qui est complementum arcus anguli *G*, ad sinum arcus *AD*, qui est complementum *AB* lateris anguli *G* – quod ex 12^a et 13^a presentis patet si vides quod super punctum *A* secant se duo circuli, qui sunt *ZAG* et *DAB*, non orthogonaliter et a puncto unius *D* super reliquum et a puncto eiusdem reliquo, quod est *G*, super priorem ducuntur duo arcus orthogonaliter. Et hoc est quod volui.

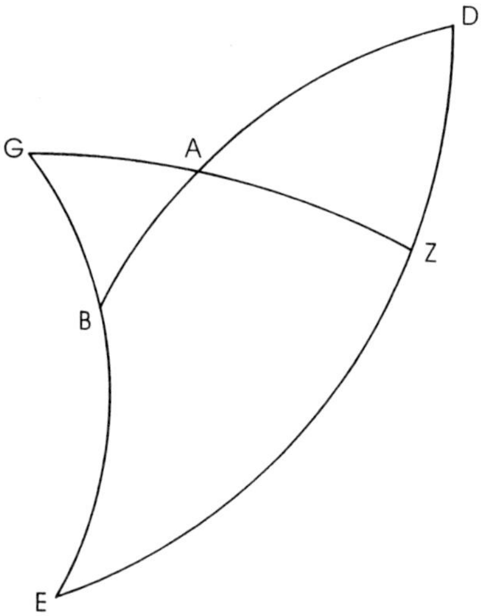

Hec 14^a propositio Geber 4^{or} modis potest variari. Unus modus est si datus triangulus duos habeat rectos et iste modus est inutilis.

Verbi gratia: sint in triangulo *ABG* duo recti scilicet *B* et *G*. Ergo ex 3^a parte 11^e huius latus *AG* 4^a est; eadem ratione, vel si maius ex 4^a parte eiusdem 11^e, latus *AB* 4^a est. Ergo tam *AG* quam *AB* est et latus et arcus sui anguli; et nec arcus nec latus habet complementum, cum sit quarta absolute. Quomodo ergo ostendetur quod proportio sinus arcus anguli *A* ad sinum arcus anguli *B* vel anguli *G* est sicut sinus complementi arcus reliqui anguli ad sinum complementi lateris eiusdem cum nullum habeat complementum, tam latus quam arcus anguli *G* sive anguli *B*? Nullius enim ad nihil nulla est proportio. Quod ergo in propositione dicitur "habente unum angulum rectum" sic debet intelligi: "id est habente unicum angulum rectum". Tunc enim, sicut videri potest ex 11ª huius, triangulus nullum habet latus quod sit 4ª.

Si vero latus oppositum recto angulo minus est 4ª, provenit utrumque laterum continentium rectum angulum sit minus 4ª et hic est secundus modus et eum probat Geber.

Tercius autem modus est si sit utrumque maius 4ª et hunc probabo. Sit ergo in triangulo *ABG* habente unicum angulum scilicet *B* rectum; latus ei oppositum *AG* minus 4ª, utrumque vero laterum *AB* et *GB* [116va] sit maius 4ª. Abscindam itaque in utroque eorum 4ᵃᵐ, et sint due 4ᵉ *BS* et *BQ*; perficiam autem ex *AB* et *GB* semicircumferentias, que sint *BAH* et *BGH*; continebunt vero necessario *H* angulum rectum ex adverso *B* recti. Sunt igitur *HA* et *HB* minores 2ᵇᵘˢ 4ⁱˢ. In triangulo igitur *AHG* constat propositum secundum probationem libri: dico autem quod sinus arcus anguli *BAG* ad sinum 4ᵉ, utpote *QS*, se habent ut sinus complementi arcus anguli *AGB* ad sinum *AQ*, qui est complementum *AB* lateris anguli *AGB*. Vides enim triangulos *AHG* et *ABG*: infer ergo ex 13ª huius quod que est proportio sinus lateris *AG* ad sinum arcus *QS* eadem est sinus lateris *AH* ad sinum arcus anguli[9] *AGH* et sinus lateris *HG* ad sinum arcus anguli *HAG* – sed et eadem est sinus lateris *AB* ad sinum arcus anguli *AGB* et sinus lateris *GB* ad sinum arcus anguli *GAB*. Ergo que est proportio sinus lateris *AH* ad sinum lateris *AB* eadem est sinus arcus anguli *AGH* ad sinum arcus anguli *BGA*. Sed sinus lateris *AH* est etiam sinus lateris *AB*, quia hii duo arcus perficiunt semicircumferentiam: ergo sinus arcus anguli *AGH* est etiam equalis sinui arcus anguli *BGA*. Patet etiam quod idem[10] arcus est complementum *AH* et *AB* laterum, scilicet arcus *AQ*; oportet quoque quod idem arcus sit complementum arcuum angulorum *HGA* et *AGB*, quia eorum arcus sicut et latera complent semicircumferentiam. Dico ergo sic: que est proportio sinus arcus anguli *HAG* ad sinum 4ᵉ *QS* eadem est sinus complementi arcus anguli *AGH* ad sinum arcus *AQ*. Sed sinus arcus anguli *AGH* est etiam sinus arcus anguli *AGB* et sinus complementi idem est quia[11] idem complementum: ergo que est proportio sinus arcus anguli *ADG*[12] ad sinum *QS* 4ᵉ eadem est sinus complementi arcus anguli *AGB* ad sinum complementi lateris *AB*, hoc est ad sinum arcus *QA*[13]. Ergo que est proportio sinus arcus anguli *GAH* ad sinum *QS* 4ᵉ eadem est sinus[14] complementi arcus anguli *AGB* ad sinum

[9]anguli: *HAG add. et del.* **MS** [10]est *add. et del.* **MS** [11]quia: *corr. ex* quod **MS**
[12]*ADG*: *corr. ex AHG* **MS** [13]*QA*: *QS* **MS** [14]sinus: *marg.* **MS**

complementi lateris *AB*, quod fuit probandum.

Sit modo ut *B* recto angulo opponatur latus 4^a longius, et sit *ZG*; datus autem triangulus sit *BZG*: erit igitur alterum laterum [116vb] continentium rectum angulum maius 4^a, et sit *BG*; reliquum vero minus 4^a, et sit *BZ*. Dico modo quod sinus arcus anguli *Z* ad sinum 4^e est sicut sinus complementi arcus anguli *ZGB* ad sinum complementi lateris *ZB*. Perficiam enim semicircumferentiam producens *ZB* et *BG* donec concurrant in puncto *H*, ubi necessario faciunt angulum rectum *H* cum *B* sit rectus. Sit autem arcus *HA* equalis arcui *ZB*, et curvabo arcum magni circuli super puncta *A* et *G*. Probabo tunc quod sinus complementi arcus anguli *HGA* est equalis sinui complementi arcus anguli *ZGB* et quod sinus arcus anguli *HAG* est sinus arcus anguli *GZB*.

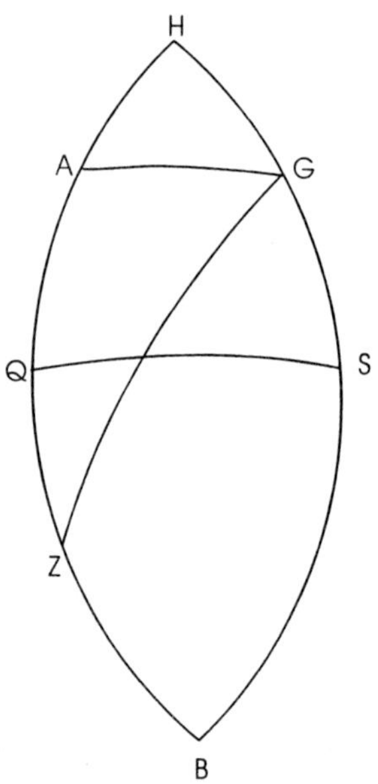

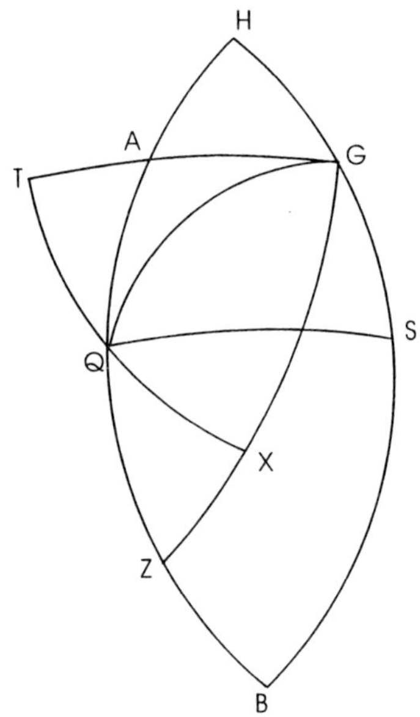

Producam enim arcum *GA* donec arcus *GT* fiat 4^a, et faciam ut prius quartam *QS* abscindentem *QB* et *SB* 4as; ducam etiam arcum magni circuli inter *Q* et *G*; deinde super *T* et *Q* puncta ducam arcum circuli magni donec in puncto *X* abscindat arcum *GZ*. Patet igitur quod polus circuli *HGB* est punctum *Q*, quia *QS* est 4^a; et oportet in triangulo *QBS* tribus 4^{is} incluso[15] tres angulos rectos esse.

[15]incluso: inclus **MS**

Hinc sequitur angulum *HGQ* rectum equalem esse angulo *QGS* recto. Similiter item necesse est punctum *G* esse polum circuli *TQX*[16] , quia *GT* et *GQ* sunt due 4ᵉ: ergo arcus *GX* est 4ᵃ; ergo arcus *TQ* et arcus *QX* sunt arcus angulorum *AGQ* et *QGZ*. Item *T* et *X* anguli recti sunt.

Inde sic: que est proportio sinus arcus *AQ* ad sinum arcus *QZ* eadem est sinus arcus *TQ* ad sinum arcus *QX* ex 12ᵃ huius, quia circuli *AQZ* et *TQX* secant se non orthogonaliter in puncto *Q* et a circulo *AQZ* demittuntur orthogonaliter *AT* et *ZX* arcus et abscindunt arcus *TQ* et *QX*. Sed sinus arcus *AQ* est equalis sinui arcus *QZ*, quia ipsi sunt arcus equales. Ergo sinus arcus *TQ* est equalis sinui arcus *QX*; et uterque arcus est minor 4ᵃ: ergo arcus *TQ* est equalis arcui *QS*. Sed hii arcus sunt arcus angulorum *AGQ* et *QGZ*: ergo oportet hos angulos equales esse. Ergo etiam equalis est angulus *HGA* angulo *ZGB*. Ergo arcus eorundem angulorum sunt equales.

Sed in triangulo *AHG* sinus lateris *HA* ad sinum arcus anguli *HGA* ita se habent ut sinus lateris *HG* ad sinum arcus anguli [117ra] *HAG* ex 13ᵃ; et similiter sinus lateris *ZB* ad sinum arcus anguli *ZGB* est sicut sinus lateris *GB* ad sinum arcus anguli *GZB*. Ex hoc et prehabitis infer quod sinus arcus anguli *GZB* ad sinum lateris *GB* est sicut sinus arcus anguli *HAG* ad sinum lateris *HG*. Sed sinus laterum *HG* et *GB* sunt equales: ergo sinus arcus anguli[17] *GZB* est equalis sinui arcus anguli *HAG*. Hiis constantibus facile est videre propositum. In triangulo enim *AHG* proportio sinus arcus anguli *HAG* ad sinum 4ᵉ est sicut proportio <sinus> complementi arcus anguli *HGA* ad sinum complementi lateris *HA*, hoc est ad ad sinum arcus *AQ*, ut probat Geber. Ergo etiam in triangulo *GZB* proportio sinus arcus anguli *GZB* ad sinum 4ᵉ est sicut proportio sinus complementi arcus anguli *ZGB* ad sinum complementi lateris *ZB*, hoc est ad sinum arcus *QZ* qui est equalis arcui *AQ*. Habes itaque plane propositum.

15. Proportio vero sinus complementi lateris subtensi recto ad sinum complementi unius duorum laterum que ipsum continent est sicut proportio sinus complementi lateris reliqui ad sinum 4ᵉ circuli.

Hoc planum est retinenti dispositionem iam factam. Circulus enim *DAB* et circulus *DZE* secant se non orthogonaliter super *D* punctum, et a punctis *A* et *B* signatis super *DB* circulum ducuntur orthogonaliter super *DZE* circulum *AZ* complementum lateris *AG* oppositi recto et *BE* complementum lateris *GB* quod est unum continentium rectum angulum: ergo proportio sinus *ZA* ad sinum arcus *BE* est sicut proportio sinus *AD* complementi lateris reliqui *AB* ad <sinum> *DB* 4ᵃᵐ circuli – quod proponebatur.

---

[16] *TQX*: *corr. marg. ex TQT* (?) **MS** [17] *HAG* ad sinum 4ᵉ *add. et del.* **MS**

In illa tamen 15^a idem 4^4 modi diversitatis occurrunt qui prius, sed primus etiam est hic inutilis scilicet si in triangulo sint duo recti anguli. Probat autem Geber propositum in triangulo cuius rectum angulum continent duo latera que ambo 4^a sint breviora ut in triangulo HAG, sed manente dispositione in precedenti facta facile est hoc videre etiam in triangulo ABG, in quo tam AB quam GB latus maius est 4^a, quia AQ [117rb] tam arcus HA quam arcus AB complementum[18] est et similiter arcus GS est complementum tam arcus GH quam arcus BG.

Ut autem videas propositum in triangulo ZBG in quo GB latus maius est 4^a et ZB minus, maneant ea que in proxima probata sunt et videbis per suppositionem quod arcus ZX, qui est complementum arcus ZG[19] , equalis est arcui AT complemento lateris AG. Cum ergo complementum lateris HA[20] est equale complemento lateris ZB et complementum lateris HG sit etiam complementum lateris GB, qui videt propositum in triangulo HAG videre etiam debet in triangulo ZBG. Habes igitur hoc totum.

In Prop. 13(3) the diagrams for the various cases are described and the theorem is proved each time, in the same way as Jābir proved it. Almost as if to obscure the unity of the reasoning, a different triangle ($\triangle ADZ$ instead of $\triangle ABG$) is taken for the third and fourth cases. This was done to save drawing another diagram. The same exasperating procedure is followed in the next proposition.

In Prop. 14 the commentator begins with a construction slightly different from that of the text ($AZ = \overline{AG}$ for $DZ \perp AG$), proves that DZ is the complement of $\hat{G}$ and then procedes as in the text. After this the commentator distinguishes four cases, or *modi*. The first of these, that of a triangle with two right angles, is *inutilis*: it leads to a ratio of nothing to nothing. The terms *modus* and *inutilis* appear to be taken over from Gerard's translation of Thābit's discussion of the compound proportion in the statement of Menelaus' theorem[36]. Here a *modus* is an expression for two of six quantities as a compound ratio of the other four – thus, in modern terms, $a : c = (b : d) \cdot (e : f)$ would be *modus* of $a : b = (c : d) \cdot (e : f)$. A *modus* is *inutilis* if no such expressiion is deducible for the chosen quantities (e.g. $a : d$ in our example). The other three *modi* of Prop. 14 are the cases arising when the side subtending the right angle $<$ or $> \frac{1}{4}\odot$, as in the series of notes considered above. In *modus* 3, in which the sides surrounding the right angle

[18]complementum: *supra* **MS** [19]vel GZ *supra* **MS** [20]*seq. ras.* **MS**

[36]See Björnbo *et al.* (1924), pp. 14–23.

are both $> \frac{1}{4}\odot$, these sides are extended to meet at H and the text result is applied to $\triangle AHG$. To convert this to apply to $\triangle ABG$, the commentator proves that Sin $A\hat{G}H$ = Sin $B\hat{G}A$ by applying the sine theorem to $\triangle$s AGH and AGB. That the sines of supplementary angles are equal is obvious from their very definition and is simply assumed in the previous theorem (penultimate sentence), but for some reason an elaborate proof is provided. The sines of the angles at A are no doubt assumed equal by analogy with those at G.

In the fourth *modus* $\triangle BZG$ is considered, in which $\hat{B}$ is rt., $ZG > \frac{1}{4}\odot$ and $BG > \frac{1}{4}\odot$. A triangle is again constructed ($\triangle HAG$) of which one side is the same as one of the sides of the chosen triangle and the other sides are the supplements of the other two sides of the chosen triangle, but the construction is not presented as such. The result is applied to $\triangle HAG$. To convert this so that it applies to $\triangle BZG$ the commentator proves that (1) $H\hat{G}A = Z\hat{G}B$ and (2) Sin $H\hat{A}G$ = Sin $G\hat{Z}B$. The demonstration for (1) is:

> Let $BQ = BS = \frac{1}{4}\odot$
> $\because \hat{B}$ is rt., Q is pole(HGB)
> So, if $GT = \frac{1}{4}\odot$, G is pole(TQX)
> From $\triangle$s QAT and QZX, $TQ = QX$[37]
> $\therefore A\hat{G}Q = Q\hat{G}Z$
> And $\because QGH = QGB$ (Q being pole(HGB)), $H\hat{G}A = Z\hat{G}B$

(2) is shown by applying the sine theorem successively to $\triangle AHG$ and $\triangle ZGB$ and using (1). Prop. 15 is treated in strictly analogous fashion.

Jābir's influence on medieval spherical astronomy[38]

The commentator of MS BN 7406 presented Jābir's book II, which is largely on spherical astronomy, in full, albeit with some changes (e.g. to II 4). However, Jābir seems to have been mostly cited as a supplementary authority or as the author of an alternative method in the solution of various problems in spherical astronomy. In a variant of al-Zarqāllu's *Canones* beginning:

> Incipit canones Magistri Gerardi Cremonensis in motibus celestium ... Quoniam cuiusque actionis ...

[37]This line is wrongly proved: Q:AT, ZX is made to yield Sin AQ:Sin QZ = Sin TQ:Sin QX.

[38]No systematic search through medieval Latin astronomical works has been made. The following selection is prejudiced by the availability of material in the Bodleian library when the research was being carried out.

we find, under the heading "Inventio arcus declivis orizontis", a demonstration "per kata coniuncta conversis proportionibus" and then[39]:

> De eodem secundum Jeber, scilicet de inventione arcus declivis orizontis. Per sequentem figuram secundum Jeber invenitur arcus orizontis ...

Towards the end of the treatise, under the heading "De kardagis declinationis inveniendis" there is again a demonstration with Menelaus' theorem, this time followed by a Jābir proposition cited by number[40], "Per 13 primi Jeber proportio ..." and then a recipe for calculation, "Secundum demonstrationem Jeber, accipe ... " The attribution to Gerard of Cremona is most probably false, but the text does serve as an example of the application of Jābir's trigonometrical method.

References to Jābir by Richard of Wallingford (c.1291–1336) were mostly of the "supplementary authority" type, but in the third part of the *De sectore* the whole of Jābir's spherical trigonometry is presented after Ptolemy's, though not quoted *verbatim*. The whole is introduced[41]:

> In tercia parte huius tractatus ponentur demonstraciones Ptholomei et Gebir commentarii de figura sectore, quibus prepono 4 conclusiones de primo *Almagesti* ... Deinde addam quasdam conclusiones Gebir suis demonstracionibus necessarias, ex quibus concludam propositum Gebir mirabiliter compendiosum et brevem ...

An interesting example of Jābir's influence on the methods of spherical astronomy is the commentary on the *Almagest* by the English Astronomer Simon Bredon[42]. From the records of Merton College, Oxford, it seems that he was a Fellow at least from 1330 to 1341. After this he apparently held various Church appointments till he died in 1372. The *Commentary on the Almagest* is extant in three manuscripts, none of which goes further than book III:

[39] MS Oxford, Bodleian, Can. misc. 51, f. 77v. The work fills ff. 73r–88v.

[40] *Ibid.* f. 88v, in a section marked (f. 87r) "Addicio".

[41] North's edition, I 175.

[42] On Bredon, see Emden I 257, Talbot. Bredon is cited in a note to a Latin copy of the *Almagest* in MS Oxford, New College 281, f. 8v, on finding the sides of the regular pentagon, decagon, etc. Other notes cite al-Farghānī (ff. 3r, 4v, 5r), al-Battānī (f. 9r) and Ibn Sīnā (f. 24r).

Oxford, Bodleian, Digby 168, ff. 21r–39r, 14c. Between ff. 21 and 22 three folios are missing and between ff. 23 and 24 two folios are missing[43].

Oxford, Bodleian, Digby 178, ff. 42r–65r, ca. 1400. Beginning missing.

Cambridge, University Library Ee III 61 (1017), ff. 43r–46r, 15c. Beginning missing.

In this elaborate work the author, while mainly concerned with Ptolemaic methods, takes a great deal from Jābir and a little from al-Battānī and others. References to relevant passages in Jābir are given for almost every theorem in the commentary on the second book, even when an extended citation is not given. Although the spherical astronomy is preceded by a full treatment of Menelaus' theorem, the Rule of Four Quantities evidently constituted the third proposition of a *preambulum* to the work, presumably in the folios of Digby 168 that are now lost, for Bredon refers back to it in his theorems comparing increments of declination and of right ascension for different arcs of the ecliptic[44]. The remaining theorems in the *preambulum* were probably of the type found in the first book of Menelaus' *Sphaerica*[45]. We may note in passing that a few congruence theorems in Jābir's book I would have eased the proofs of his book II considerably.

3. Plane Trigonometry

In contrast to his spherical trigonometry, Jābir's plane trigonometry is presented as a series of procedures rather than theorems. The procedures are not new — they are to be found in the *Almagest* —, but his presentation shows systematically how any triangle may be solved. Gerard's translation of his treatise laid the foundation of this branch of mathematics in the West.

The problems are reduced to an appeal to the definition of chord: for the standard diameter — and therefore, by proportion, for a given diameter —

[43]The numbers of missing folios have been estimated from an old foliation. In his *Early Science in Oxford* (II 52) Gunther claimed that MS Dibgy 168 was an autograph, but since he also thought that the "Quinque conclusiones de numeris quadratis" in MS Digby 178, ff. 12v–13r, was also an autograph, his judgement must be treated with reservation, for the writing is different in the two manuscripts.

[44]MS Digby 168, f. 23r.

[45]*Ibid*. Th. 6 was evidently a congruence theorem; theorems 1 and 2 involved inequalities in triangles.

the chord YZ (see Fig. 3) is known from the angle X that it subtends, and *vice versa*. Accordingly, triangles are divided into right triangles in which two of three things are known: hypotenuse, side and angle opposite the side.

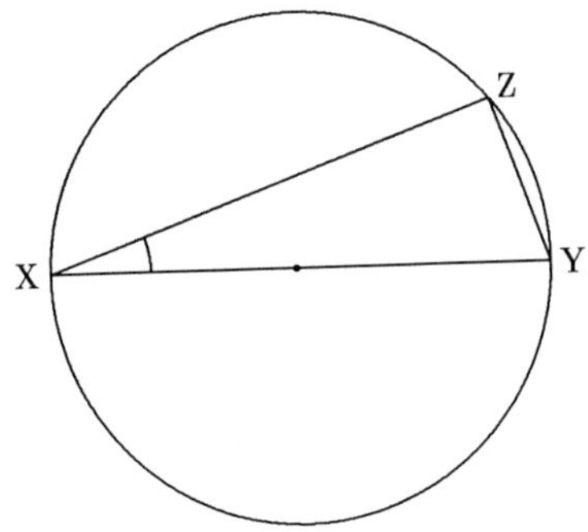

Fig. 3

When in $\triangle ABG$ (see Fig. 4) two sides AB, AG and the included angle B are known[46], a perpendicular AD is dropped from A to BG; the right triangle ABD is solved to give AD, BD. Then in the right triangle AGD the sides are known, and hence also the angles. A similar procedure applies to the case in which two sides and one of the non-included angles are known (Fig. 5). As Jābir points out, one must know in this case if angle G is acute or obtuse.

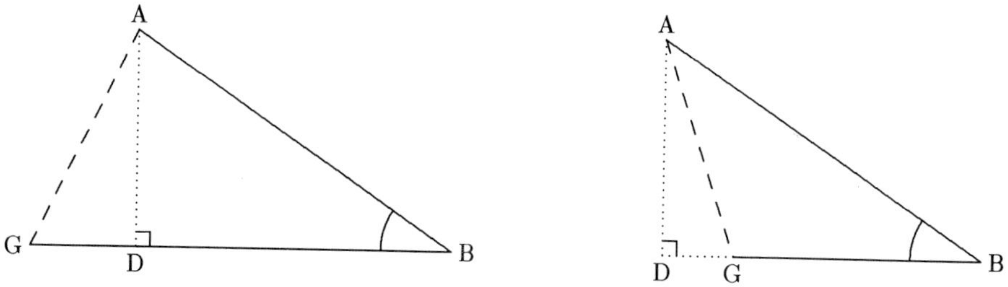

Fig. 4

[46]In Figs. 4, 5 and 6 known sides are drawn solid, unknown sides as broken lines and construction lines as dotted lines; known angles are marked with a small arc.

If three angles are known, then the ratio of the diameter of the circle to each side, considered as a chord of the circle, is known, and so the ratio of any two sides is known. Therefore, if one of the sides is known, they are all known. This is tantamount to a statement of the sine theorem for plane triangles.

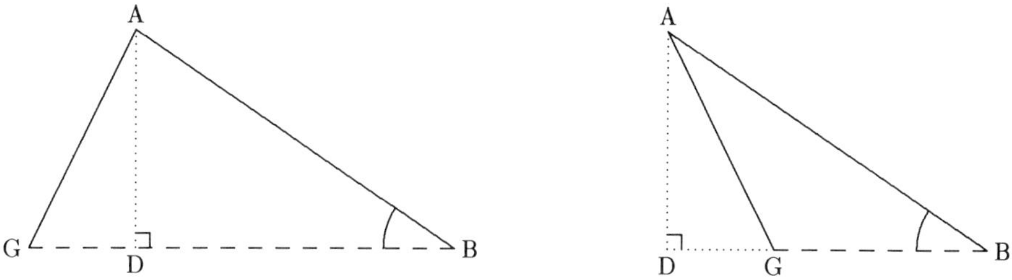

Fig. 5

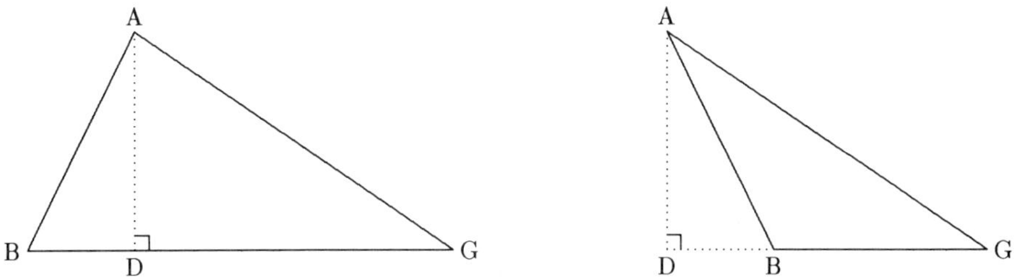

Fig. 6

Just before considering triangles with three known angles, Jābir deals with the more interesting case of three known sides. If $AG = AB$ (see Fig. 6), the matter is of course trivial. Otherwise, let AB be the smaller. Then

$$AG^2 - AB^2 = GD^2 - BD^2 \;\ldots\ldots\ldots\ldots\ldots\ldots \text{(i)}$$

and if either side of the equation is divided by BG, half the difference of the quotient and BG yields BD. For

$$\frac{AG^2 - AB^2}{BG} = \frac{GD^2 - BD^2}{GD + BD} = GD - BD \ldots \text{(ii)}$$

for an acute-angled triangle. Now $\triangle ABD$ may be solved as before.[47]. As is

[47]Ptolemy applies the procedure in *Almagest* VI 7 and in *Planisphaerium*, section 8. See *Opera omnia* I 514 and II 238. The passage equivalent to *Almagest* VI 7 is found in Jābir's book V (printed edition, p. 79).

stated in a Latin manuscript of perhaps the late fourteenth century[48] Jābir uses this in Book VII, Proposition 14.

Further, if $(AG^2 - AB^2)/BG > BG$, then the [acute] angle B, when found as usual, will be the supplement of $\hat{B}$ in the triangle, because in this case AD falls outside the triangle. There is a disturbance in the text here: we follow Gerard; MS Berlin 5653 (f. 17r) has "less than" instead of "greater than" and MS Cambridge add. 1191 correctly gives the two cases.

In this passage the two groups of Arabic manuscripts diverge: the Berlin and Cambridge manuscripts carry the text translated by Gerard but in the two Escurial manuscripts (910 and 930) there is a different recension[49]. In the treatment of the case of three known sides the Escurial manuscripts only say:

> Since BG is known and is divided into unequal parts at D and the difference of the squares of the parts is known, it is necessary that each of them be known.

In the margin of Escurial 930 is a note (very hard to read) that gives an arithmetic procedure to find the larger and smaller of the unequal parts, a method that involves an unnecessary operation of squaring. In modern terms, if a is the larger part and B the smaller,

$$\frac{a^2 - b^2 + (a+b)^2}{2(a+b)} = a$$

We may notice in passing that in this recension the point from from which the perpendicular is dropped is chosen as the one opposite the greatest side – a precaution that obviates the question of angle B's being acute or obtuse.

Another difference between the two recensions is the explicit treatment in the Escorial manuscripts of the case of two known angles and one known side. In this case the triangle is solved by drawing a perpendicular from one end of the known side and solving the two resulting triangles as usual. As we

[48] Lyon, Bibliothèque de la Ville 328, f. 78v.

[49] After an earlier version of this section was given as a paper at the Fifth International Symposium for the History of Arabic Science, Professor G. Saliba suggested in the discussion that the author himself may have produced the two recensions at different times. It may be added that an example of such a phenomenon is al-Bīrūnī's writing and rewriting of the *Istikhrāj al-awtār*; I am grateful to Dr. A. Abdul-Latif (Amman) for pointing this out to me.

have seen, the other group, which is followed by Gerard, treats the case as as appendage to the case of three known angles.

These procedures in plane trigonometry attracted comments in the manuscripts of the Latin translation – here the most interesting are Madrid BN lat. 10006 and British Library, Harley 625 –, in separate collections of notes – particularly those in Madrid BN 10010, ff. 51r–v (also Paris BN lat. 7377B, ff 39v–41r) – and in the commentary on Jābir's astronomy in Paris BN lat 7406. In one of the Madrid-Paris notes on Jābir's plane trigonometry, his first book is referred to as "liber 30 figurarum"[50]. This may indicate some connection with the commentary in MS Paris BN lat. 7406, which has the phrase in its *incipit*. Some of these notes – the second in the Madrid 10010 series and the Paris commentary – go over the same results as Jābir's text, but for obtuse-angled triangles. The last note in the series, the one it has in common with Digby 168, investigates the condition for the perpendicular to fall inside or outside the triangle by finding (Fig. 4) BD and comparing it with BG[51]. As the note points out, AB is needed for this as well as angle B and side GB , as Jābir said (here the Berlin manuscript and the Latin translation agree).

The majority of the notes on the case of three known sides are justifications of (ii) from (i) by "geometric algebra": Euclid II 12 and 13 (equivalent to the cosine formula for acute- and obtuse-angled triangles) is the most direct[52], but II 4 (in modern terms $(a+b)^2 = a^2 + b^2 + 2ab$)[53] and II 6 (difference of squares) are also applied[54]. Sometimes the "geometric algebra" was done by hand, after the manner of Euclid, as in the Paris commentary, where, incidentally, extraordinary letters are used for the extra points needed in the diagram, letters that do not come at the beginning of either the Arabic or Latin alphabet[55]. To help with the geometric algebra in the case of acute-angled triangles, an extra point is sometimes added on line BG so that it and B are symmetrically placed about point D[56]. We may notice that (i) itself appears elsewhere, e.g. in Pappus' *Collectio* VII 120 and in the *Book*

[50] Madrid 10010, f. 51r; Paris 7377B, f. 40v.

[51] Madrid 10010, f. 51r–v; Paris, BN lat. 7377B, ff. 40v–41r; Oxford, Bodleian, Digby 168, f. 126v (alt. pag. 127v).

[52] First note in Ma 10010.

[53] Ma 10006, on f 13r.

[54] British Library, Harley 625, f 16v. Lyon 328, f 79r quotes *arismetica Jordani* for the difference of squares formula.

[55] This is also true of the first comment in the Madrid 10010 series.

[56] E.g. in the second note in the Madrid 10010 series.

of Assumptions attributed in the two known manuscripts to Archimedes and to a certain Aqāṭun[57].

Two unusual notes on the case of three known sides are of some interest. The first is a numerical example[58]:

> Potest hoc ostendi in numeris. Sit latus *AB* 8 et latus *AG* 6 et latus *GB* 7; quadratum *AB* id est 8 superat quadratum *AG* id est 6 in 28, que si dividantur per 7 remanent 4. Superfluitas de 7 ad 4 est 3, que si dividantur per medium remanet unum et dimidium: et ibi est punctus perpendicularis ex parte linee *AG* id est 6. Quod multiplica in seipsum et erit 2 et $\frac{1}{4}$; quod trahas de quadrato linee *AG* id est 6 et remanent 33 et $\frac{3}{4}$: et illud est quadratum linee perpendicularis id est linea *AD*. Patet etiam hoc: si accipiatur quadratum residui de 7 ab uno et de[1] id est 5 et de, et addatur quadrato perpendicularis, et fiet quadratum linee *AB*, id est quadratum 8.

The second invokes Hero's formula for the area of a triangle[59]:

> glossa[2]. Alio modo potest cognosci perpendicularis *AD*, nam cum 3ª latera *AB BG AG*[3] sint nota, si accipiatur medietas ipsorum et differentie 3 ipsius medietatis ad singula latera primaque in 2ᵃᵐ ducatur et in productum 3ᶜⁱᵃ iterumque in productum ipsa medietas, radix producti dabit aream trianguli, quod probatum est alibi. Si duplum illius aree dividatur per latus *BG* in quod cadit perpendicularis, egredietur quantitas perpendicularis *AD* quod proponebatur.

Perhaps the most interesting of all these notes is the fourth in the Madrid 10010 series, an alternative method of solving triangle *ABG* (see Fig. 7) with three known sides (Madrid 10010, f. 51r; BN 7377B, f. 40v):

[1] Here and in the next occurrence = $\frac{1}{2}$ [2] glossa: *marg.* **MS** The gloss is to th. 26 in the printed edition (here 27). [3] *AG*: *corr. ex AB* **MS**

[57] See Dold-Samplonius (1976), p. 26. (i) is also the subject of a note in MS Erlangen, Universitätsbibliothek 836 (786), f. 95r.

[58] MS Madrid 10006, f. 13r.

[59] MS Lyon 328, f. 79r.

Sit triangulus AGB cuius tria latera sint[4] nota: dico igitur quod ipse est notus. Probacio eius. Quia si duo latera AG AB sunt equalia, tunc manifestum est quod ipse est notus. Sed si latus AB[5] fuerit maius latere AG, describam super centrum A[6] cum longitudine AG circulum $DGSH$, et producam lineam BA usque ad punctum D. Ergo multiplicatio DB in BH est equalis multiplicationi GB in BS. Ergo hic sunt quatuor linee proportionales, quarum prima que est DB est nota quia AB est nota, et AH et AD sunt note[7] quia utraque est equalis AG note; ergo DH est nota, quare remanet BH quarta nota; et GB[8] secunda est nota; quare tercia SB[9] est nota quare et GS est nota. Cum ergo producetur perpendicularis AE, erit et eius medietas ES nota, ergo tota BE est nota. Et AB est nota, ergo perpendicularis AE est nota. Ergo triangulus ABG est notus.

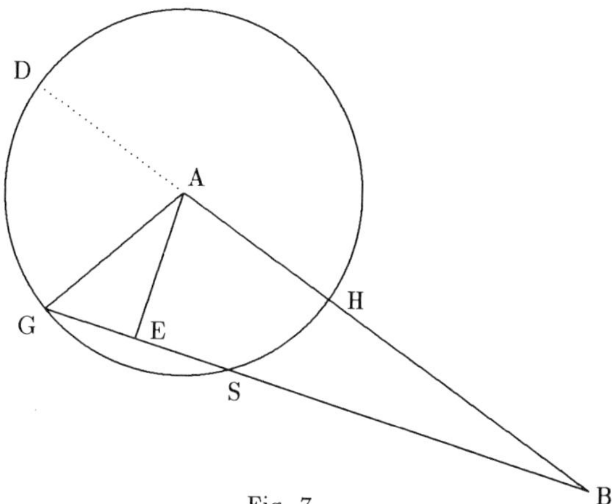

Fig. 7

About point A a circle of radius AG [the smallest side] is drawn. Now

$$DB \cdot BH = GB \cdot BS \text{ [by Euclid III 36]}$$

In the usual modern symbols

$$(g + b)(g - b) = a \cdot BS$$

[4]sint: sunt **P** [5]AB: *supra* **P** [6]A: *supra* **M** [7]sunt note: est nota **M** [8]GB: BG **P** [9]SB: BS **P**

Thus BS, and so SG and its half GE, are found. Of course, for purposes of calculation, this is equivalent to (ii). This theorem was translated by Gerard of Cremona, in quite different words, and appears in MS Digby 168 as well as in the celebrated codex Paris BN lat. 9335. It was later incorporated into the anonymous *Liber de triangulis Jordani*[60].

There were other sources for the development of early western plane trigonometry. We may mention the short treatise by Levi b. Gerson (14c.) translated into Latin under the title *De Sinibus, chordis et arcubus, Item Instrumento Revelatore Secretorum*[61], in which plane trigonometry and Jacob's staff are treated. But by this time the tradition had been established. Again the procedure for finding BD in Fig. 6 is to be found already in the *Geometria incerti autoris* in a numerical example (curiously, a right-angled triangle is taken)[62]. But it is suggested here that the *Iṣlāḥ al-Majasṭī*, with its systematic solution of plane triangles, was the basis of this branch of trigonometry.

The anonymous text *De tribus notis*[63], possibly of the fourteenth century, is virtually a commentary on Jābir's plane trigonometry. Using Jābir's methods and often mentioning him by name, taking the cases in almost the same order and using chords only, the author presents a long-winded account of all possible cases and subcases. For the case of one known side and two known angles, for instance, he gives ten sub-cases without mentioning that they could be reduced to one by first calculating the third angle. Not yet investigated is a series of numbered propositions on plane trigonometry in MS Cracow, Bib. Jag. 619, ff 1r–v. The first $3\frac{1}{2}$ sections are missing and the text begins abruptly with "6 partes et 4 [?] minutorum de partibus quibus AB est 7 partes ... ". Jābir is mentioned in at least one proposition (no. 8, line 8)[64].

All these results, including the alternative method for solving a triangle of known sides, are to be found in book I, Props. 44–53 of Regiomontanus' *De triangulis*, written in the 1460's, but printed only in 1533. Finally, in the

[60]Proposition IV 25. See Clagett, pp. 328-329, 424-425, 475, 600-601, 658.

[61]Edited by Curtze (1898).

[62]Gerbert, p. 344. According to M. Folkerts (private communication) this text is probably of the ninth century at the latest.

[63]Edited by Curtze (1900), pp. 380–390.

[64]There are other series of propositions on plane trigonometry in MS Erfurt, Amplon. F 375, 137v, "Omnis trianguli, cuius latera sunt nota, anguli sunt noti tam super centrum quam super circumferentiam ... ", and in MS Erfurt, Amplon. Q 349, f. 71v *et seq.*, "In omni triangulo orthogonio, noto uno angulorum ... "; but they are relatively late and do not appear to have any connection with Jābir.

first book of the *De revolutionibus* (1543), Copernicus gave a concise account of the solution of plane triangles, with chords only, and using the methods described above. He, too, includes the alternative method for the case of three known sides, describing it as "commodius forsitan".

4. After Regiomontanus

According to Ricciolus' *Almagestum novum*, Andreas Stiborius [d. 1515] wrote an epitome of "Geber"[65]:

> *Andreas Stiborius* Boius, Philosophus, Theologus, & Astronomus acutissimus, Viennensis Canonicus, scripsit Epitomen Albategni, Almagesti, Gebri & alia multa, floruit sub Maximiliano Imper.

If this is so, the work does not appear to have survived. But Jābir was still cited well into the sixteenth century, for instance by Snell[66], Stevin[67] and Maurolycus[68]. No doubt many more such references could be found, but in general Jābir's critique of the *Almagest* was becoming out of date and by the sixteenth century his trigonometry had been superseded by more systematic and complete treatments of the subject. Thus Cardanus, who, at eighteen, imbued with an inextinguishable desire for an immortal name after his father had died, started to write a book on finding distances of places from longitude and latitude:

> Erat enim mihi liber quidam antiquus de triangulis Hebri Hispani, cuius auxilio, ut dixi, libellum de diagnoscenda locorum distantia conscribere adortus sum: quem tamen perficere non potui, quaedam enim deerant, quae postea Monteregium docuisse animadverti.

The book on triangles by Heber Hispanus had, after all, to be supplemented by Regiomontanus' work.

[65] *Chronicon*, which prefaces the work, part 2, p. xxx.
[66] On the complexity of Ptolemy's method of finding the Moon's apogee, *Appendix*, p. 106.
[67] Stevin, III 70.
[68] In the preface to his *Sphaerica*, f. 45v

5. *Bibliography*

The Latin Translation of Anaritius' Commentary on Euclid's Elements of Geometry, Books I–IV, edited by P.M.J.E. Tummers. *Artistarium*, supplementa IX. Nijmegen, 1994.

Al-Bīrūnī, *Kitāb maqālīd 'ilm al-hay'a*, La Trigonométrie sphérique chez les Arabes de l'Est à la fin du Xe siècle. Édition et traduction pat M.-Th. Debarnot, Damascus 1985.

A. Björnbo, "Thabits Werk über den Transversalensatz (liber de figura sectore)" mit Bemerkungen von H. Suter. Herausgegeben und ergänzt von ... H. Bürger und K. Kohl, *Abhandlungen zur Geschichte der Naturwissenschaften und der Medizin* 7, Erlangen 1924.

A. von Braunmühl, "Nassîr Eddîn Tûsi und Regiomontan", *Abhandlungen der Kön. Leopold. Akad.* 81 (1897). .

A. von Braunmühl, *Vorlesungen über Geschichte der Trigonometrie*, 2 vols., Leipzig 1900.

H. L. L. Busard, "Der Traktat *De isoperimetris*, der unmittelbar aus dem Griechischen ins Lateinische übersetzt wordem ist", *Medieval Studies* 42 (1980) 61–88.

Hieronymus Cardanus, *De subtilitate*, Nuremberg 1550.

Hieronymus Cardanus, *Opera omnia*, 10 vols., Lyons 1663.

M. Clagett, *Archimedes in the Middle Ages*, vol. 5 (2 parts), Philadelphia 1984.

M. Curtze, "Die Abhandlung des Levi ben Gerson über Trigonometrie und den Jacobstab", *Bibliotheca mathematica* NF 12 (1898) 97–112.

M. Curtze, "Urkunden zur Geschichte der Trigonometrie im christlichen Mittelalter", *Bibliotheca mathematica* 3F 1 (1900) 321–416.

M. Curtze, "Der Briefwechsel Regiomontans mit Giovanni Bianchini, Jakob von Speier und Christian Roder", *Urkunden zur Geschichte der Mathematik im Mittelalter und der Renaissance, 1. Theil* (= *Abhandlungen zur Geschichte der mathematischen Wissenschaften* ... 12 (1902)) 185–336.

Debarnot: *v.* Bīrūnī.

Y. Dold-Samplonius, *Kitāb al-mafrūdāt li-Aqātun*, Heidelberg 1977.

A. B. Emden, *A Biographical Register of the University of Oxford to A.D. 1500*, 3 vols., Oxford 1957–1959.

The Thirteen Books of Euclid's Elements, translated with introduction and commentary by Sir Tho. L. Heath, Cambridge 1926 (repr. 1956).

Gebri filii Affl'a Hispalensis, de astronomia libri IX, Nuremburg 1543. [copy used: Bodleian, Savile X3]

Gerberti postea Silvestri II papae Opera Mathematica, ed. N. Bubnov, Berlin 1899, repr. Hildesheim 1963.

A. Gerl, *Trigonometrisches Rechnen kurz vor Copernicus. Der Briefwechsel Regiomontanus-Bianchini*, Stuttgart 1989. Annotated German translation of Curtze, "Der Briefwechsel ..." above.

B. R. Goldstein, "The Survival of Arabic Astronomy in Hebrew", *Journal for the History of Arabic Science* 3 (1979) 31–39.

R. T. Gunther, *Early Science in Oxford*, Oxford 1923.

N. G. Hairetdinova, "On the oriental sources of the Regiomontanus' trigonometrical treatise", *Archive Internationale d'Histoire des Sciences* 90–91 (1970) 61–66.

N. G. Hairetdinova, "On Spherical Trigonometry in the Medieval Near East and in Europe", *Historia Mathematica* 13 (1986) 136–146.

Ibn Muʿādh: *v.* Villuendas.

W. Knorr, "The Medieval Tradition of a Greek Mathematical Lemma", *Zeitschrift für Geschichte der Arabisch-Islamischen Wissenschaften* 3 (1986) 230–264.

P. Kunitzsch, "Letters in Geometrical Diagrams. Greek - Arabic - Latin", *Zeitschrift für Geschichte der Arabisch-Islamischen Wissenschaften* 7 (1991/92).

R. Lorch, "Some remarks on the *Almagestum parvum*, in *Amphora. Festschrift für Hans Wussing*, Basel etc. 1992, 407–437, = item V in this volume.

R. Lorch, "The Transmission of Theodosius' *Sphaerica*", in the *Tagungsband* for the *Arbeitsgespräch* "Mathematische Probleme im Mittelalter", 25pp. (forthcoming – ?1995).

F. Maurolycus, *Sphaerica*, Messina 1558.

Claudii Ptolemaei opera quae exstant omnia, ed. J. L. Heiberg, Leipzig 1898 (vol. I) and 1907 (vol. II).

Regiomontanus on Triangles, *De trinagulis omnimodis* by ... Regiomontanus, translated by Barnabas Hughes, O.F.M., with an Introduction and Notes, Madison etc. 1967.

G. B. Ricciolus, *Almagestum novum*, Bononiae 1651.

Richard of Wallingford. An edition of his writings with introductions, English translation and commentary by J. D. North, 3 vols., Oxford 1976.

W. Snell, *Appendix ad observationis Hassiacas*, 1618.

M. Steinschneider, *Die hebraeischen Übersetzungen des Mittelalters und die Juden als Dolmetscher*, Berlin 1893 (repr. Graz 1956).

The Principal Works of Simon Stevin, 5 vols., Amsterdam 1955–1968.

H. Suter, "Zur Trigonometrie der Araber", *Bibliotheca Mathematica* 10 (1909-10) 156–160.

C. H. Talbot, "Simon Bredon (c.1300–1372), Physician, Mathematician, and Astronomer", *British Journal for the History of Science* 1 (1962) 19–30,

M. V. Villuendas, *La trigonometría europea en el siglo XI. Estudio de la obra de Ibn Muʿāḏ* El Kitāb maŷhūlāt, Barcelona 1979.

Sr. M. C. Zeller, "The Development of Trigonometry from Regiomontanus to Pitiscus", Ph.D. Dissertation, University of Michigan, 1944.

E. Zinner, *Leben und Wirken des Joh. Müller von Königsberg genannt Regiomontanus*, 2nd ed., Osnabrück 1968.

E. Zinner, *Regiomontanus: His Life and Works*, Amsterdam etc., 1990. *Studies in the History and Philosophy of Mathematics*, vol. I. An English translation of the above.

IX

Abū Kāmil on the Pentagon and Decagon

The Text

Abū Kāmil Shujā' b. Aslam lived after al-Khwārizmī, whom he quotes at the beginning of his *Algebra* and elsewhere, and before the *Fihrist* was compiled in the late tenth century[1]. His most famous work, the three-part *Algebra*, which is known in only one Arabic manuscript[2], Istanbul, Kara Mustafa Paşa 379 (dated 651/1253)[3], had considerable influence in the medieval Arabic world[4]. Of the Latin translation there is only one known manuscript, Paris, Bibliothèque Nationale, lat. 7377A, 71v–97r (14c.), which breaks off shortly after the beginning of the third part, on indeterminate equations. There was a further transmission into Latin in the form of the *Practica geometriae*, which was based on Arabic sources; it contains an elaborated version of the second part of the *Algebra*, on the pentagon and decagon[5]. A few citations of Abū Kāmil's procedures by the author of the *Liber Mahameleth* probably represent yet another independent transmission from the Arabic[6]. The whole treatise was translated into Hebrew in the fifteenth century by Mordecai Finzi; and of this translation there is a complete manuscript (Munich, cod. heb. 225) and a fragment in Paris (Bib. Nat., heb. 1029,7)[7].

In this paper an edition of the Latin translation of the second part[8] of the *Algebra* is presented with a note on the terminology of the translator. Since there are several modern translations of the work[9], only a translation of the

[1]For general references to Abū Kāmil see Sezgin 277-281, Levey (DSB), the introduction to Abū Kāmil (Hebrew). His dates are usually given as ca. 850 – ca. 930.

[2]Dr. Jacques Sesiano informs me (private communication) that the Meshed manuscript of the *Algebra* mentioned by Sezgin (p. 281) in fact contains another work. MS Teheran, Sanā 2672/6, ff. 189r–303r, *Misāḥat al-arāḍīn*, suggested *ibid* as being the treatise on the pentagon and decagon, has an *incipit* (see *Nashriye* II, 241) that does not occur in the Istanbul manuscript of the treatise; it has not been investigated.

[3]Published in facsimile by the Institut für Geschichte der Arabisch-Islamischen Wissenschaften in 1986.

[4]See e.g. Karpinski.

[5]For a detailed comparison with the Abū Kāmil text as transmitted through Hebrew, see Suter, pp. 38-42.

[6]See Sesiano, 73–74, 95–96. The first of these citations, at the beginning of the treatise, MS Paris BN lat. 7377A, f. 101r, quotes the titel of Abū Kāmil's book in transliteration, not in translation as in our text.

[7]The first part was translated from Hebrew into German by Josef Weinberg for his doctoral dissertation and was published by him in 1935; a Hebrew edition and English translation was published by Martin Levey in 1966.

[8]An edition of the Latin translation of the first and third part is to be found in the contribution of Dr. J. Sesiano in this volume.

[9]Sacerdote (from the Hebrew), Suter (from Sacerdote) and Yadegari and Levey (from

preface from the Arabic (for comparison with the Latin) and a summary of the content of the rest will be given here. The preface[10]:

> Abū Kāmil Shujāʿ said: since we have expounded what needs explanation of what was hard to understand in the calculation [called] *al-jabr* and *al-muqābala* for those advanced in the science and its application and the skilled geometers who are experts in the book of Euclid and other [books], and [since] we have explained and commented on that in this book of ours, we shall expound the amount of each of the sides of the equilateral and equiangular pentagon and decagon, each of which is contained in a known circle or contains a known circle, and the amount of the diameter of the circle that contains or is contained by a known equilateral and equiangular pentagon or decagon, and the amount of the chord of a fifteenth of the circumference of a known circle and the amount of each of the sides of the equilateral and equiangular pentagon and decagon of known area and the amount of the sides of triangles of known area when they are in an equilateral and equiangular pentagon or decagon, and other things that we have commented upon in this book of ours, the finding of much of which has been impossible for those of the calculators and geometers of our time whom we have met and for those advanced in the science of calculation and geometry of whom report has reached us. God (be He mighty and glorious) make easy for me to reach what remained inaccessible to them of it and to achieve what was difficult for them to understand of it. Praise, thanks, blessings and eternity [belong] to God, who has no companion.

In the following summary of the mathematical content of the treatise, the application of algebra to geometrical problems, the division into twenty propositions presented by Suter (and taken over from Sacerdote) will be retained; some propositions are subdivided. All polygons are regular and the figures referred to are those in the Latin text unless otherwise specified. x means the *res*, or thing sought; s_5 and S_5 are the sides of the inscribed and circumscribed regular pentagon, and similarly for the other regular figures; Ptol.($ABGD$) means Ptolemy's theorem applied to the cyclic quadrilateral $ABGD$; Pyth.(ABG) means Pythagoras' theorem applied to triangle ABG. Juxtaposed fractions are to be understood as added together. Numbers enclosed in square brackets after an enunciation indicate the previous propositions used in the problem. Only the essence of the argument is given, not necessarily in the order of the mathematical steps in the text. All references to book and proposition number of Euclid's *Elements* ("Eu.") and most of

the Arabic). Suter's translation, though from the Italian, translated from the Hebrew, which was translated probably direct from the Arabic, is the best, but is sometimes a paraphrase.

[10][Abū Kāmil], p. 134. It is a pleasure to thank Professor Paul Kunitzsch for considerable help with this translation and with many other parts of this paper.

the justifications for mathematical steps are supplied by the editor. The use of Ptolemy's theorem is justified only in a marginal note in the Latin translation: where it is first used (Prop. 1) "Ptolomeus in libro de arcu et corda" (presumably *Almagest* I 10) is quoted.

1. To find the side (x) of the pentagon inscribed in a circle of diameter 10[11]. (See p. 234)

   $EL = \frac{1}{10}x^2$ since $ED^2 = HE \cdot EL$

   Pyth.(ELD) : $GL = \sqrt{x^2 - \frac{1}{10}\cdot\frac{1}{10}x^2\,x^2}$

   Ptol.($ABDG$): $AB \cdot GD + AB^2 = GD^2$

   Now $GD^2 - AB^2 = 3x^2 - \frac{1}{25}x^2$

   $(3x^2 - \frac{1}{5}\cdot\frac{1}{5}x^2x^2) \div x = GD = \sqrt{4x^2 - \frac{1}{5}\cdot\frac{1}{5}x^2x^2}$

   Square, *Oppone*, $\div x^2$: $5 + \frac{1}{625}x^2x^2 = \frac{1}{5}x^2\,x^2$

   Restaura to *census census*[12] ($\times 625$), solve by formula:

   $x = \sqrt{62\frac{1}{2} - \sqrt{781\frac{1}{4}}}.$

2. To find the side of a decagon (x) inscribed in a circle of diameter 10. [1] (See p. 236)

   Ptol.($AGDH$): $10x + s_5^2 = GH^2$; and $GH^2 + s_5^2 = 100$

   $125 + 10x - \sqrt{3125} = 100$

   Oppone, $\div 10$: $x = \sqrt{31\frac{1}{4}} - 2\frac{1}{2}.$

3. To find the side (x) of a pentagon circumscribed about a circle of diameter 10. [1] (See p. 236)

   Pyth.(ELT): $EL^2 = 9\frac{1}{4}\frac{1}{8} + \sqrt{48\frac{1}{2}\frac{1}{4}\frac{5}{8}\cdot\frac{1}{8}}$

   Similar $\triangle$s ETB, ELT: $x^2 : 25 = s_5^2 : EL^2$

   Cross-multiply, *reduc* to *census* ($\times\ \frac{1}{5}\frac{1}{25} - \sqrt{\frac{4}{125}}$):

   $x^2 = 375 + \sqrt{15625} - \sqrt{78125} - \sqrt{28125} = 500 - \sqrt{200000}.$

4. To find the side (x) of a decagon circumscribed about a circle of diameter 10. [2] (See p. 238)

   Similarly: $x^2 : s_{10}^2 = 25 : EL^2$. Hence $x^2 = 100 - \sqrt{8000}.$

[11] 10 was often used as a token value. It is clear from Prop. 16 that Abū Kāmil could work just as easily with an unknown instead of 10.

[12] These terms are explained below. Throughout the summary the Latin terms and the Latin diagrams are used.

5. To find the diameter of the circle in which a pentagon of side 10 is inscribed. (See p. 238)

   Put $x = EG$. Ptol.$(ABGE)$: $x \cdot 10 + 10^2 = x^2$

   Solve by formula: $x = 5 + \sqrt{125}$

   Pyth.(EDL): $DL^2 = 62\frac{1}{2} - \sqrt{781\frac{1}{4}}$

   Put $x = CD$.

   Similar $\triangle$s CDE, EDL: $x^2 : 10^2 = 10^2 : DL^2$

   Cross-multiply, *reduc* to *census* $(\times\ \frac{1}{50} + \sqrt{1\frac{1}{4} \cdot \frac{1}{15625}})$:

   $x^2 = 200 + \sqrt{8000}$.

6. To find the diameter of the circle circumscribed by a pentagon of side 10. [5] (See p. 239)

   $AM^2 = 50 + \sqrt{500}$

   Pyth.(AMC): $CM^2 = 25 + \sqrt{500}$

   (diam. inner $\odot)^2 = 100 + \sqrt{8000}$

   Or: (diam. outer $\odot)^2 - S_5^2 =$ (diam. inner $\odot)^2$ and similarly for other regular figures.

7. = Prop. 5 by another method. [1]

   Eu. within XII 2: chords of similar arcs $\propto$ diameters of the circles in which they stand; put $x =$ diam.:

   $100 : 62\frac{1}{2} - \sqrt{781\frac{1}{4}} = x^2 : 100$ etc., as in 5.

8. = Prop. 6 by another method. [3]

   $100 : 500 - \sqrt{200000} = x^2 : 100$

   Cross-multiply, *reduc* to *census* $(\times\ \frac{1}{10} \cdot \frac{1}{10} + \sqrt{\frac{1}{125} \cdot \frac{1}{100}})$:

   $x^2 = 200 + \sqrt{8000}$.

9. To find the diameter (x) of a circle when a decagon of side 10 is inscribed in it. [2]

   $x : 10 = 10 : \sqrt{31\frac{1}{4}} - 2\frac{1}{2}$

   Cross-multiply, *restauremus* with $2\frac{1}{2}x$: $100 + 2\frac{1}{2}x = \sqrt{31\frac{1}{4}x^2}$

   $(100+2\frac{1}{2}x)^2 = 10000 + 6\frac{1}{4}x^2 + 500x = 31\frac{1}{4}x^2$

   Opponamus, $\div 25$, and solve: $x = 10 + \sqrt{500}$.

10. *i.* To find the diameter (x) of a circle when a decagon of side 10 is circumscribed. [4]

    $100 - \sqrt{8000} : 100 = 100 : x^2$

    Cross-multiply, *reduc* to *census* $(\times\ \frac{1}{2} \cdot \frac{1}{10} + \sqrt{\frac{1}{100} \cdot \frac{1}{5}})$

    $x^2 = 500 + \sqrt{200000}$.

10. *ii.* If AG is the side of a regular polygon and CY and AD are the diameters of the inscribed and circumscribed circles, $AG^2 + CY^2 = AD^2$. (See p. 242)

    $\triangle AML \equiv \triangle DMZ : \quad LG = DZ;$

    and since $A\hat{L}M = M\hat{Z}D, \quad ZD \parallel LG$

    So $ZL \parallel$ and $= DG$

    The result follows by Pyth.(AGD).

11. *i.* To find the side of the 15gon inscribed in a circle of diameter 10. [2] (See p. 243)

    $BD = s_{10} = \sqrt{31\frac{1}{4}} - 2\frac{1}{2}, BG = s_6 = 5,$ so that $GD = s_{15}; AD$ and AG are found by Pyth.(ABD, ABG)

    Ptol.($ABDG$): $AD \cdot BG - AG \cdot BD = AB \cdot GD$

    Multiply out, $\div AB$ i.e. 10:

    $GD = \sqrt{15\frac{5}{8} + \sqrt{48\frac{3}{4}\frac{5}{8}\cdot\frac{1}{8}}} + \sqrt{4\frac{1}{2}\frac{1}{8}\frac{1}{2}\cdot\frac{1}{8}} - \sqrt{23\frac{3}{8}\frac{1}{2}\cdot\frac{1}{8}}.$

11. *ii.* Approximately, $GD = 2; 4, 44, 48$ in sexagesimals.

12. If the area of an equilateral triangle + its altitude = 10, to find its altitude (x). (See p. 244)

    $DG = \sqrt{\frac{1}{3}x^2}$ and so area $= \sqrt{\frac{1}{3}x^2x^2}$

    $x + \sqrt{\frac{1}{3}x^2x^2} = 10$

    Restaura $\sqrt{\frac{1}{3}}$(*census census*) to $\sqrt{}$(*census census*) $(\times\sqrt{3})$:

    $\sqrt{3x^2} + x^2 = \sqrt{300}$

    Solve: $x = \sqrt{\frac{3}{4} + \sqrt{300}} - \sqrt{\frac{3}{4}}$.

13. If an equilateral triangle has side 10, to find the base of a rectangle of area 10 constructed within it[13]. (Propositions 13 and 14 are spoilt in Latin. The Arabic diagrams, with letters according to the system of

[13]There is a problem similar to Props. 13 and 14 in al-Khwārizmī: Rosen's edition, Arabic pp. 33–34, English pp. 84-85. It is not in the Latin translations by Robert of Chester and Gerard of Cremona.

transliteration used by the translator, are reproduced here. The Latin text jumps to "So area of △" at the end of Prop. 14.)

$x = HE$. Then $HB = \sqrt{\frac{1}{3}x^2}$, etc., much as in Prop. 12.

$x = \sqrt{18\frac{1}{2}\frac{1}{4} - \sqrt{75}} + \sqrt{18\frac{1}{2}\frac{1}{4}}$.

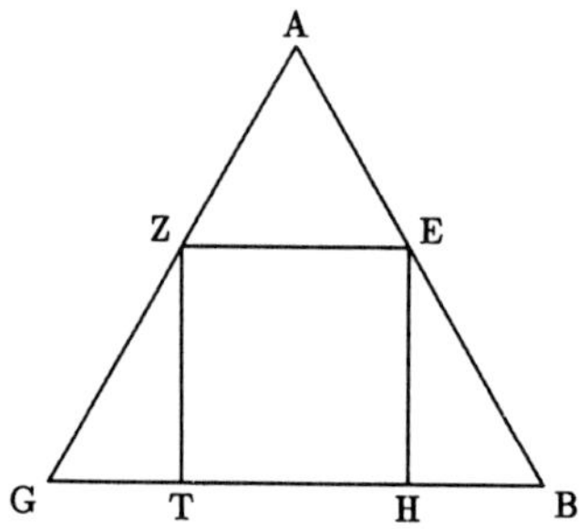

Diagram for Prop. 13 in Arabic

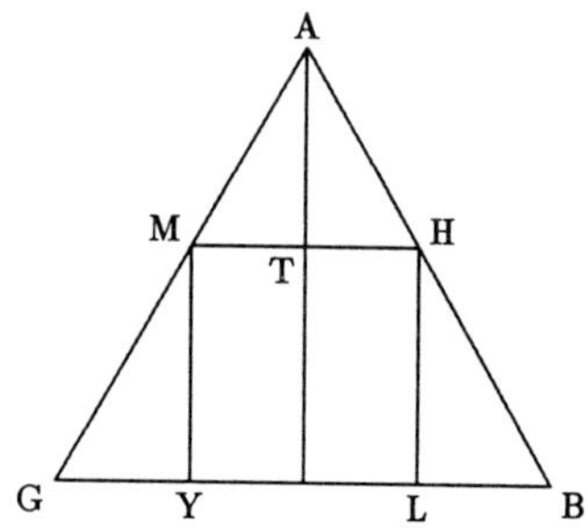

Diagram for Prop. 14 in Arabic

14. If the area of an equilateral triangle + the area of the square constructed within it is 10, to find the side (x) of the square.

$\Delta MYG + \Delta HLB = \sqrt{\frac{1}{3}x^2x^2}$; $\Delta HMA = \sqrt{\frac{1}{8}x^2x^2 + \frac{1}{2}\frac{1}{8}x^2x^2}$

$\sqrt{\frac{1}{3}x^2x^2} + \sqrt{\frac{1}{8}x^2x^2 + \frac{1}{2}\frac{1}{8}x^2x^2} = 10 - 2x^2$

$x^2 = \sqrt{45\frac{1395}{20449}} - \sqrt{11\frac{10261}{20449}}$

So area of $\Delta = 10 - x^2 = 3\frac{41}{143} + \sqrt{11\frac{10261}{20449}}$.

15. To find the side (x) of an equilateral pentagon constructed within a square of side 10. (See p. 245)

$BE = 10 - x$

From ΔHGZ, $GH = \sqrt{\frac{1}{2}x^2}$; and so $HB = 10 - \sqrt{\frac{1}{2}x^2}$

$200 + 1\frac{1}{2}x^2 - 20x - \sqrt{200x^2} = BE^2 + HB^2 = x^2$

Oppone and solve: $x = 20 + \sqrt{200} - \sqrt{200 + \sqrt{320000}}$.

16. To find the side (x) of a pentagon of area 50. [5 or 7] (See p. 246)

From Prop. 5 or 7: for side x, diam.$^2 = 2x^2 + \sqrt{\frac{4}{5}x^2x^2}$

Pyth.(MEH): $MH^2 = \frac{1}{4}x^2 + \sqrt{\frac{1}{2}\frac{1}{10}x^2x^2}$

$\frac{1}{2}\frac{1}{8}x^2x^2 + \sqrt{\frac{1}{320}x^2x^2x^2x^2} = 100$

Restaura $\frac{1}{2} \cdot \frac{1}{8}$ *census census* to *census census* (× 16):

$x^2x^2 + \sqrt{\frac{4}{5}x^2x^2x^2x^2} = 1600$

Reduc to *census census* $(\times\ 5 - \sqrt{20})$:

$x^2x^2 = 8000 + \sqrt{51200000}$.[14]

17. To find the side (x) of a decagon of area 100. [9] (See p. 247)

From Prop. 9: diam. $= x + \sqrt{5x^2}$

Eu. XIII 9 and 5[15]:

$ZM(s_6), ZU(s_{10})$ and their sum are in e.m.r.[16]

Eu. XIII 3: $(ZM - x + \frac{1}{2}x)^2 = 5(\frac{1}{2}x)^2$

$ZM^2 = 1\frac{1}{2}x + \sqrt{1\frac{1}{4}x^2x^2}$

Pyth.(MZS): $MS^2 = \quad 1\frac{1}{4}x^2 + \sqrt{1\frac{1}{4}x^2x^2}$

$MS^2 \cdot ZS^2 = \frac{1}{4}x^2x^2 + \frac{1}{2}\frac{1}{8}x^2x^2 + \sqrt{\frac{5}{64}x^2x^2x^2x^2} = 100$

Restaura $\frac{1}{4} + \frac{1}{2} \cdot \frac{1}{8}$ *census census* to *census census*: $(\times 3\frac{1}{5})$:

$x^2x^2 + \sqrt{\frac{4}{5}x^2x^2x^2x^2} = 320$

Reduc to $x^2x^2(\times \quad 5 - \sqrt{20}): \quad x^2x^2 = 1600 - \sqrt{2048000}.$

18. If $\triangle BDE$ is 10, to find the side (x) of the pentagon. (See p. 249)

Eu. XIII 8: if M divides BE in e.m.r., $ME = DE$.

Eu. XIII 3: if L bisects ME, $LM^2 = \frac{1}{5}LB^2$

$BL = \sqrt{1\frac{1}{4}x^2}$ and so $BE = \frac{1}{2}x + \sqrt{1\frac{1}{4}x^2}$

Pyth.(BEH): $BH^2 = 1\frac{1}{4}x^2 + \sqrt{1\frac{1}{4}x^2x^2}$

$BH^2 \cdot EH^2 = \frac{1}{4}x^2x^2 + \frac{1}{2}\frac{1}{8}x^2x^2 + \sqrt{\frac{5}{64}x^2x^2x^2x^2} = 100$

Restaura $\frac{1}{4} + \frac{1}{2} \cdot \frac{1}{8}$ *census census* to *census census* $(\times 3\frac{1}{5})$:

$x^2x^2 + \sqrt{\frac{4}{5}x^2x^2x^2x^2} = 320$

Reduc to *census census* $(\times\ 5 - \sqrt{20})$:

$x^2x^2 = 1600 - \sqrt{2048000}$.

---

[14]This should be $8000 - \sqrt{51200000}$.

[15]XIII 9 and 5 are quoted *verbatim* and XIII 3 is later used. But the whole passage is not necessary to the argument.

[16]e.m.r. means "extreme and mean ratio".

19. If $\triangle ABE$ is 10, to find BE (x). (See pp.249, 251)
    Eu. XIII 1 and 8: $(ME + \frac{1}{2}BE)^2 = 5(\frac{1}{2}BE)^2$
    Let $BE = x$. $AE = \sqrt{1\frac{1}{4}x^2} - \frac{1}{2}x$
    Pyth.(AEY): $AY^2 = 1\frac{1}{4}x^2 - \sqrt{1\frac{1}{4}x^2x^2}$
    $AE^2 \cdot YE^2 = \frac{1}{4}x^2x^2 + \frac{1}{2}\cdot\frac{1}{8}x^2x^2 - \sqrt{\frac{5}{64}x^2x^2x^2x^2} = 100$
    Reintegra $\frac{1}{4} + \frac{1}{2} \cdot \frac{1}{8}$ *census census* to *census census* $(\times 3\frac{1}{5})$:
    $x^2x^2 - \sqrt{\frac{4}{5}x^2x^2} = 320$
    Reintegra so that *census census* becomes whole
    $(\times\ 5 + \sqrt{20})$: $x^2x^2 = 1600 + \sqrt{2048000}$.
20. If $\triangle CTH$ is 10, to find CH (x). [5 or 7] (See p. 252)
    As in Prop. 16, $(\text{diam.})^2 = 2x^2 + \sqrt{\frac{4}{5}x^2x^2}$
    Eu. XIII 10: $s_5^2 = s_6^2 + s_{10}^2$ and so $CT^2 = \frac{1}{2}x^2 - \sqrt{\frac{1}{2}\cdot\frac{1}{10}x^2x^2}$
    Pyth.(CTM): $TM^2 = \frac{1}{4}x^2 - \sqrt{\frac{1}{2}\frac{1}{10}x^2x^2}$
    $CM^2 \cdot TM^2 = \frac{1}{2}\cdot\frac{1}{8}x^2x^2 - \sqrt{\frac{1}{320}x^2x^2x^2x^2} = 100$
    Reintegra $\frac{1}{2} \cdot \frac{1}{8}$ *census census* to *census census* $(\times 16)$:
    $x^2x^2 - \sqrt{\frac{4}{5}x^2x^2x^2x^2} = 1600$
    Reintegra to *census census* $(\times\ 5 + \sqrt{20})$:
    $x^2x^2 = 8000 + \sqrt{51200000}$.

As Suter has pointed out[17], the obtrusion of Euclid XIII 3 in Prop. 17 in the transmitted form of the text may well be due to a copyist. We may note that in the Latin translation of Prop. 18, which relies on XIII 8 and 3, XIII 5 is cited as well: this may well represent more interference of the same kind in the Arabic manuscript from which the translator was working. But we are concerned here only with the translation, not with the formation of the text itself.

The Latin Translation

The Latin terms used in the above summary should now be explained in modern terms. *Res*, *census*, *cubus* (Arabic: *shay*ʾ, *māl*, *kaʿab*) are used for the unknown, here called x, and its square and cube. Higher powers of the unknown are expressed by combinations of them, e.g. *census census*. *Restaurare*, *opponere* (Arabic: *jbr*, *qbl* III) and their associated abstract nouns are

[17]Suter, p. 30, n. 6.

used in their tradition senses: *restaurare* in Prop. 9 means to take a negative term from one side of the equation to become a positive term on the other; *opponere* means to collect like terms and to express the result as one term, which is usually positive (there is an exception near the end of Prop. 2). But *restaurare* usually occurs as a translation of *kml* IV in the sense of making the coefficient of a specified term (here always *census census*) unity by multiplying all terms of the equation by a suitable factor, which may be a number or a root. By *reducere* (Arabic *rdd*) a similar reduction of a specified term (here always *census* or *census census*) is meant, but this time the factor is a number plus (or minus) a root. These reduction factors Abū Kāmil produces without calculation or explanation. In fact, to reduce a coefficient $a + \sqrt{b}$ to unity, we must multiply by $\frac{a-\sqrt{b}}{a^2-b}$ or, in the form given by Abū Kāmil, $\frac{a}{a^2-b} - \sqrt{\frac{b}{(a^2-b)^2}}$. A complication arises in the last two propositions: *reintegrare* appears in the text four times for *kml* IV in the sense of reducing coefficients to unity, twice when the factor is a number and twice when it is a number plus a root. Finally, *integra* is twice (Props. 17,18) written over *restaura* as an interlinear note, presumably to convert it to to *reintegra*.

The diagram letters are translated as follows:

ا *A*, ب *B*, ج *G*, د *D*, ه *E*, و *U*, ز *Z*, ح *H*, ط *T*, ك *C*, ل *L*, م *M*, ع *Y*, س *S*.

In Props. 4 and 20 the correspondence is not exact. The three letters *E*, *F*, *U* stand for the four letters ع, س, ص, ق in Prop. 4; and in Prop. 20 *Z*, *I* stand for the single letter و in the Arabic text and for و, ز in the diagram. But it is only a matter of letters mentioned once as the vertices of a decagon and not mentioned again. The representation of ع by *Y* is striking — Gerard of Cremona, for instance,usually has *Q* and Adelard of Bath in at least part of his translation of Euclid has *i* or *j* —, but Hermann of Carinthia has *Y* in his translations of both the *Elements* and the *Planisphaerium*[18]. The representation of ك by *C* seems to be very unusual.

In the translation the diagrams themselves seem to have been reconstructed from the text. The fact that many are reversed, so that the vertices of a pentagon, say, will usually run anticlockwise in Arabic and clockwise in Latin, does not count for much. But the diagrams in the Latin manuscript certainly cannot have been copied from an Arabic manuscript such as Kara Mustafa Paşa 379, in which the diagrams in general correspond well with the text. In Prop. 3, for instance, we find that in the Arabic *TH* is parallel to *AB*; but this is not so in the Latin. Similar observations apply to Prop. 4. Further, the Latin manuscript has two diagrams for Props. 18 and 19, but in Arabic there is one composite diagram. Finally, both in the spoilt text of Props. 13–14 and in the explanatory note the figure is made to suit the new problem formulated in the note and does not exist in Arabic. Of course, these

[18]Kunitzsch, pp. 8, 15, 16.

differences and the implied reconstruction of the diagrams from the text may have occurred within the Arabic tradition.

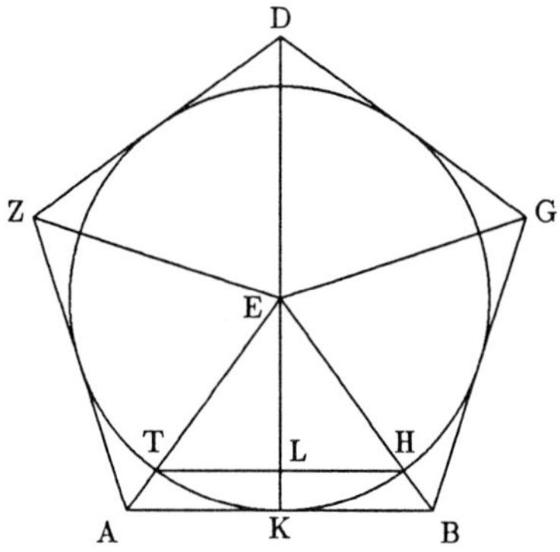

Diagram for Prop. 3 in Arabic

Numbers, which are written out in words in Arabic, are usually expressed in the Latin translation by Hindu-Arabic numerals, simple fractions usually appearing as in the modern notation. There are also some curious fractions, with two numerators and two denominators, which are also to be found in the works of Leonardo of Pisa. $\frac{e\ 1}{8\ 2}$ means $\frac{1}{2} \cdot \frac{1}{8}$. In such cases the notation usually follows the Arabic, which would in this case read "half of an eighth". There are exceptions; e.g. in Prop. 1 $\frac{e\ 1}{5\ 5}$ appears several times for the Arabic equivalent of $\frac{2}{5} \cdot \frac{1}{10}$. The "*e*" may be explained as representing the Latin word-ending (here of *octave*)[19]. $\frac{a\ b}{c\ d}$ at least once means $\frac{a}{c} + \frac{b}{c \cdot d}$, for $\frac{1\ \ \ \ 1}{15625\ 4}$ appears in Prop. 5 as equivalent of the Arabic for $1\frac{1}{4} \cdot \frac{1}{15625}$. In a similar place in Prop. 7 the fraction has been corrected. Of the two occurrences (Props. 8,10) of fractions of form $\frac{1\ 0}{c\ d}$ meaning $\frac{1}{c \cdot d}$ at least one has been corrected. Accordingly, it is reasonable to suppose that the writer of MS BN 7277A was using the same convention as Leonardo, only reversed. For Leonardo has $\frac{5\ e}{8\ 8}$ for $\frac{5}{64}$ and $\frac{a\ b}{c\ d}$ for $\frac{b}{d} + \frac{a}{c \cdot d}$, and he commonly writes his numbers "backwards", i.e. fractions first[20]. Since these fractions also appear in Arabic writings[21], the question arises whether this notation was taken over from the Arabic manuscript used by the translator or whether it was introduced at translation stage or later. If it was not in the Arabic manuscript, the

[19] I am indebted to Dr. Jacques Sesiano for this explanation. It is supported by the appearance of $\frac{e}{10}$ for $\frac{1}{10}$ both in our manuscript (e.g. in Prop. 10*i*) and in Leonardo (e.g. in II 212).

[20] See Karpinski. p. 53, referring to *Liber abaci*, I p. 447, for $\frac{1\ e}{8\ 8}$ and to I pp. 24-25 for Leonardo's explanation of the system.

[21] See Vogel, pp. 3, 18; Suter (1901), pp. 19ff.

translator or some later copier must have found it in some other Arabic-based source. Unfortunately, no help is afforded by the Hebrew translation. For although $\frac{6}{8}\frac{5}{8}$ occurs for the value $\frac{53}{64}$ several times in Suter's translation of Props. 3 and 4, it is clear that he himself inserted most of them to patch up the text. There are nine places in these propositions where the value $48\frac{53}{64}$ occurs and in each of them both Arabic and Hebrew have the equivalent of $\frac{1}{2}+\frac{1}{4}+\frac{5}{8}\cdot\frac{1}{8}$ or $\frac{3}{4}+\frac{5}{8}\cdot\frac{1}{8}$ for the fraction[22].

Before leaving the matter of numerical notation we should notice that, although the "4" in most occurrences is the normal single loop usual in manuscripts of the fourteenth century, a form resembling the Arabic "4" occurs on f. 94r in the fraction $\frac{4}{125}$ and in the marginal note taken from Leonardo at the bottom of the same page[23].

The whole text is liberally supplied with interlinear notes, some giving extra information (e.g. "que est 10" not far from the beginning of Prop. 1), but most giving alternative renderings of the text. Of the latter, some bring the text nearer to the Arabic and some make the Latin easier to understand. Again, some are accompanied by crossings out or underscoring of the text and some are not. Further, there are several marginal notes. Almost all of these additions seem to be in the same hand as the text. A lengthy note in the bottom margin of f. 94r is an extract from Leonardo of Pisa and another — not in the same hand — on f. 94v appears to be a reworking of another such extract[24]. Some of the notes carries the identification"g."; in the first part of the algebra there are several notes with this sign and others with more of the name "Guilelmus" written out[25].

A note in the bottom margin of f. 95v is an attempt to repair a large omission in Props. 13–14. In the text the problem is stated for Prop. 13, but just as rectangle (*quadratum longum*) $EZHT$ is being described, the text jumps to the end of Prop. 14 at the words "the square $HMLY$", thus producing "quadratum longum $GMLY$", H (ح) being mistaken for G (جـ or ـجـ), no doubt within the Arabic transmission or at translation stage[26]. In the note an entirely different problem about the figure $GMLY$ (which becomes *rumbus*) constructed inside triangle ABG is posed and solved; and then a new problem bearing some similarity to the last few lines of the text and introduced by the same formula, "Et si volueris, dic ..." is stated, no doubt

---

[22]The places in the Hebrew manuscript, Munich, cod. heb. 225, are, with the corresponding passages in the Arabic manuscript in the pagination of the facsimile: 1. 156v12-13 [i.e. lines 12–13 of f. 156v] (137:4 [i.e. line 4 of p. 137]); 2. 156v15 (137:16); 3. 156v20 (137:20-21); 4. 156v22 (138:2); 5. 157r2 (138:7-8); 6. 157r22 (139:8); 7. 157r24 (139:10); 8. 157v2-3 (139:15); 9. 157v5-6 (139:10). There are omissions in the Hebrew, e.g. of $\frac{1}{2}$ in no. 1, and other mistakes, e.g. no. 4 reads "$\frac{6}{8}$ of $\frac{1}{8}$".

[23]For this form of "4" in early western manuscripts see Lemay, 455-456.

[24]The note on f. 94r is from Leonardo II 209, lines 31–39; the note on f. 94v is a reworking of II 211, lines 12–18.

[25]I am grateful to Dr. Jacques Sesiano for this information.

[26]In the facsimile of the Arabic text, everything between p. 147, line 19, word 2 and p. 149, line 16 (end) is omitted.

intended as an emendation[27].

Rather disappointing is the "g." note to Prop. 12 on the left margin of the same page. Abū Kāmil stated simply that if $AD = x$ (in modern terms), then $DG = \sqrt{\frac{1}{3}x^2}$. In fact this must follow by applying Pythagoras' theorem to $\triangle AGD$: $x^2 + DG^2 = (2DG)^2$ — a procedure that involves two unknowns. But the note assumes that $AD \cdot \frac{1}{3}AD = DG^2$ and appears to explain only the appearance of *census*.

No conclusion has been reached on the identity of the translator. In the main the translation is literal; and the few passages of great deviation are probably to be explained by the vagaries of transmission. This might suggest Gerard of Cremona[28] as translator — and indeed the vocabulary seems in general more like that of Gerard of Cremona than that of Robert of Chester in their translations of al-Khwārizmī's *Algebra*. But a similar vocabulary is also to be found in the works of Leonardo of Pisa; and neither the inconsistencies of terminology nor the translation of diagram letters favours Gerard as translator. Identification of the translator by style must await detailed analysis; and only a brief note on terminology is added here.

Terminology

Before we come to the technical vocabulary, a few generalities about the translation and about non-technical words may be made. First, we may observe that although the translated Latin usually mimics the original Arabic — e.g. by translating قد regularly as *iam* , ثمّ as *deinde* or (more ludicrously) باب by *porta* — and although fixed phrases in Arabic are rendered by (more or less) fixed phrases in Latin — so فإن قيل لك becomes *et si dicemus tibi* —, the translator has evidently felt at liberty not only to translate several different Arabic words by the same Latin word — e.g. أراد and شاء by *velle*; قال, ذكر (also translated as *rememorare*) and وصف (which is also translated by *exponere*, *rememorari* and *memorare*) by *dicere*, or ضلع and جانب by *latus* — but even to translate one Arabic word by two or more Latin words. Examples of this are إذا , which becomes *cum*, *quando* or even *si*, and جعل in the phrase جعل ... شيئًا, which is usually translated by *facere*, but occasionally by *ponere*. The pronouns هو, ذلك etc. and the suffixes ـه, ـها are all rendered by *is*, *ille*, etc., apparently at will. The conjunction أن is usually rendered by *quia*, though typically by *ut* in such phrases as وهو أن, but *quod* occasionally appears for it. The particle فـ is sometimes represented by *et*, *igitur*

[27]Note that $6\frac{82}{123} - \sqrt{11\frac{1681}{15129}}$ is simply $6\frac{2\cdot41}{3\cdot41} - \sqrt{11\frac{41^2}{41^2\cdot9}}$.

[28]In Steinschneider's discussion (pp. 585, 588) of the text he mentions the suppositions that Gerard of Cremona or Abraham bar Ḥiyya translated it.

or similar, but is often suppressed, particularly when coupled with a form of the imperfect translated as an imperative — so فيسقط *prohice*, فتقسم *divide*, etc. It should be noticed that a Latin imperative sometimes corresponds to an Arabic imperative with فـ, so فرد *reduc*, فأكمل *restaura*, فاضرب *duc*.

فيكون[29] introducing the result of a calculation or operation is usually translated as *et proveñ*[30] or *et erit/erunt*. تبيّن (verb) or بيّن (adjective) is rendered by *manifestum est* (typically فقد تبين أن *iam manifestum est quia*). The second form is usually translated *exponere* (so وقد كنا بيّنّا أن *et iam exposuimus quia* and similar, occasionally with *quod*, often without كنا). In the "Q.E.D." formula at the end of most of the propositions we find وذلك ما أردنا أن نبين *et hoc est quod voluimus exponere*.

As for the commonalty of non-technical terms we might mention بدأ *incipere*, علة *causa* in the sense of "general principle", علم and عرف *scire* (and معرفة *scientia*), فهم *intelligere*, تقدّم *premittere*, كل *unusquisque* or *quisque* meaning each of a specified and limited number and *omnis* in a more general sense, كله after a noun *totus*, يجب *necesse est*, موضع *locus*.

The *termini technici* are here listed alphabetically according to the root of the word (or of the principal word in a phrase). Not all variants are given, but it is intended that the principal Latin equivalents are sufficiently indicated. References to specific examples are only given when they are unique or are needed to point up some difference of usage.

أخذ *accipere*; مأخوذ جذره *accepta radice eius* (prop. 11*i*), but *accipiamus radicem eius* (Prop. 11*i*).

إلا in contexts where English would use "minus" is expressed by means of *diminutus*. Although this word is seldom written out, but abbreviated to *dim̄* or *dimiñ*, it is clear that it is constructed as an ablative absolute, e.g. إلا جذر *diminuta radice*, "minus the root of [what follows]".

وبرهان ذلك أن *cuius demonstratio est quod* (Prop. 10(ii)).

بقي *remanere*, typically فيبقى *et remane(n)t*, *et remanebit*, etc.; فما بقي *residuum*.

تام *integer* (Prop. 20).

---

[29]The prefix letter in an imperfect form of the verb is sometimes deliberately left without dots, so that the verb can be either masculine or feminine. In the manuscript the masculine form is usual in the present case. A similar convention is also used when it is not desired to specify second person singular or first person plural.

[30]The word appears like this in the manuscript. From similar passages at the beginning of the whole work it appears that the future is probably meant (I am indebted to Jacques Sesiano for this information).

مثلث triangulus.

جبر: فنجبره بـ restauremus cum (Prop. 9). حساب الجبر ومقابلة computatio restaurationis et oppositionis (Intro.).

جذر *radix*.

جزء "part" used to express fractions translated by symbols, *pars*, *portio* (Prop. 7, in quotation from Euclid).

جمع *agregare*; ما اجتمع *agregatum*; من جميع ذلك *ex toto hoc*.

جانب *latus*.

مجهول *qui non est scitus* (Prop. 7), *ignotus* (Prop. 8).

حساب see جبر.

حفظ *servare*.

أحاط بـ *circumdare in* + ablative. The Latin verb always appears as a participle. A typical example is الذى يحيط به *circumdans in eo* (note the pedantic, and misleading, rendering of the preposition بـ). Sometimes this is rendered as *circumdans* + accusative (e.g. in Prop. 3), more often in the suprascript emendations than in the text, though on two occasions (Prop. 3,4) the text with the accusative is "corrected" into the *in* construction. Other examples are تحيط به الدائرة *circumdans circulum* (Prop. 1), *circumdata a circulo* (Prop. 11*i*); التى يحيط بها هذا المعشر *circumdans hunc decagonum* (Prop. 10(i)); يحيط بها المخمس *contentus a pentagono* (Prop. 8). Another translation of أحاط بـ is *continere* + accusative or (Prop. 4) with *in* + ablative. محيط (n.) *circumferentia*, *circumdans*.

خرج *exire* (فيخرج *exibit*) for the result of an operation. خارج *extra* (of figures) (Prop. 10(i)). أخرج *producere*, *ducere* (lines). استخراج *exitus* corrected into *inventio* (Intro.), *inventio* (Prop. 1), *exiens* (Prop. 11(ii)).

خط *linea*.

مخمس *pentagonus*. Sometimes *pentagonus* appears in a phrase translating an Arabic phrase containing خمس.

داخل *infra* (of figures).

درهم *dragma* (almost always abbreviated to *drag*).

دور: يدير (of drawing a circle), *facere*, *addere* (Prop. 2,3) and *adiungere* (Prop. 4), presumably by confusion with نزيد. دائرة *circulus*.

مربع *quadratum* (in algebraic sense or four-sided figure); مربع مستطيل *quadratum longum* (Prop. 13-14).

ردّ: فردّ *reduc*.

رسم *facere* in geometrical constructions; رسمنا *diximus* (Prop. 7).

مركز *centrum*.

زاوية *angulus*; على زاوية قائمة *secundum angulum rectum.*

زاد ... على *addere ... ad* (+ dat. in Prop. 11*i*).

مسدس *exagonus.*

أسقط ... من *prohicere ... ex.*

مساوٍ لـ *equalis* + dative; متساوى الأضلاع *equilaterus* (Prop. 15), متساوى الأضلاع والزوايا *equilaterus et equiangulus* (up to Prop. 12) and *equalium laterum et angulorum* (frequent from Prop. 13-14 on).

شكل *figura*; شكل له خمس عشرة زاوية *figura quindecagona.*

شيء *res*, i.e. the unknown quantity to be found, x.

ضرب ... بـ (vb.) *ducere ... in* (multiply), *multiplicare ... in*; ضرب ... بـ (n.) *multiplicatio ... in*; مضروب *ductus* (usual), *productum* (Prop. 6).

ضعّف *duplare.*

طرف *extremitas*; and see نسبة.

مطلوب *quesitus.*

طول *longitudo.*

من العدد *ex numero* usually, but also *numero* (Prop. 2,3).

عدل (vb.) *est equalis, equatur*, in sentences of structure "there results A *equal to* B" usually *equalis.*

معشر *decagonus.* Sometimes *decagonus* appears in a phrase translating an Arabic phrase containing عشر.

معلوم *scitus, notus*; معلوم أن *scitum est quia.*

عمل ... على *facere ... super* of circumscribing one figure about another and similarly with في for inscribing.

قبل: فنقابل بـ *oppone in* (Prop. 1,15), *opponamus* (Prop. 2,9); مقابلة *oppositio* (Prop. 2 and see جبر).

قدر *quantitas* (Prop. 1,4), *mensura* (Prop. 2); كمية قدر *quantus* (Prop. 3) and there are other such variants. قدر ... من *proportio ... ad* (usual) or *proportio ... ex* (Prop. 4); قدر ... عند *proportio ... ad* (Prop. 7,8). *v.* كم.

بالتقريب *in propinquitate* (Prop. 11(ii)), *secundum propinquitatem* (Prop. 11(ii)), *fere* (Prop. 11(ii)).

قسم ... على *dividere ... per.* انقسم dividere (various constructions) (Prop. 17).

القسم الأعظم *maior pars* (of a divided line).

قطر *diameter.*

قاعدة *basis.*

قائمة *rectus* (of angles). قائم الزوايا *rectorum angulorum* (of figures). على استقامة *secundum rectitudinem*.

قياسه أن *exemplum, verbi gratia*.

كعب *cubus*, x^3.

كل شيء معك and كل ما معك *totum quod habes*.

كم *quantus*; كم مقدار *quantus* (Prop. 1).

كمل: فأكمل ... حتى يكون *restaura ... ut sit* (Prop. 1,12,16,17,18), *reintegra ... ut sit* (Props. 19,20); وإكمالك إياه ... ان *et restaurabis illud si ...* (Prop. 12).

كمية see قدر.

ألقى ... من *prohicere ... ex*.

مثل in the sense of "equal" *similis*; في مثله in multiplication *in se*. مثال ذلك *exemplum* (Prop. 10(ii)), *verbi gratia* (Prop. 17).

ماسّ *contingere*.

مساحة *mensura*.

مال *census*, the square of the unknown, x^2.

على نسبة ذات وسط وطرفين *per proportionem per medium et 2 partes* corrected into *secundum proportionem medii et duo extrema*, *secundum proportionem medii et extrema*, *secundum proportionem habentem medium et duo extrema*, ditto with *duoque* (all Prop. 17)., *secundum proportionem medii et extremorum* (Prop. 17,18,19).

نصف *medietas, medium* (of a line), $\frac{1}{2}$. نصّف *dimidare*.

في نفسه in multiplication *in se*.

نقص ... من *subtrahere ... ex*; منقوص *subtractus* (Prop. 15), *sublatus* (Prop. 9,11*i*,11*ii*); نقصته *minus eam* (Prop. 17).

وتر *corda*. وتّر *est corda* etc.

وقع *cadere* (of one figure inside or outside another).

موازٍ... لـ *equidistans* + dative.

وصل ... بـ *coniungere*; وصل فيما بين *coniungere quod est inter*.

The Manuscript

The contents of MS Paris, BN lat. 7377A may be itemized as follows:

1. ff. 1r–33v: Muḥammad ibn 'Abd al-Bāqī's commentary on Euclid's *Elements*, book X, in the translation by Gerard of Cremona; ed. Curtze, 1899, in his edition of Anaritius, pp. 252–384.
2. ff. 34r–43v: al-Khwārizmī's *De algebra et almuchabala* in the translation by Gerard of Cremona; ed. Hughes, 1986.
3. ff. 43r–56v: Abū Bakr's *Liber mensurationum* in the translation of Gerard of Cremona; ed. Busard, 1968.
4. ff. 56v–57v: *Liber Saydi Abuothmi* on mensuration; ed. Busard, 1969.
5. ff. 57v–58v: *Liber Aderameti* on mensuration; ed. Busard, 1969.
6. ff. 58v–68r: *Liber augmenti et diminucionis* (anon.); ed. Libri, 1838, pp. 304–376[31].
7. ff. 68r–70v: Pappus' commentary on Euclid's *Elements*, book X, probably translated by Gerard of Cremona; ed. Junge 1936.
8. 71v–97r: Abū Kāmil, *Algebra.*
9. 99r–203r: *Liber Mahameleth*, a Latin compilation of problems in arithmetic and algebra, based on Arabic sources[32].

Most, if not all, of the first seven items are translations by Gerard of Cremona. Items 2–6 appear in the same order in MSS Paris, BN lat. 9335 and other manuscripts[33]. The writing appears to be of the fourteenth century.

Editing conventions

For the purposes of the edition it has been assumed that the literal style of the translator was smoothed in the process of copying and that both literal and smooth readings often appear in text and interlinear note (or in note and text), as will often happen when a scribe has two manuscripts before him[34]. It is intended that the edited text be as near to the original translation as the evidence allows. Where appropriate, the Arabic has been used to decide on readings. Occasionally an Arabic reading in transliteration or translation is given in the apparatus; diagram letters are quoted according to the translator's system (see above). Emendations, notice of omission, trivial corrections made in the manuscript and the usual small editorial corrections

[31]See also Tannery.

[32]See Sesiano.

[33]Busard (1969), 163–164.

[34]For a completely different view of the matter, see the article by Dr. Jacques Sesiano in this volume.

appear in the first apparatus. The second apparatus is to accommodate the numerous interlinear and marginal notes.

In either apparatus, if a word or phrase in the accepted text is commented upon, it is written in the apparatus followed by a square bracket,]. A word followed by a shallow angle-bracket, ⟩, is used simply to indicate the position for the note that follows; additions in the text come after the word so indicated. In the case of a phrase the line-numbers given in the apparatus refer to its last word. Text added by the editor appears in angle-brackets, < >. Text between square brackets is not legible because of physical damage to the manuscript[35]; difficult readings in the note to Prop. 3 on f. 94r, have been resolved with the help of Boncompagni's edition of Leonardo's *Practica geometriae*.

Sometimes the text confuses the diagram letters G and H. Since this mistake can reasonably only be made in Arabic or by the translator, the "incorrect" reading has been retained, as being what the translator wrote. But to help the reader a dash is put on the letter. Thus G', H' respectively mean G, H in the Latin manuscript, which stand for the equivalent of H, G in Arabic.

Several abbreviations have been left unexpanded: *dim̄* or *dimin̄*, *drag'*, *proven̄*, *rad'* for a form of *diminutus*, *dragma*, *provenire*, *radix*, respectively. With the exception of *drag'*, which usually has no sign of abbreviation, these resemble the writing of the manuscript.

The diagrams have been redrawn much as they are in the manuscript, albeit a little idealised. The two Es in each of the diagrams for Props. 3 and 4, as well as the corresponding mistakes in the text, have been left (in each case the E referred to in the mathematical argument is the centre of the circle).

Bibliography

[Abū Kāmil] Abū Kāmil Shujā' ibn Aslam, *The Book of Algebra. Kitāb al-Jabr wa 'l-muqābala*, Frankfurt am Main 1986.

[Abū Kāmil (Hebrew)] *The* Algebra *of Abū Kāmil*, Kitāb fī al-jabr wa'l-muqābala *in a Commentary by, Mordecai Finzi: Hebrew Text, Translation, and Commentary with Special Reference to the Arabic Text*, edited with introduction, by Martin Levey, Madison etc. 1966.

[Anaritius] M. Curtze, *Anaritii in decem libros priores Elementorum Euclidis*, in *Euclidis Opera Omnia*, ed. I. L. Heiberg et H. Menge, *Supplementum*, Leipzig 1899.

H. L. L. Busard, "L'algèbre au moyen âge: le 'Liber mensurationum d'Abū Bekr'", *Journal des Savants*, (1968), 65–214.

H. L. L. Busard, "Die Vermessungstraktate *Liber Saydi Abuothmi* und *Liber Aderameti*", *Janus*, 16 (1969), 161-174.

[35] Dr. Jacques Sesiano very kindly eliminated a large class of uncertain readings – words that were caught in the binding and could not be read on the film available to me – by direct inspection of the manuscript. It is a pleasure to acknowledge my debt to him.

Barnabus Hughes, "Gerard of Cremona's Translation of al-Khwārizmī's *al-Jabr*: a critical edition", *Medieval Studies*, 48 (1986), 211-263.

Barnabus Hughes, *Robert of Chester's Latin Translation of al-Khwārizmī's al-Jabr*, Stuttgart 1989.

G. Junge, "Das Fragment der lateinischen Übersetzung des Pappus-Kommentars zum 10. Buch Euklids", *Quellen und Studien zur Geschichte der Mathematik, Astronomie und Physik*, Abt. B, 3, Studien (1936), 1–17.

L. C. Karpinski, "The algebra of Abu Kamil Shoja' ben Aslam", *Bibliotheca mathematica*, 3. F., 12 (1911-12), 40–55.

Paul Kunitzsch, "Letters in Geometrical Diagrams, Greek – Arabic – Latin", *Zeitschrift für Geschichte der Arabisch-Islamischen Wissenschaften*, 7 (1991/92), 1–20.

[al-Khwārizmī] *The Algebra of Mohanned ben Musa*, ed. and trans. Frederic Rosen, London 1831 (repr. Hildesheim-Zurich-New York 1986).

Richard Lemay, "The Hispanic Origin of Our Present Numeral Forms", *Viator*, 8 (1987), 435–462 + 15 pp. of ilustrations.

[Leonardo of Pisa] *Scritti di Leonardo Pisano matematico del secolo decimoterzo*, ed. Baldassarre Boncompagni, 2 vol., Rome 1862.

Martin Levey, "Abū Kāmil Shujā' ibn Aslam ibn Muḥammad ibn Shujā'", *Dictionary of Scientific Biography*, I, 30-32, New York 1970.

Libri, *Histoire des sciences mathématiques en Italie depuis la renaissance des lettres jusqu'à la fin du dix-septième siècle*, 4 vol., Paris 1838-1841.

Nashriya II (1340-1341).

G. Sacerdote, "Il trattato del pentagono e del decagono di Abu Kámil Shogia' ben Aslam ben Muhammed", *Festschrift zum achtzigsten Geburtstag Moritz Steinschneiders*, Leipzig 1896, 169–194.

J. Sesiano, "Le Liber Mahameleth, un traité mathématique latin composé au XIIe siècle en Espagne", Premier Colloque Internationale sur l'Histoire des Mathématiques Arabes, *Actes*, Algiers 1988.

Fuat Sezgin, *Geschichte des arabischen Schrifttums*, V, Leiden 1974.

M. Steinschneider, *Dir hebraeischen Übersetzungen des Mittelalters und die Juden als Dolmetscher*, 1893 (repr. Graz, 1956).

[Suter 1901] Heinrich Suter, "Das Rechenbuch des Abū Zakarījā", *Bibliotheca mathematica*, 3.F 2 (1901), 12-40.

[Suter] Heinrich Suter, "Die Abhandlung des Abū Kāmil Šoǵā ben Aslam 'Über das Fünfeck und Zehneck', *Bibliotheca mathematica*, 3. F, 10 (1909-1910), 15–42.

P. Tannery, "Sur le 'Liber augmenti et diminutionis' compilé par Abraham", *Bibiotheca mathematica*, 3.F, 2 (1901), 45–47.

Josef Weinberg, *Die Algebra des Abū Kāmil Šoǵā ben Aslam*, Munich 1935.

Kurt Vogel, "Zur Geschichte der Stammbrüche und der aufsteigenden Kettenbrüche", *Sudhoffs Archiv*, 66 (1982), 1–19.

Mohammad Yadegari and Martin Levey, "Abū Kāmil's 'On the Pentagon and Decagon'", *Japanese Studies in the History of Science*, Supplement 2 (1971), 54 pp.

Dixit Abucamel ssagia filius abrahim agregator istius libri: postquam exposuimus quod indiget expositione in eo quod est occultum ex computatione restaurationis et oppositionis ex modis antiquorum in scientia et in aggravatione in eo et intelligentium ex geometris et sapientibus in libro euclidis et aliis et glosavimus illud et discooperuimus in libro nostro isto, an nobis est nunc ad exponendum quantitatem cuiusque lateris de lateribus cuiusque figure pentagoni et decagoni equilateri et equianguli descripti intra circulum notum vel extra circulum notum, et quantitatem omnis diametri circuli circumdantis pentagonum vel decagonum scitos equilateros et equiangulos vel circumdati ab eis, et mensuram corde partis ex 4 partibus circumferentie circuli sciti, et mensuram cuiusque lateris de lateribus decagoni equilateri et equianguli cum scitur mensura eius et quantitatem laterum triangulorum mensura scitorum cum fuerint in pentagonis vel decagonis equilateris et equiangulis, et alia ex eis que exponemus in libro nostro isto ex illo quod est exitus multus ex eo. Et ego iudico in bonum quos vidi compotistas et geometras nostri temporis huius et antiquos de quibus audivi sapientes in numero et mensura. Creator tamen erexit me ad posse applicare ad id quod eis fuit impossibile contingere.

< 1 > Incipiamus in inventione quantitatis corde $\frac{1}{5}$ circuli ex diametro eius. Et illud est ut faciamus circulum scitum $ABDEG$ cuius diameter sit 10 ex

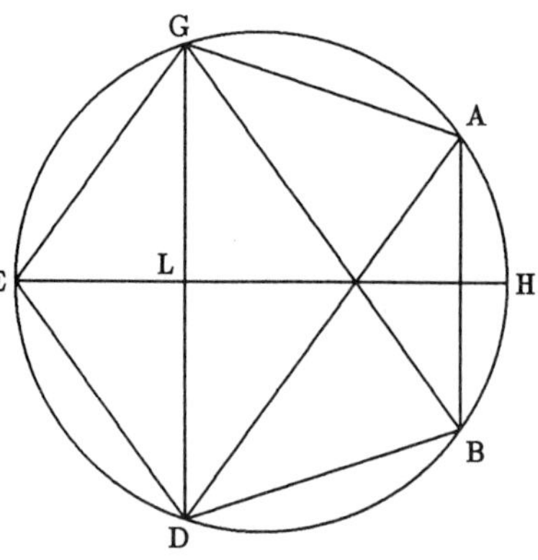

Prop. 1

5 glosavimus] *corr. ex* eglosavimus 9 equilateros] equilatos 9 vel⟩ c'ade *add. et del.* 12 eius⟩ s *add. et del.* et mensuram corde ... mensura eius *om. Ar.* 15 Et ego ... bonum] *Ar. yata'adhdharu*

3 ex modis] *corr. in* respicientium 6 an nobis est] *corr. in* pervenimus 6 quantitatem⟩ vel mensuram *supra* 7 pentagoni] *supra* 11 equilateri] *corr. ex* equilateris 14 illo⟩ per *supra* 15 exitus multus ex eo] *corr. in* inventio multa 18 contingere⟩ et ad id quod occultum fuit eis, et Ipsi sit gloria et honor Qui est unum et secundum non habet *marg.*

numero, sitque linea HE, et in eo sit pentagonus equilaterus et equiangulus circumdans circulum sitque pentagonus $ABGD$. Cum autem voluerimus scire quantum est quodlibet latus de lateribus huius pentagoni, producemus lineam DLG, que est corda $\frac{2}{5}$ circuli, et faciamus lineam ED rem. Scitum est quia linea EL est $\frac{1}{10}$ census, quia multiplicatio ED in se est sicut multiplicatio HE in EL; ergo linea GL est radix census dimiñ $\frac{1}{10}$ decime census census; et linea GL est sicut linea LD: linea GD est radix 4 censuum dimiñ $\frac{e\ 1}{5\ 5}$ census census. Et scitum est quia multiplicatio AB in GD et AB in se est sicut multiplicatio GD in se, quia multiplicatio AB in GD et AG in BD est sicut multiplicatio AD in BG, et AG in BD est sicut multiplicatio AB in se et AD in BG sicut GD in se. Et GD in se est 4 census dimiñ $\frac{e\ 1}{5\ 5}$ census census. Prohice ex eo multiplicationem AB in se, que est census, et remanebunt 3 census dimiñ $\frac{e\ 1}{5\ 5}$ census census equalis AB in GD. Divide 3 census dimiñ $\frac{e\ 1}{5\ 5}$ census census per lineam AB, que est res, exibit linea GD 3 res dimiñ $\frac{e\ 1}{5\ 5}$ cubi. Et iam exposuimus quia linea GD est radix 4 censuum dim̃ $\frac{e\ 1}{5\ 5}$ census census. Ducamus 3 res dimiñ $\frac{1}{25}$ cubi in se, et proveñ 9 census et $\frac{1}{625}$ cubi cubi dimiñ $\frac{6}{25}$ census census equalis 4 censibus dimiñ $\frac{e\ 1}{5\ 5}$ census census. Oppone in ipsis et erit $\frac{1}{5}$ census census equalis 5 censibus et $\frac{1}{625}$ cubi cubi. Divide totum quod habes per censum, et exibunt 5 drag′ et $\frac{1}{625}$ census census equalis $\frac{1}{5}$ census. Restaura ut sit census census, quod est ut ducas eum in 625, et ducas totum quod habes in 625, et proveñ census census et 3125 drag′ equalis 125 censibus. Dimidia ergo census, et exibit 62 et $\frac{1}{2}$; duc ea in se et proveñ 3906 et $\frac{1}{4}$; ex quibus prohice 3125, et remanebit 781 et $\frac{1}{4}$; radix eius subtracta ex 62 et $\frac{1}{2}$; accepta rad′ residui est linea ED, que est unum ex lateribus pentagoni. Et hoc est quod voluimus exponere.

< 2 > Et quando voluerimus scire mensuram quantitatis corde $\frac{1}{10}$ circuli sciti et iam scivimus $\frac{1}{5}$ eius cordam, quod addemus circulum super quem sint $ABGDEG'$ et faciemus diametrum eius 10 numero, que sit linea AG', et faciamus in medietate eius 5 cordas equales, que sint AB et BH' et linea $H'D$ et linea DE et linea EG'. Manifestum est quia linea DG' est corda $\frac{1}{5}$ huius circuli; et iam exposuimus quia ducta in se est 62 et $\frac{1}{2}$ diminuta radice 781 et $\frac{1}{4}$; et linea AH' sicut ipsa. Et multiplicatio AG in DG' et GD in AH est sicut AD in GH. Set AD sicut GH et AG sicut DH, multiplicatio $G'D$

---

21 equilaterus] equilatus *26 GL*] *Ar. DL* *31* in se[1]] *repet. et del.* *32* in] *repet.* *33* dimiñ[1]⟩ *seq. ras.* *33 GD*] *BG* *33* dimiñ[2]⟩ $\frac{e\ 1}{10\ 5}$ census census *add.* *46* scire⟩ quantit *add. et del.* *48* et⟩ 10 *add. et del.* *49* faciamus⟩ i[a] *add.* *50* $\frac{1}{5}$⟩ seq. ras.

---

22 circumdans⟩ intra *supra* *22 ABGD*] *GD del., corr. in ABDEG* *24 DLG*] *supra* *25* se⟩ que producit censum *supra* *26 HE*⟩ que est 10 *supra* *26 EL*⟩ quam $\frac{1}{10}$ census esse ab hoc necesse est *supra* *26* est] *supra* *26* decime] *supra* *27* linea[3]⟩ igitur *supra* *30 BG*⟩ ex eo quod dicit ptolomeus in libro de arcu et corda cum fuerit quadrangulum in circulo *marg.* *31* $\frac{e\ 1}{5\ 5}$ census] *corr. ex* 25. censibus *34* census census] *corr. ex* censibus *35* cubi] *corr. ex* cubis *36* cubi] *corr. ex* cubis *37* census[3]] *corr. ex* censibus *41* ducas eum in 625] *is underlined in the MS, but is present in Arabic* *46* quantitatis] *del. in textu* *46* corde] *supra* *47* cordam] *supra* *48 ABGDEG′*] *corr. in ABHDEG, Ar. ABGDEH* *48* faciemus] *supra* *48* 10] *supra* *52 DG′*] *corr. in DH* *53* sicut[2]] *corr. in* in *53* et *AG* sicut *DH*] *del. in textu, H* in *D...G* (?) *add. et del. supra*, est sicut *GH* in se Et *supra*

| in AH' est sicut AH' in se: ergo multiplicatio DH in AG et AH' in se est sicut multiplicatio GH in se. Et linea $H'D$ est corda $\frac{e}{10}$ huius circuli; faciamus eum rem. Et ducamus eam in lineam AG', que est diameter circuli et est 10, et proveñ 10 res; et ducamus lineam DH in se et proveñ 62 et $\frac{1}{2}$ dimiñ radice 781 et $\frac{1}{4}$. Ergo multiplicatio linee GH in se 62 et $\frac{1}{2}$ et 10 res dimiñ radice 781 et $\frac{1}{4}$. Addamus ei multiplicationem AH' in se, que est 62 et $\frac{1}{2}$ dimiñ radice 781 et $\frac{1}{4}$: erit 125 drag' et 10 res dimiñ radice 3125 drag' equalis multiplicationi AG' in se, que est 100. Opponamus ergo in ipsis, et erit post oppositionem radix 3125 drag' dim̃ 25 drag' equalis 10 rebus ergo res equatur radici 31 et $\frac{1}{4}$ dim̃ 2 et $\frac{1}{2}$, et hoc est linea GD que est corda $\frac{1}{10}$ circuli. Et hoc est quod voluimus exponere. *94r*

< 3 > Cum voluerimus scire quanta sit corda unius lateris pentagoni equilateri et equianguli circumdantis in circulo scito, addamus circulum scitum, et faciamus diametrum eius 10; et faciamus super ipsum pentagonum equilaterum et equiangulum, qui sit pentagonus $ABGDE$. Manifestum est quia corda TH corda pentagoni circumdantis circulum, et iam exposuimus quia ducta in se est 62 et $\frac{1}{2}$ dim̃ rad' 781 et $\frac{1}{4}$ et quadratum linee TL 15 et $\frac{5}{8}$ dim̃ rad' 48 et $\frac{1}{2}$ et $\frac{1}{4}$ et $\frac{e\ 5}{8\ 8}$. Prohice ergo ipsum ex quadrato linee TE, que est $\frac{1}{2}$ diametri, quod est 25, et remanebit quadratum linee EL 9 et $\frac{1}{4}$ et $\frac{1}{8}$ et radix 48 et $\frac{1}{2}$ et $\frac{1}{4}$ et $\frac{e\ 5}{8\ 8}$. Deinde faciamus lineam AB, que est una ex

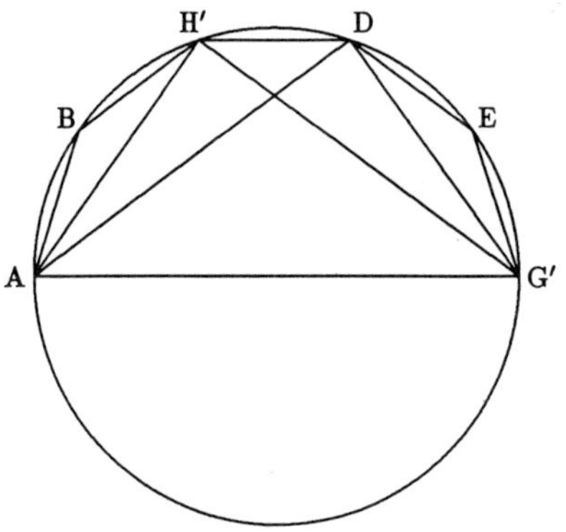

Prop. 2

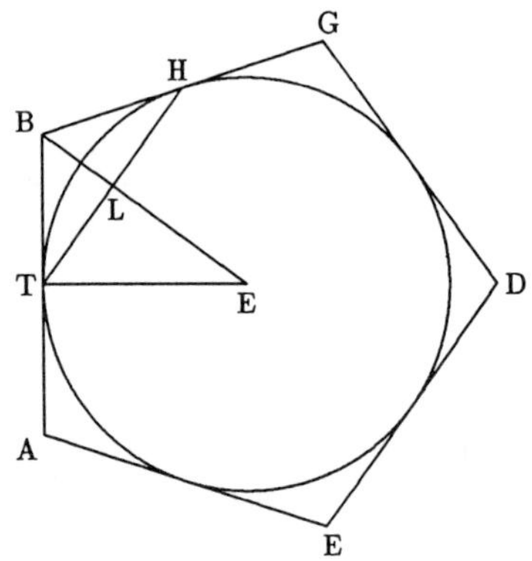

Prop. 3

---

54 multiplicatio GD in ... se] *om. Ar.* *56* in⟩ se *add. et del.* *65* Cum] Et non *71* rad'⟩ 8 *add. et del.*

---

55 et AH' in se est] *supra* *57* DH] *corr. in* GD *59* 781] *corr. ex* 581 *60* 781] *corr. ex* 981 *63* radici] *supra* *66* addamus] *corr. in* faciamus *69* circumdantis⟩ vel continentis *supra* *69* circulum⟩ intra *supra* *72* et[2]] *supra* *73* $\frac{1}{4}$ et $\frac{1}{8}$] $\frac{e}{8}$ (*corr.*) $\frac{1}{4}$

lateribus pentagoni circumdantis in circulo, rem et quadratum eius censum, et proportio quadrati AB, quod est census, ad quadratum TE, quod est 25, sicut proportio quadrati TH, quod est corda $\frac{1}{5}$ circumdantis circulum, quod est 62 et $\frac{1}{2}$ dim̄ radice 781 et $\frac{1}{4}$, ad quadratum EL, que est 9 et $\frac{1}{4}$ et $\frac{1}{8}$ et radix 48 et $\frac{1}{2}$ et $\frac{1}{4}$ et $\frac{e}{8}\frac{5}{8}$. Duc censum in 9 et $\frac{3}{8}$ et radix 48 et $\frac{1}{2}$ et $\frac{1}{4}$ et $\frac{e}{8}\frac{5}{8}$, et proveñ 9 census et $\frac{3}{8}$ census et radix 48 census census et $\frac{3}{4}$ census census et $\frac{e}{8}\frac{5}{8}$ census census: hoc est equalis quadrato TE in quadratum TH, quod est 1562 et $\frac{1}{2}$ dim̄ radice 488281 et $\frac{1}{4}$. Reduc totum quod habes ad censum, quod est ut ducas totum quod habes in $\frac{1}{5}$ et $\frac{e}{5}\frac{1}{5}$ dim̄ radice $\frac{4}{125}$ drag′. Duc 9 census et $\frac{3}{8}$ census et radicem 48 census census et $\frac{3}{4}$ census census et $\frac{e}{8}\frac{5}{8}$ census census <in $\frac{1}{5}$> et $\frac{e}{5}\frac{1}{5}$ dimiñ rad′ $\frac{4}{125}$ unius, et proveñ census. Deinde duc 1562 et $\frac{1}{2}$ dim̄ radice 488281 et $\frac{1}{4}$ in $\frac{1}{5}$ et $\frac{e}{5}\frac{1}{5}$ dim̄ radice $\frac{4}{125}$ unius, et proveñ hoc 375 drag′ et rad′ 15625 drag′ et dim̄ rad′ 78125 drag′ et dim̄ rad′ 28125 drag′, et est 500 dim̄ rad′ 200000. Ergo census equatur 500 dim̄ rad′ 200000; accipe radicem eius, AB, que est unum de lateribus pentagoni equilateri et equianguli circumdantis in circulo scito cuius diameter est 10 numero. Iam manifestum est quia quadratum linee AB, que est unum de lateribus pentagoni, est 500 dim̄ rad′ 200000. Et hoc est quod voluimus exponere.

< 4 > Et si dicemus tibi, nos volumus scire quanta sit mensura corde decagoni equilateri et equianguli circumdantis in circulo scito, adiungamus circulum et faciamus diametrum eius 10 ex numero, et faciamus super eo circumdantem in ipso decagonum $ABGDEFUZM$; et volumus scire quanta est AB, que est unum ex lateribus decagoni. Manifestum est quia linea TH facta est corda decagoni continentis circulum. Et iam exposuimus quia ipsa est radix 31 et $\frac{1}{4}$ dim̄ 2 et $\frac{1}{2}$, et linea TL est $\frac{1}{2}$ eius et est radix 7 et $\frac{1}{2}$ et $\frac{1}{4}$ et $\frac{e}{8}\frac{1}{2}$ dimiñ 1 et $\frac{1}{4}$; quadratum eius 9 et $\frac{1}{4}$ et $\frac{1}{8}$ dim̄ radice 48 et $\frac{1}{2}$ et $\frac{1}{4}$ et $\frac{e}{8}\frac{5}{8}$. Prohice illud ex quadrato TE, quod est 25: remanebit quadratum EL 15 et $\frac{5}{8}$ et rad′ 48 et $\frac{1}{2}$ et $\frac{1}{4}$ et $\frac{e}{8}\frac{5}{8}$. Et quadratum TH 37 et $\frac{1}{2}$ dim̄ radice 781 et $\frac{1}{4}$. Deinde fac lineam AB rem, que est unum de lateribus decagoni circumdantis in circulo, et quadratum eius censum. Et erit proportio census ex quadrato TH, quod est 37 et $\frac{1}{2}$ dim̄ radice 781 et $\frac{1}{4}$, sicut proportio quadrati CE, et est

---

75 TE] *Ar. KE* *80 TE*] *Ar. KE* *89* circumdantis] circumdatis *97* facta] ? *101* 37] 39 *104* est²⟩ *seq. ras.*

---

74 in circulo] *corr. ex* circulum *75* quadratum] *corr. ex* quadrati *77* ad] *supra* *77* quadratum] *corr. ex* quadrato *79* census⁴] *supra* *80 TE*⟩ ducto *supra* *82* radice] *supra* *88* eius⟩ que est *supra* *91* exponere⟩ [Aliter] promtius quia in orthogonio trigono ETB ab angulo recto producta est cathetus [TL erit prop]ortio quadrati BE ad quadratum ET sicut quadratum ET ad quadratum [EL Quare multip]licatio quadrati BE in quadratum EL est sicut multiplicatio quadrati ET [in se quod est 6]25 Quare ponam BE rem et ducam eam in se exibit census quem du[cam in qua]dratum EL quod est $\frac{3}{8}$9 radix $\frac{5}{8}\frac{6}{8}$48 venient (census *add. et del.*) $\frac{3}{8}$9 et ra[dix $\frac{5}{8}$]$\frac{6}{8}$48 census census que equatur 625 dragmis. Reduc hec ad censum [hoc est multip]lica 625 per $\frac{6}{25}$ dragme diminuta radice $\frac{4}{125}$ ue[nient] 150 minus radice 12520 que equatur censui hoc est quadrato BE [de quo si au]feratur ... [quad]rato [T]B ... *marg.* *95* decagonum] *corr. ex* pentagonum *95 ABGDEFUZM*⟩ *N add. supra* *97* decagoni] *corr. ex* pentagoni *100 EL*] ? *corr. in TL*; *T supra* *101* et⁴] *supra* *103* in circulo] *corr. ex* circulum *103* Et erit proportio] *supra*; rad′ *add. et del.* *103* ex quadrato] *corr. supra in* ad quadratum ; eius *add. et del.*

25, ex quadrato EL, et est 15 et $\frac{5}{8}$ et rad′ 48 et $\frac{1}{2}$ et $\frac{1}{4}$ et $\frac{e\ 5}{8\ 8}$. Ducamus eum in quadratum AB, quod est census, et provẽn 15 census et $\frac{5}{8}$ census et rad′ 48 census census et $\frac{3}{4}$ census census et $\frac{e\ 5}{8\ 8}$ census census equalis quadrato CE in quadatum TH, et est 937 et $\frac{1}{2}$ dim̃ radice 488281 et $\frac{1}{4}$. Reduc totum quod habes ad censum, quod est ut ducas totum quod habes in $\frac{e\ 2}{5\ 5}$ unius dim̃ rad′ $\frac{e\ \ 4}{625\ 5}$ unius, et erit census equalis 75 drag′ et rad′ 625 drag′ dim̃ radice 3125 drag′ et dim̃ rad′ 1125 drag′. Et illud est 100 drag′ dim̃ rad′ 8000, et illud est quadratum linee AB, quod est unum de lateribus decagoni continentis in circulo scito cuius diameter est 10 ex numero. Et hoc est quod voluimus exponere.

| < 5 > Et cum voluerimus scire quanta est diameter circuli circumdan- *94v*
tis in pentagono scito equilatero et equiangulo, faciamus pentagonum notum $ABGDE$ et faciamus unumquodque laterum eius 10 ex numero et circulum super eo circumdantem in eo cuius diameter linea CL, et scientia diametri eius est ut coniungamus EG et ponamus eum rem. Manifestum est quia multiplicatio eius in AB, et est 10, et AB in se est sicut multiplicatio EG in se. Sed multiplicatio EB, que est res, in GD, que est 10, provẽn 10 res; et DG in se producit 100; et EG in se producit censum. Erit census equalis 10 rebus et 100 drag′. Ergo res equatur 5 et rad′ 125, que est linea EG; medietas eius 2 drag′ et $\frac{1}{2}$ et rad′ 31 et $\frac{1}{4}$, et hoc est linea EL. Duc eam in

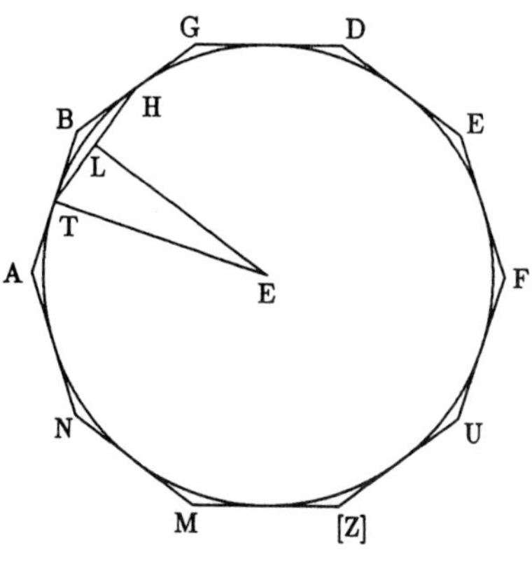

Prop. 4

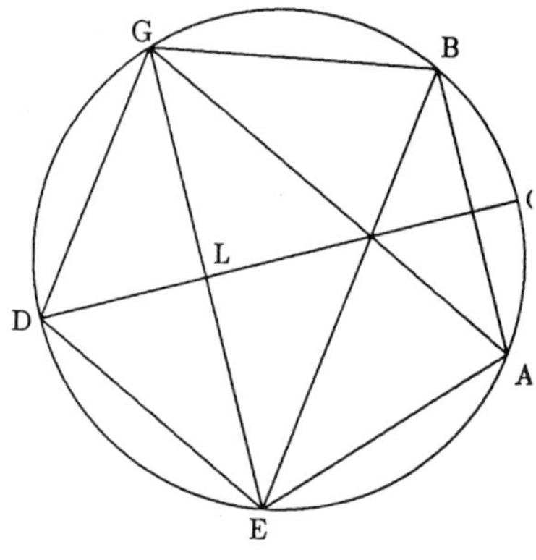

Prop. 5

106 et²⟩ ig ? *add. et del.* *109* rad′⟩ $\frac{4}{5}$ partis *add.* *110* et¹] *supra* *121* se⟩ erit 10 res *add. et del.*

105 ex quadrato] *corr. supra in* ad quadratum *112* continentis⟩ circumdantis *supra* *115* cum] ?, *corr. in* si *116* in pentagono ... equiangulo] *corr. in* pentagonum scitum equilaterum et equiangulum *117* 10] *corr. ex* 16 *118* circumdantem] *corr. ex* circumdantis *118* in eo] *corr. in* eum *119* *EG*] *BH corr. in BE, Ar. EG corr. ex* ? *120* *AB*¹] *corr. in DG* *120* 10⟩ res *supra* *120* *AB*²] *corr. in BG* *121* Sed] Sed *supra* *121* provẽn] *corr. in* producit *122* censum⟩ cum sit equalis *EB marg.*

se et provenit 37 et $\frac{1}{2}$ et radix 781 et $\frac{1}{4}$. Prohice eum ex ED in se, quod est 100, et remanebit 62 et $\frac{1}{2}$ dim̄ rad′ 781 et $\frac{1}{4}$, et hoc est quadratum linee DL. Deinde faciamus diametrum circuli, que est linea CLD, rem et quadratum eius censum. Proportio census ad quadratum DE, quod est 100, est sicut proportio quadrati DE, quod est 100, ad quadratum DL, quod est 62 et $\frac{1}{2}$ dim̄ rad′ 781 et $\frac{1}{4}$. Duc eum in censum, et proveñ 62 census et $\frac{1}{2}$ census dim̄ rad′ 781 et $\frac{1}{4}$ census census equalis multiplicationi quadrati ED in se, quod est 10000. Reduc totum quod habes ad censum, quod est ut ducas totum quod habes in $\frac{e\ 1}{10\ 5}$ unius et rad′ $\frac{1\quad\ 1}{15625\ 4}$ unius, et proveñ census equalis 200 et rad′ 8000. Iam manifestum est quia quadratum diametri circuli, que est CLD, 200 et radix 8000. Et hoc quod voluerimus exponere.

< 6 > Et si voluerimus scire diametrum circuli contenti a pentagono $ABGDE$ noto equilatero et equiangulo cuius quodlibet latus est 10 ex numero, et centrum circuli M < ... > et medietas diametri circuli contenti a pentagono linea AM. Et iam exposuimus quod quadratum diametri eius 200 et radix 8000. Ergo quadratum linee AM, quod est $\frac{1}{2}$ diametri eius,

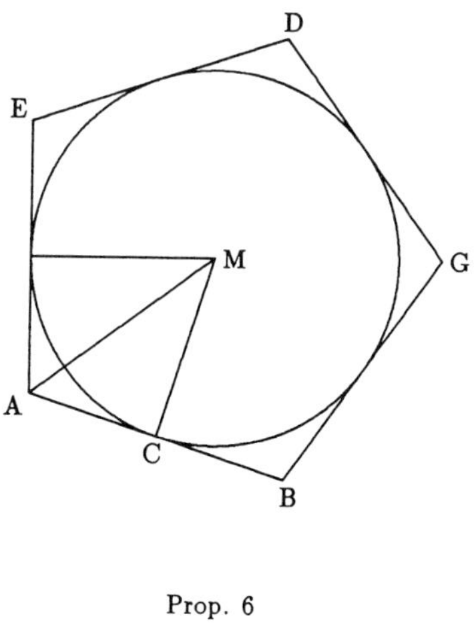

Prop. 6

129 DE] *corr. ex D* *133* $\frac{e\ 1}{10\ 5}$⟩ uni *add. et del.* *135* exponere⟩ Aliter promtius multiplica *EB* que est 5 et radix de 125 in se quia est corda anguli pentagonici erunt 150 diminutis [ ] et 10 radices de 125 quibus adde quadratum lateris pentagonici quod est 100 erunt 250 et 10 radices de[ ] quorum quinta pars que est 50 et 2 (*ras.* radicis *ras.* de 125 est quadratum lateris exago[ ] silicetmedietatis diametri ut in quartodecimo libro euclidis habetur quadruplum vero eius silicet200 Et 8 radices de 125 que sunt radix 8000 est quadratum diametri *DC* ut superius in ... *marg.* *138* < ... >] *Ar. wa-niṣf quṭrihā khaṭṭ KM* *140* radix⟩ 8 quadra *add. et del.*

139 contenti in pentagono] *corr. in* tan<g>entis pentagonum *139* eius⟩ est *supra* *140* Ergo] *supra*

50 et radix 500. Prohice ex eo quadratum AC, quod est 25, et remanebit quadratum CM, quod est $\frac{1}{2}$ diametri circuli cadentis in hoc pentagono, 25 et radix 500, quorum quadruplum scilicet 100 et radix 8000 est quadratum totius diametri . Et hoc est quod voluimus exponere.

Et si volueris, duc unum de lateribus pentagoni in se et prohice illud ex producto diametri circuli circumdantis cadentis < ... > infra pentagonum. Et hec porta ducitur in omni figura equilatera et equiangula. Et intellige hoc.

< 7 > Et quando voluerimus scire quanta sit diameter circuli continentis in pentagono noto equilatero et equiangulo cuius quodlibet latus sit 10 ex numero, secundum modum operationis quam exposuimus in illo. Et est manifestum ex eo quod dixit euclides quia omnis corde portionis circuli proportio ad cordam similis portionis alterius circuli est sicut proportio diametri illius ad diametrum illius. Et iam exposuimus quia circuli cuius diameter est 10 numero pentagoni lateris quadratum est 62 et $\frac{1}{2}$ dimin̄ rad′ 781 et $\frac{1}{4}$. Manifestum est ex eo quod diximus quod proportio centum ad 62 et $\frac{1}{2}$ dim̄ rad′ 781 et $\frac{1}{4}$ est sicut proportio census, qui est quadratum diametri circuli qui non est scitus circumdantis in pentagono cuius quodlibet latus est 10, quem fecimus rem, ex 100 et illud est ut quadratum diametri circuli noti circumdantis in pentagono noto. Ducamus 100 in 100, et proven̄ 10000. Deinde ducamus censum in 62 et $\frac{1}{2}$ dim̄ radice 781 et $\frac{1}{4}$ et proven̄ 62 census et $\frac{1}{2}$ census dim̄ rad′ 781 et $\frac{1}{4}$ census census: hoc est equale 10000. Reduc totum quod habes ad censum, et illud est ut ducas totum quod habes in $\frac{1}{50}$ unius et radicem $\overline{\frac{1}{15625}\,\frac{1}{4}}$ unius, et proven̄ census equalis 200 drag′ et radici 8000, et hoc est quadratum diametri circuli continentis pentagonum quando erit unumquodque latus eius 10.

< 8 > Et sic si voluerimus facere in pentagono noto equilatero et equiangulo cuius quodlibet latus est 10 circulum contentum a pentagono — et iam exposuimus quod circuli, cuius diameter est 10 ex numero, quadratum cuiusque lateris pentagoni circumdantis in ipso est 500 dim̄ rad′ 200000 — dicemus proportio 100, que sunt quadratum unius laterum pentagoni, ad 500 dim̄ rad′ 200000 est sicut proportio census, qui est quadratum circuli ignoti

---

141 AC] *corr. ex ABC* *145* lateribus⟩ eius *add. et del.* *146* diametri] diametro *146* circuli cadentis < ... > infra pentagonum] *Ar. al-dā'ira allatī taqa'u 'alā hādhā al-mukhammas fī mithlihi wa-khudh jidhr mā baqiya fa-mā kāna fa-huwa quṭr al-dā'ira allatī taqa'u dākhil al-mukhammas* *147* ducitur] duc̄ *150* secundum modum operationis] *Ar. bi-ghayr al-'amal* *154* quadratum] *corr. ex* quadraztum (?) *157* circumdantis] circumdatis *157* in⟩ circulum *add. et del.* *158* 100] 10 *166* pentagono] petagono *corr. ex* pedtagono

---

141 50] *corr. ex* $\frac{1}{5}$ *144* scilicet 100 ... diametri] *supra For* 8000 *MS has* 800; quorum quadruplum ... totius diametri] *Ar. fa-murabba' quṭr al-dā'ira allatī taqa'u fī hādhā 'l-mukhammas mi'a wa-jidhr thamāniyat ālāf* (*MS 'lf*) *146* cadentis] productum ex diametro circuli ca *marg.*, -dentis *in textu* *146* infra⟩ ipsum *supra* *148* continentis] *corr. in* circumdantis *152* cordam] *supra* *152* circuli] *corr. ex* circulus *153* quia⟩ diameter *add. et del.* *153* circuli] *corr. ex* circulus *154* pentagoni⟩ ei contenti ab eo *supra* *158* ex] *corr. in* ad *158* et⟩ et ut ? *supra* *158* illud] *supra* *158* ut] *supra* *162* $\frac{1}{50}$] *corr. ex* ? *162* $\overline{\frac{1}{15625}\,\frac{1}{4}}$] ? *corr. in* $\overline{\frac{1}{3125}\,\frac{1}{4}}$, *Ar.* $1\frac{1}{4}\,\frac{1}{15625}$ *163* 200] *supra*; *ras. in textu* *171* circuli⟩ diametri *supra*

quesiti, ad quadratum diametri circuli sciti, quod est 100. Duc 100 in 100 et proveñ 10000. Deinde duc censum in 500 dim̄ rad′ 200000, et proveñ 500 census dim̄ rad′ 200000 census census equalis 10000 drag′. Reduc totum quod habes ad censum, quod est ut ducas eum in $\frac{e\ \ 1}{10\ 10}$ unius et rad′ $\frac{1\ \ \ \ 0}{100\ 125}$ unius, et proveñ census equalis 100 et rad′ 8000, | et illud est quadratum diametri *95r*
circuli cadentis in pentagono cuius quodlibet latus est 10.

< 9 > Et similiter si voluerimus facere super decagono equilatero et equiangulo cuius quodlibet latus sit 10 circulum circumdantem in eo. Et iam exposuimus quod circuli cuius diameter est 10 corda decagoni cadentis in eo est radix 31 $\frac{1}{4}$ dim̄ 2 et $\frac{1}{2}$ faciamus diametrum circuli rem. Deinde dicemus proportionem rei ad 10 sicut est proportio 10 ad radicem 31 et $\frac{1}{4}$ dim̄ 2 et $\frac{1}{2}$. Ducamus igitur 10 in 10, et proveñ 100. Deinde ducamus rem in radicem 31 et $\frac{1}{4}$ dim̄ 2 et $\frac{1}{2}$, et proveñ rad′ 31 census et $\frac{1}{4}$ census sublatis ex eo 2 rebus et $\frac{1}{2}$, quod equatur 100. Restauremus cum 2 rebus et $\frac{1}{2}$ et addamus illas ad 100, erunt 100 drag′ et 2 res et $\frac{1}{2}$ rei equalis radici 31 et $\frac{1}{4}$ census. Duc 100 drag′ et 2 res et $\frac{1}{2}$ rei in se, et proveñ 10000 drag′ et 6 census et $\frac{1}{4}$ census et 500 res equales 31 censibus et $\frac{1}{4}$ census. Opponamus ergo in ipsis, et erit res 10 et radix 500, quod est diameter circuli circumdantis in pentagono cuius quodlibet latus est 10.

< 10*i* > Et similiter si volueris facere in decagono noto equilatero et equiangulo cuius quodlibet latus est 10 circulum circumdantem decagonum. Et iam exposuimus quod circuli cuius diameter est 10 corda $\frac{e}{10}$ cadentis super eo est 100 dim̄ radice 8000. Dicamus quia proportio 100 dim̄ rad′ 8000 ad 100 est sicut proportio 100 ad censum. Duc 100 in 100, et proveñ 10000. Deinde duc 100 dim̄ rad′ 8000 in censum, et proveñ 100 census dim̄ rad′ 8000 censuum census equalis 10000. Reduc totum quod habes ad censum, et illud est ut ducas eum in $\frac{1}{2}$ $\frac{e}{10}$ unius et rad′ $\frac{1\ \ \ 0}{5\ 100}$ unius, et erit census equalis 500 et radici 200000, et hoc est quadratum diametri circuli cadentis infra decagonum cuius quodlibet latus est 10.

Et si volueris, duc diametrum circuli cadentis extra hunc decagonum in se, et iam exposuimus quod <est> 10 et rad′ 500, erit 600 et rad′ 200000. Prohice itaque ex eo 100, quod est multiplicatio lateris decagoni in se, et remanebit 500 et rad′ 200000. Radix eius est diameter circuli circumdantis hunc decagonum.

< 10*ii* > Et ego ponam causam huius porte cum adiutorio dei. Omnis figure equilatere circumdantis circulum vel circumdate in ipso circulo multiplicatio

---

175 $\frac{1\ \ \ \ 0}{100\ 125}$] *corr. ex* ?, *Ar.* $\frac{4}{5}\frac{1}{10000}$ *175* unius] unus *177* in⟩ circulo (?) *add.* *181* radix⟩ unius *add.* *186* radici⟩ unius *add. et del.* *189* circuli⟩ conteñ *add. et del.* *193* corda] *lege* quadratum corde *193* cadentis] cadetis *193* super⟩ s *add. et del.* *198* $\frac{e}{10}$] $\frac{e\ \ 1}{10\ 12}$ *198* rad′⟩ *seq. ras.* *198* $\frac{1\ \ \ 0}{5\ 100}$] *manu altera?Ar.* $\frac{4}{5} \cdot \frac{1}{400}$

---

174 equalis] *corr. ex* equale *179* circulum circumdantem] *corr. ex* circulus circumdans *180* corda] *corr. ex* cordan *181* $\frac{1}{4}$] *supra* *182* proportio] *corr. in* proportionem *186* 100^1⟩ et *supra* *186* et $\frac{1}{4}$ census] *om.*, $\frac{1}{4}$ censuum *supra*; Restauremus etiam censum et $\frac{1}{4}$ census *add. et del.* *192* circumdantem⟩ ia (?) *supra* *194* 8000^1] *corr. ex* 8 *207* in] *supra*; ab *add.*

lateris in se et diametri circuli circumdantis in ipso in se agregate sunt sicut multiplicatio diametri circuli circumdantis figuram in se. Exemplum: faciamus triangulum equilaterum et equiangulum sitque triangulus ABG, et in ipso circulum circumdantem triangulum sitque HZL, extra triangulum vero circulum circumdantem in ipso sitque ABG. Et diameter eius sit AD, diameter vero circuli circumdantis triangulum sit linea CY. Dico quia multiplicatio AG in se et CY in se est sicut multiplicatio AD in se. Cuius demonstratio est quod producamus super lineam AG ex puncto L, qui est locus in quo contingit circulus lineam AG, lineam secundum angulum rectum. Manifestum est quod in ipsa est centrum duorum circulorum, et applicent secundum extremitates ipsius ad circumferentiam circuli HZL sitque linea LZ: linea LZ est diameter circuli HZL que est equalis linee CY. Et coniungemus punctum Z cum puncto D cum linea ZD, et punctum D cum puncto G cum linea DG. Linea AM est sicut linea MD et linea ML est sicut linea MZ, et due linee AM et ML sunt sicut 2 linee DM et MZ; et angulus AML est sicut angulus DMZ: et basis DZ est sicut basis AL. Et AL est sicut LG: ergo LG est sicut DZ et triangulus AML est sicut triangulus DMZ et anguli eius sicut anguli illius. Angulus ALM sicut angulus MZD; et linea ZD equidistans linee LG et iam exposuimus quia ipsa equalis illi; et iam coniunctum fuit quod est inter duas quasque extremitates ipsarum lineis ZL et DG: linea ZL

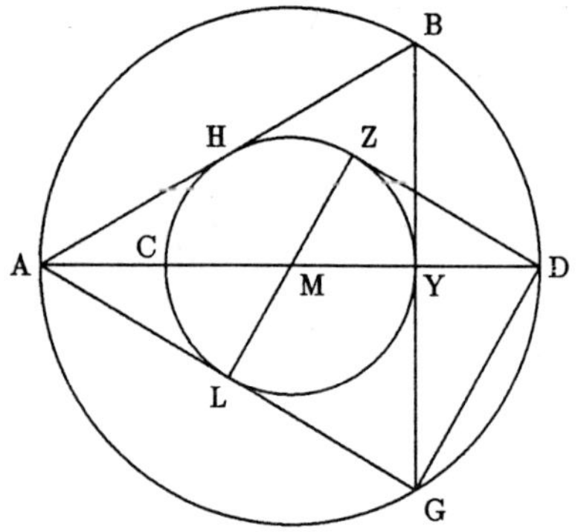

Prop. 10*ii*

---

208 circuli⟩ cont *add. et del.* 212 in ipso] *corr. in* ipsum 212 sitque] t *supra* 215 quod⟩ sciamus *add. et del.* 221 Linea⟩ *DG add. et del.*

---

209 circumdantis] circumdatis, continentis *supra* 211 triangulum[1]] *underlined* 212 circumdantem] *corr. ex* circumdantes 213 triangulum] *corr. in* in triangulo 216 contingit⟩ eum *add. et del.*; applicat ad *supra* 221 *AM*] *corr. ex AG* 225 angulus] *supra* 227 quod est inter duas extremitates ipsarum] *corr. ex* cum duabus extremitatibus ipsas, *Ar. fīmā bayna aṭrāfihimā*; in *add. et del.* 227 *DG*⟩ ergo *supra*

equidistans et equalis linee DG. Sed multiplicatio AG in se et GD in se est sicut multiplicatio AD in se, quia angulus AGD est rectus; et DG est sicut LZ et LZ est sicut CY. Ergo multiplicatio AG in se et CY in se est sicut multiplicatio AD in se. Et hoc est quod voluimus exponere.

< 11*i* > Si fuerit figura quindecagona circumdata a circulo et diameter eius sit 10 sitque equilatera et equiangula et voluerimus scire unumquodque latus eius, faciemus circulum notum AGB, cuius diameter sit AB, et faciemus in eo latus decagoni, quod erit corda partis et $\frac{1}{2}$ partis, sitque linea BD. Et iam exposuimus quia ipsa est radix 31 et $\frac{1}{4}$ dim̄ 2 et $\frac{1}{2}$ corda AD est corda $\frac{2}{5}$ circuli et est 62 et $\frac{1}{2}$ et radix 781 et $\frac{1}{4}$ accepta radice eius. Et faciamus in ea cordam $\frac{1}{6}$, sitque BG, que corda 2 partium et $\frac{1}{2}$ et <est> 5, et linea AG est corda $\frac{1}{3}$ et est radix 75. Ducamus lineam AG in lineam BD et proveñ rad′ 2343 et $\frac{1}{2}$ et $\frac{1}{4}$ dim̄ rad′ 468 et $\frac{1}{2}$ et $\frac{1}{4}$. Deinde prohice eum ex multiplicatione AD in lineam BG, | que est 1<562 et $\frac{1}{2}$ et rad′ 488>281 et $\frac{1}{4}$ (accipiamus 95*v*
radicem eius), remanebit 1562 et $\frac{1}{2}$ et radix 488281 et $\frac{1}{4}$ accepta rad′ eius. Addemus ei rad′ 468 et $\frac{1}{2}$ et $\frac{1}{4}$, sublata ex toto hoc rad′ 2343 et $\frac{1}{2}$ et $\frac{1}{4}$: et hoc est sicut multiplicatio linee AB in lineam GD. Dividamus per lineam AB, que est 10, et exibit linea GD 15 et $\frac{5}{8}$ <et> rad′ 48 et $\frac{3}{4}$ et $\frac{e}{8}\frac{5}{8}$ accepta rad′ eius et rad′ 4 drag′ et $\frac{1}{2}$ et $\frac{1}{8}$ et $\frac{e}{8}\frac{1}{2}$ dim̄ rad′ 23 et $\frac{3}{8}$ et $\frac{e}{8}\frac{1}{2}$.

< 11*ii* > Scietur exiens corda GD ad intelligendum in acceptione radicis 48 et $\frac{3}{4}$ et $\frac{e}{8}\frac{5}{8}$ in propinquitate, exitus eius 6 et 59 minuta et 15 secunda et

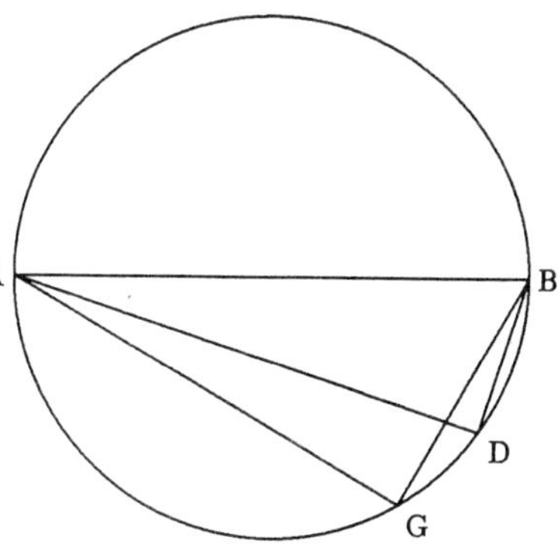

Prop. 11*i*

236 2] 62 *236* corda] corde *236 AD*⟩ seq.ras. *238 AG*⟩ 8 que (?) *add. et del.* *239* Ducamus ... *BD*] *om. Ar.* *242* accepta] acceptam *242* eius²] *corr. ex* oeius *243* 468] 462 *245* 15] 25 *247* Scietur ... acceptione] *Ar. fa-nqrb istikhrāj watar GD li-tafham kayfa ya'khudhu* *248* 59] *Ar.* 56

228 equalis⟩ est *supra* *234 AB*] *corr. ex AG* *238* sitque] *corr. in* et est *240* 468] *corr. ex* 488 *244* Dividamus] *corr. ex* divisa

46 tertia. Addimus ea ad 15 et $\frac{5}{8}$, et erit 22 et 36 minuta et 45 secunda et 46 tertia; quorum accipe radicem et invenies eam fere 4 et 45 partes et 19 secunda et $\frac{1}{3}$ unius; cui adde radicem 4 et $\frac{5}{8}$ et $\frac{e\ 1}{8\ 2}$ fere, et erit 2 et 9 minuta et 54 secunda et 12 tertia, et erit 56 et $\frac{2}{5}$ partes et 13 secunda et 13 tertia. Sublata ex eo rad′ 23 et $\frac{3}{8}$ et $\frac{e\ 1}{8\ 2}$ secundum propinquitatem, quod est 4 drag′ et 50 minuta et 28 secunda et 25 tertia, et remanebunt 2 et 4 minuta et 44 secunda et 48 tertia, et hoc corda *GD* secundum propinquitatem. Et hoc est quod voluimus exponere.

< 12 > Et si dicemus tibi trianguli equilateri et equianguli mensura cum perpendiculari ipsius est 10 ex numero, quanta est perpendicularis. Faciamus triangulum *ABG* cuius perpendicularis *AD*. Cum autem voluerimus scire

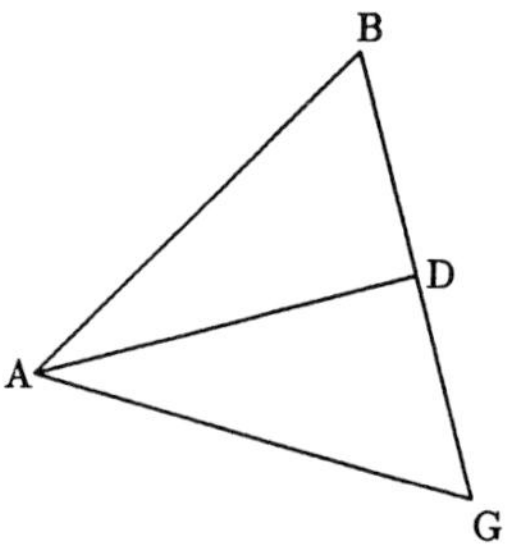

Prop. 12

quanta sit linea *AD*, que est perpendicularis trianguli *ABG*, faciamus ipsum rem. Igitur linea *DG* radix $\frac{1}{3}$ census, et quodlibet latus trianguli *ABG* radix census et $\frac{1}{3}$ census, et mensura trianguli *ABG* est radix $\frac{1}{3}$ census census, et perpendicularis eius res: erit res et radix $\frac{1}{3}$ census census equalis 10. Restaura radicem $\frac{1}{3}$ census census ut sit radix census census, et restaurabis illud si multiplicabis illud in radicem 3: ergo multiplica totum quod habes in radicem 3, et proveñ census et rad′ 3 censuum equalis radici 300. Dimidia igitur radicem 3 censuum, et exibit radix $\frac{3}{4}$; quam duc in se. Et proveñ $\frac{3}{4}$; quas adde ad radicem 300, et erunt $\frac{3}{4}$ et radix 300; quorum accipe radicem et prohice ex

---

250 invenies] *corr. ex insuenies* *250* et[2]⟩ $\frac{1}{5}$ et *add. et del.* *251* et $\frac{1}{3}$ unius] *lege* et tertium unum, *Ar wa-thālitha wāḥida* *252* et[1]⟩ 4 *add. et del.* *252* 56 et $\frac{2}{5}$ partes] *lege* 6 et 55 minuta, *Ar. sitta wa-khams wa-khamsīn daqīqa* *262* *ABG*⟩ radix *ABG* *add. et del.* *267* proveñ⟩ *seq. ras.*

---

252 et[3]] *supra* *257* mensura⟩ est *supra* *258* perpendiculari⟩ cum (?) *supra* *260* *AD*] *supra*, *DG in textu* *261* *DG*⟩ *DG supra* *261* census⟩ g. quia tanta erit multplicatio linee *AD*, qui est res, in tertiam sui, que multiplicatio est tertia census, quanta (est *supra*) multplicatio linee *DG* in se. Ergo *DG* est radix tertie census. Intellige. *263* erit] *corr. in* ergo *263* equalis⟩ sunt *supra* *267* censuum] *supra* *267* quas] *corr. ex* que

eo rad′ $\frac{3}{4}$, et residuum est linea AD, que est perpendicularis trianguli ABG. Et hoc est quod voluimus exponere.

< 13 & 14 > Si dicemus tibi trianguli equalium laterum et angulorum unumquodque latus est 10 et in ipso est quadratum longum rectorum angulorum cuius mensura est 10, quanta est longitudo quadrati. Exemplum: faciamus triangulum triangulum ABG et quadratum longum < ... > $GMLY$, quod est in triangulo ABG. Prohice ergo eum ex 10 et quod remanet est mensura trianguli ABG. Et si volueris, dic: mensura quadrati $6\frac{102}{143}$ unius dim̄ rad′ 11 drag′ et $\frac{10261}{20449}$ par<tis> drag′, remanebit mensura trianguli 3 et $\frac{41}{143}$ unius et rad′ 11 et $\frac{10261}{20449}$ unius.

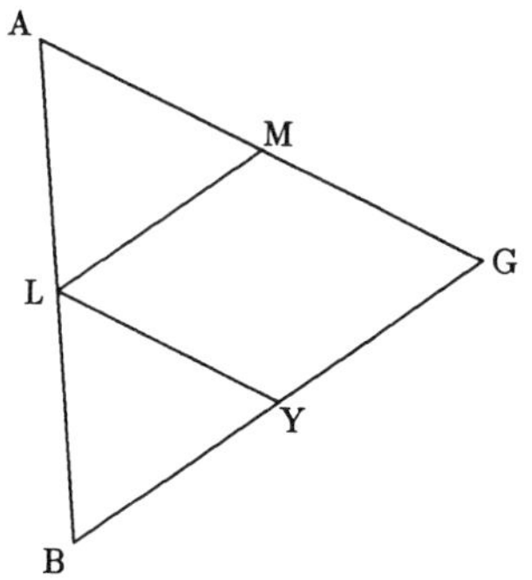

Props. 13-14

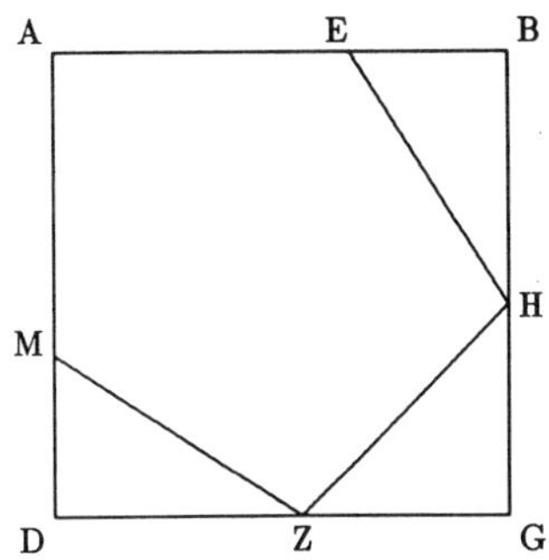

Prop. 15

< 15 > Et si dicemus tibi quadrati equilateri et rectorum angulorum quodlibet latus est 10, sitque quadratum $ABGD$, in quo fecimus pentagonum secundum hanc figuram, et illud pentagonum est $AEHZM$, scientia cuiusque laterum eius pentagoni est ut faciamus unum de lateribus pentagoni, et est linea AE, rem. Remanebit linea EB 10 dim̄ re. Et linea GH est rad′ $\frac{1}{2}$ census; remanebit linea $G'B$ 10 diminuta radice $\frac{1}{2}$ census. Et linea BE 10 diminuta re. Duc unumquodque ipsorum in se et agrega ea, et erunt 200 drag′

---

276 $\frac{102}{143}$] $\frac{101}{143}$ *277* et[1]⟩ 20461 ***add. et del.*** *277* $\frac{10261}{20449}$] $\frac{20261}{20449}$ *277* mensura⟩ quadrati ***add. et del.*** *278* $\frac{10261}{20449}$] $\frac{20261}{20449}$; 20449 ***add. et del. supra*** *282* eius] ***del. in textu***

---

271 < 13 & 14 >⟩ G (?) mihi videtur quod sic debeat esse lecta. Si dicemus tibi trianguli equalium laterum et angulorum et quad[rati] longi id est rumbi et in triangulo contenti mensura est 10 ex numeris tunc quanta est mensura rumbi predicti et trianguli. verbi gratia sit triangulus equilaterus ABG et rumbus intra ipsum rumbus $GMLY$ fac et invenies rumbum esse medietatem trianguli. quare ipse erit $\frac{1}{3}$ de 10 id est $3\frac{1}{3}$ trianguli vero 6 et $\frac{2}{3}$ vel si volueris dic quod si (?) rumbum est 6 (de rad′ ... que est $3\frac{1}{3}$ ***add. et del.***) et $\frac{82}{123}$ diminuta rad′ 11 et $\frac{1681}{15129}$ que est [ ] et radix [ ] et $\frac{1681}{15129}$ ***marg.*** *273* quadrati⟩ guli ***supra*** *275* eum] ***supra*** *275* quod[2]] ***supra*** *276* dic] ***supra*** *276* quadrati⟩ est ***supra*** *278* 11] ***corr. ex*** unius *280* in quo] ***supra*** *283* est[2]] ***supra*** *284* census[1]⟩ .g. quia GH et GZ sunt equales ***supra*** *284* $\frac{1}{2}$⟩ tis ***supra*** et census et $\frac{1}{2}$ census dim̄ 20 partibus et dim̄ radice 200 censuum equalis census. Oppone in ipso sicut exposui tibi: et exibit linea *AE*, que est unum de lateribus pentagoni, 200 et rad′ 320000 accepta radice illorum subtracta ex 20 et radice 200. Et hoc est quod voluimus exponere.

< 16 > Et si dicemus tibi pentagoni equilateri et equianguli mensura est 50 drag′, quantum est quodlibet latus eius. Verbi gratia faciamus pentagonum pentagonum *ABGDE* et centrum circuli circumdantis in eo pentagonio M; et exeant linee *AM MB GM DM ME*. Manifestum est quia iam dividitur triangulis equalibus, qui sunt triangulus *AMB* et *BMG* et *GDM* et *DME* et *EMA*: triangulus *EMD* 10 ex numero. Et iam exposuimus in eo quod

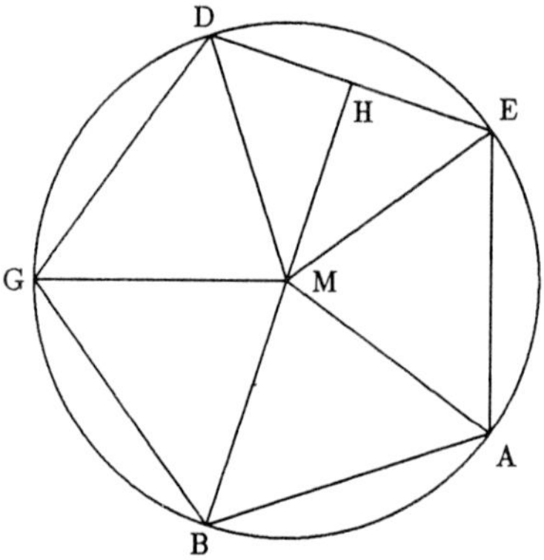

Prop. 16

premisimus ex libro nostro hoc quia, quando erit pentagonus equilaterus et equiangulus quodlibet latus eius est 10 ex numero, quadratum diametri circuli circumdantis ipsum est 200 et radix 8000. Manifestum est ex eo quod diximus quia, quando volueris scire diametrum circumdanti a pentagono noto equalium laterum et angulorum, duces unum de lateribus pentagoni in se, deinde duplabis eum et servas eum; deinde ducis etiam similiter unum de lateribus pentagoni in se, deinde agregatum in se, deinde accipias $\frac{4}{5}$ eius et id quod iam accepisti radicem eius. Quod exit addes | ad id quod servasti, et *96r*
quod habebis accipe radicem, et quod exibit est diameter circuli. Et postquam est hoc ita, faciamus lineam *ED* rem. Et illud est unum de lateribus pen-

---

292 circumdantis] circumdatis *293 MB*⟩ et *B add. et del.*

---

288 accepta radice ipsorum] *corr. ex* accipiamus radicem eius *292* M] *supra* *293 DM*] *corr. ex DC* (?) *295 EMD*⟩ est *supra* *297* equiangulus⟩ et *supra* *301* servas] *corr. ex* serva *303* et id quod iam accepisti] *Ar. fa-mā kāna akhadhta*; ad accipionem *supra* *303* et⟩ illius *supra* *304* postquam⟩ ita *add. et del.*; fuit *supra*

tagoni: quadratum diametri circuli circumdantis in eo pentagono, sicut iam exposuimus, est 2 census et rad′ $\frac{4}{5}$ census census, et quadratum $\frac{1}{2}$ diametri eius, que est linea EM, est $\frac{1}{2}$ census et radix $\frac{e}{10}\frac{1}{2}$ census census. Deinde produc perpendicularem trianguli EMD — manifestum quia ipsa cadet super medium linee ED — et ipsa est linea MH. Linea EH est $\frac{1}{2}$ rei: duc eam in se, et proveniet $\frac{1}{4}$ census; prohice ergo eam ex quadrato linee EM, que est $\frac{1}{2}$ et radix $\frac{1}{2}$ decime census census, et remanebit quadratum linee MG' $\frac{1}{4}$ census census et radix $\frac{e}{10}\frac{1}{2}$ census census. Duc eum in lineam EG', que est $\frac{1}{2}$ rei, et proveñ $\frac{e}{8}\frac{1}{2}$ census census et radix $\frac{1}{320}$ census census census census equalis 10 ductis in se, que sunt 100. Restaura $\frac{e}{8}\frac{1}{2}$ census census ut sit census census, quod est ut ducas eum in 16. Duc ergo totum quod habes in 16, et proveñ census census et radix $\frac{4}{5}$ census census census census equalis 1600. Reduc totum quod habes ad censum census, et illud est ut ducas totum quod habes in 5 dim̃ rad′ 20, et proveñ census census equalis 8000 et radici 51200000. Rad′ eius est quodlibet latus pentagoni.

< 17 > Et si dicemus tibi decagoni equalium laterum et angulorum mensura est 100 ex numero, quantum est unumquodque latus eius. Verbi gratia faciamus decagonum $ABGDEUZHTL$ et centrum circuli circumdantis in ipso decagono M. Et manifestum est quia ipse dividitur 10 triangulis equalibus; triangulus ZMU 10. Iam exposuimus in eo quod premisimus ex verbis

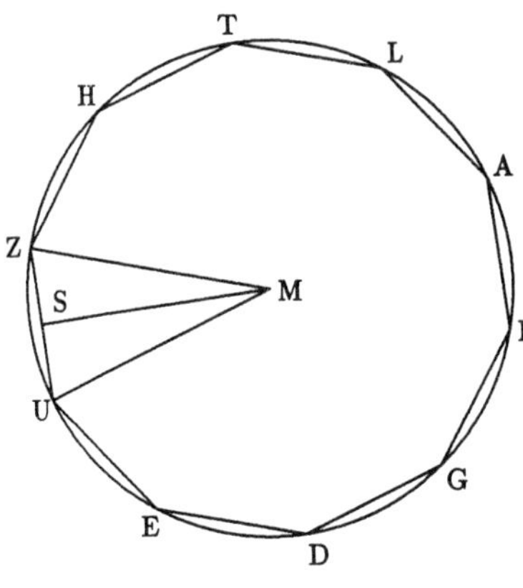

Prop. 17

308 Deinde⟩ fa *add. et del.* *309* trianguli⟩ *EM add. et del.* *310* est¹⟩ linea *END* et est *add. et del.* *314* radix⟩ 3 *add. et del.* *314* 10⟩ censibus *add. et del.* *317* et radix $\frac{4}{5}$ census census] *om. Ar.* *321* tibi⟩ pen *add. et del.* *323* circumdantis] circumdantes *325* equalibus⟩ trian e *add. et del.*; similibus *marg.*

311 $\frac{1}{2}$⟩ census *supra* (*māl supra in Ar.*) *312* *MG′*] *corr. in MH* *317* census⁶] *supra* *323* decagonum] *corr. ex* pentagonum *325* exposuimus⟩ autem *supra* nostris quia, quando erit decagonus equilaterus et equiangulus unumquodque latus cuius sit 10, diameter circuli circumdantis eum 10 et radix 500. Manifestum est ex eo quod exposuimus quia, quando voluerimus scire diametrum circuli circumdantis in decagono noto, ducas unum de lateribus decagoni in se, deinde in 5 deinde; accipias radicem agregati et adde illud ad unum de lateribus decagoni, et quod exierit erit diameter circuli. Et postquam manifestum est ex eo quod diximus quod faciamus lineam ZU rem, et illud unum de lateribus decagoni. Necesse est quod sit diameter circuli circumdantis in eodem decagono, sicut exposuimus, res et radix 5 censuum; et $\frac{1}{2}$ diametri eius, que est linea ZM, $\frac{1}{2}$ rei et radix census et $\frac{1}{4}$ census et manifestum est etiam similiter ex hoc quod diximus ex eo quod rememoravit euclides, et est quia, quando dividetur linea per proportionem per medium et 2 partes et additum in longitudine linee sicut maior pars eius, tota linea addita divisa erit secundum proportionem medii et duo extrema maiorque pars eius est linea prima. Manifestum est ex eo quod diximus quia quando dividetur linea secundum proportionem habentem medium et duo extrema et dividetur maior pars secundum proportionem habentem medium duoque extrema, maior pars linee sic divise erit minor pars linee totius scilicet prime. Et dixit in alio loco: quando coniungetur latus exagoni cum latere decagoni unius circuli secundum rectitudinem, tota linea divisa est secundum proportionem medii et extremorum maiorque pars eius est linea exagoni. Set linea ZU est latus decagoni et linea ZM est latus exagoni. Et quando fuerint hee 2 linee linea una, iam divisa est secundum proportionem medii et extremorum et maior pars est linea ZM; linea igitur ZM quando dividetur secundum proportionem medii et extremorum sicut linea ZU. Et quando facies ZU rem et minues eam ex ZM, deinde addes ad residuum $\frac{1}{2}$ linee ZU et duces agregatum in se: erit sicut 5 ad quadratum medietatis linee ZU, sicut exposuit euclides. Duc medietatem linee ZU in se, deinde in 5, et accipe radicem eius et adde ad $\frac{1}{2}$ linee ZU, que est $\frac{1}{2}$ rei: exibit linea ZM. Iam manifestum est quia linea ZM $\frac{1}{2}$ rei et radix census et $\frac{1}{4}$. Deinde produc perpendicularem trianguli ZMU, que est MS; manifestum est quia ipsa cadit super medium linee ZU, quod est S. Deinde duc lineam ZM in se, que est $\frac{1}{2}$ rei et radix census et $\frac{1}{4}$: erit census et $\frac{1}{2}$ et radix census census et $\frac{1}{4}$ census census. Deinde duc lineam ZS, que est $\frac{1}{2}$ rei, in se, et provenī $\frac{1}{4}$ census. Prohice ipsam ex eo, et remanebit census et $\frac{1}{4}$ et radix census census et $\frac{1}{4}$ census census. Rad′ huius est linea MS: duc eam in lineam ZS que est $\frac{1}{2}$ rei, et provenī $\frac{1}{4}$ census

---

330 deinde⟩ dic *add. et del.* *336* quod[2]⟩ dixit e *add. et del.* *343* sic divise] *om. Ar.* *343* erit⟩ linea *add. et del.* *346* linea⟩ *ZU add. et del.* *347* *ZM*⟩ est latus *add. et del.* *349* est linea *ZM*; linea igitur *ZM*] *lege* linee *ZM*, *Ar. min khaṭṭ ZM* *356* *ZMU*] *corr. ex ZM* *359* ipsam] ipsa

---

337 per[1]⟩ secundum *supra* *337* partes⟩ medii (*corr. ex* mediu) et duo extrema *supra* *338* et⟩ erit *supra* *338* in⟩ se *marg.* *338* eius⟩ quia *supra* *338* tota] *corr. ex* totam *339* secundum⟩ item *supra* *343* linee[1]⟩ illius *supra* *345* rectitudinem⟩ quia *supra* *350* extremorum⟩ est maior pars *supra* *352* ad quadratum] *corr. ex* cum quadrato *357* *ZU*] *corr. ex HT* *357* quod est *S*] *corr. ex* quod est *ZH*, *Ar. wa-huwa* ZS; in *supra* *359* census⟩ et *supra*; extrap (?) *add. et del.* *360* eo et] eo (?) et *supra*, et *in textu*

census et $\frac{e\ 1}{8\ 2}$ census census et radix $\frac{5}{64}$ census census census census equalis 100 drag′. Restaura $\frac{1}{4}$ census census et $\frac{1}{2}$ octave census census ut sit census census, quod est ut ducas eum in 3 et $\frac{1}{5}$: duc ergo totum quod habes in 3 et $\frac{1}{5}$, et provẽ census census et radix $\frac{4}{5}$ census census census census equalis 320 drag′. Reduc totum quod habes ad censum census, et illud est ut ducas totum quod habes in 5 dim̄ rad′ 20, et erit census census equalis 1600 dim̄ rad′ 2048000, quorum accipe radicem, et hec est linea ZU, et illud est unum de lateribus decagoni. Et hoc est quod voluimus exponere.

| < 18 > Si dicemus tibi pentagoni $ABGDE$ equalium laterum et an- 96v
gulorum mensura trianguli BDE est 10, quanta est linea DE, que est basis trianguli, unum de lateribus pentagoni. Exemplum: faciamus lineam DE rem. Et iam exposuit euclides quod linea BE quando dividetur secundum proportionem medii et extremorum, maior eius pars erit equalis linee DE; et exposuit etiam quia, quando dividetur linea secundum proportionem medii et

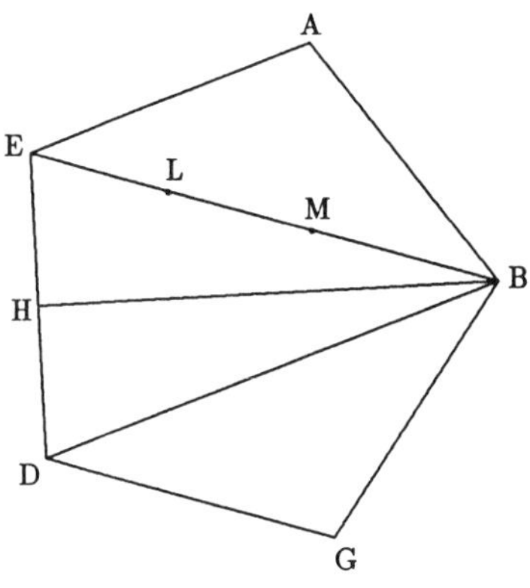

Prop. 18

extremorum et addetur toti linee sicut maior pars eius, tota addita divisa erit secundum proportionem eandem et maior pars erit linea prima; et similiter exposuit quod, quando dividetur linea secundum proportionem medii et extremorum et addetur minori parti medietas maioris et multiplicabitur addita in se, erit quadratum productum 5^{plum} quaddrati medietatis linee longioris. Ergo dividamus lineam BE secundum proportionem medii et extremorum in

377 et exposuit etiam ... prima] *om. Ar.* *381* extremorum⟩ sm *add. et del.*

363 Restaura] staura *underlined*, integra *supra* *365* census[5] census[6]] *supra* *367* erit⟩ provenit *supra* *371* mensura] *supra* *371* *BDE*] *E supra* *372* trianguli⟩ et est *supra*

punctum M, et faciamus maiorem partem lineam ME: linea ME equalis est linee DE. Et dividamus lineam ME in duas partes in puncto L: erit multiplicatio linee LM in se $\frac{1}{5}$ multiplicationis LB in se. Et multiplicatio LE in se est $\frac{1}{4}$ multiplicationis DE in se, ergo est $\frac{1}{4}$ census, quia fecimus lineam DE rem et lineam ME rem et LE $\frac{1}{2}$ rei. Ergo multiplicatio BL in se est census et $\frac{1}{4}$; ergo linea BL est radix census et $\frac{1}{4}$. Et linea LE est $\frac{1}{2}$ rei: linea BE $\frac{1}{2}$ rei et radix census et $\frac{1}{4}$. Exibit perpendicularis trianguli BED, que est BH: <linea EH est> $\frac{1}{2}$ rei. Duc igitur eam in se, et proveñ $\frac{1}{4}$ census; et prohice eum ex multiplicatione BE in se, quod est census et $\frac{1}{2}$ et radix census <census> et $\frac{1}{4}$ census census, et remanebit census et $\frac{1}{4}$ et radix census census et $\frac{1}{4}$ census census; radix huius est linea BH. Duc eam in lineam EH, que est $\frac{1}{2}$ rei, et proveñ $\frac{1}{4}$ census census et $\frac{1}{2}$ octave census census et radix $\frac{5}{64}$ census census census <census> equalis 100 dragmis. Restaura $\frac{1}{4}$ census ex censu tuo et $\frac{e\;1}{8\;2}$ census census tui ut sit census census, quod est ut ducas eum in 3 et $\frac{1}{5}$: duc ergo totum quod habes in 3 et $\frac{1}{5}$, et proveñ census census et radix 4 censuum census census <census> equalis 3 censibus et 20 drag′. Reduc totum quod habes ad censum census, quod est ut ducas eum in 5 dim̄ rad′ 20, et proveñ census census equalis 1600 dim̄ rad′ 2048000; cuius radix est linea ED. Et manifestum est quod triangulus BED ex isto pentagono cuius rememorati sumus similis trianguli ZM ad decagonum quem memoravimus in figura que est superius ab hac.

< 19 > Si dicemus tibi mensura trianguli ABE est 10 ex numero, quanta est linea BE. Verbi gratia faciamus lineam BE rem. Et iam exposuit euclides quia, quando dividetur linea secundum proportionem medii et extremorum, pars maior quando addetur in longitudinem ipsius sicut medietas linee totius et ducetur in se, erit quadratum illius 5^{plum} quadrati quod erit ex $\frac{1}{2}$ linee in se. Faciamus igitur lineam BE rem, et cum ducetur $\frac{1}{2}$ eius in se, deinde in 5, proveñ census et $\frac{1}{4}$. Manifestum est ex eo quod diximus quia linea AE est radix census et $\frac{1}{4}$ dim̄ $\frac{1}{2}$ rei. Exibit perpendicularis super lineam BE, et est AY: ducamus lineam AE in se, et proveñ census et $\frac{1}{2}$ dim̄ rad′ census census et $\frac{1}{4}$ census census. Deinde duc lineam EY in se et est $\frac{1}{2}$ rei et proveñ $\frac{1}{4}$ census; prohice ergo eum ex eo, et remanebit census et $\frac{1}{4}$ dim̄ rad′ census census et $\frac{1}{4}$ census census: erit census census. Radix huius est linea AY, que est trianguli ABE perpendicularis. Duc ergo eam in lineam EY, que est $\frac{1}{2}$ rei, et proveñ $\frac{1}{4}$ census census et $\frac{e\;1}{8\;2}$ census census dim̄ rad′ $\frac{5}{64}$ census census census: radix huius equatur triangulo ABE, que est 10. Duc ergo 10 in

384 erit multiplicatio ... in se] *Ar. fa-ḍarb BL fī mithlihi amthāl ḍarb LE fī nafsihi* *385* est $\frac{1}{4}$ ... ergo] *om. Ar.* *387* et⟩ c̄ *add. et del.* *396* 4] *Ar.* $\frac{4}{5}$ *401* similis] similitudo *409* linea] repet. *414* erit census census] this appears misplaced, but is present here in the Arabic

383 dividamus] *corr. ex* divisa est *385* $\frac{1}{4}^{1}$] *corr. ex* quadruplum *386* LE] *corr. ex* LNE *387* BE] *corr. ex* ZE *389* rei⟩ exibit ... rei *underlined*; medietas autem linee ED est $\frac{1}{2}$ rei et est comunis *supra* *394* Restaura⟩ integra *supra* *394* censu] *corr. ex* censui *403* dicemus] *corr. in* dicit *404* exposuit] *corr. ex* exposuimus *406* pars maior quando] *del.*; et *supra* *406* ipsius⟩ parti maiori *supra* *410* Exibit⟩ exit *supra* *410* super⟩ per *supra* *415* perpendicularis] *marg.*

se, et proveñ 100 equales $\frac{1}{4}$ census et $\frac{e}{8}\frac{1}{2}$ census census dim̃ rad′ $\frac{5}{64}$ census census census <census>. Reintegra $\frac{1}{4}$ census census tui et $\frac{e}{8}\frac{1}{2}$ census census tui ut sit census census, quod est ut ducas eum in 3 et $\frac{1}{5}$: duc ergo totum quod

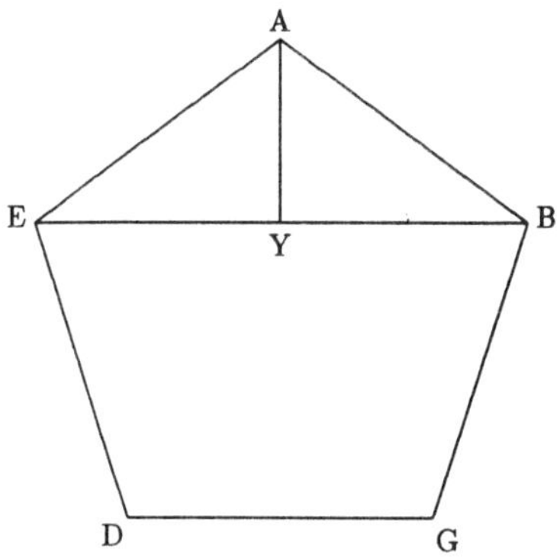

Prop. 19

habes in 3 et $\frac{1}{5}$, et proveñ census census dim̃ rad′ $\frac{4}{5}$ census census equalis 320. Reintegra ergo illud ut sit census census integer, quod est ut ducas eum in 5 et radicem 20: duc ergo totum quod habes in 5 et rad′ 20, et proveñ census census equalis 1600 drag′ et rad′ 2048000; cuius radix est radix linee BE. Et hoc est quod voluimus exponere.

< 20 > Et si dicemus tibi decagoni $ABGDEZIHTC$ equilateri et equianguli mensura trianguli CTH est 10 ex numero, quanta est linea CH, que est latus decagoni $\frac{1}{5}$ circuli circumdantis hunc decagonum. Exemplum: faciamus lineam CH rem et iam exposuimus quod quando erit corda $\frac{1}{5}$ circuli res, erit quadratum diametri eius 2 census et radix $\frac{4}{5}$ census census: ergo est quadratum $\frac{1}{2}$ diametri eius $\frac{1}{2}$ census et radix $\frac{e}{10}\frac{1}{2}$ census census. Prohice ergo illud ex quadrato corde $\frac{1}{5}$, quod est census, et remanebit quadratum CT $\frac{1}{2}$ census dim̃ radice $\frac{e}{10}\frac{1}{2}$ census census. Et quod cogit nos facere hoc est quia euclides iam exposuit quia corda $\frac{1}{5}$ potest super cordam $\frac{1}{6}$ et $\frac{1}{10}$, quando erit circulus unus. Prohiciamus ex eo quadratum CM in se quod est $\frac{1}{4}$ census et remanebit quadratum TM, que est perpendicularis trianguli

---

421 equalis] *corr. ex* equales *424* radix[2]] *Ar. om.* *424* cuius radix ... BE] *lege* cuius radix radicis est linea BE ? *432* quadratum] *lege* quarta

---

431 ergo est quadratum] *corr. ex* ex *431* diametri⟩ ergo est semidiametrum *supra* *435* unus⟩ continens eos *supra* *436* $\frac{1}{4}$] *supra, lege* quadratum *436* que] *corr. ex* quod

CTE, quadratum census dim̃ rad′ census et $\frac{1}{2}\ \frac{e}{10}$ census census. Duc ergo eum in quadratum CM, quod est $\frac{1}{4}$ quadrati census: | erit $\frac{e\ 1}{8\ 2}$ census census 97r
dim̃ rad′ $\frac{1}{320}$ census census <census census>, cuius radix equatur mensure MTH, que est 10 ex numero. Duc ergo 10 in se, et proven̄ 100 equalis $\frac{e\ 1}{8\ 2}$ census census dim̃ rad′ $\frac{1}{320}$ census census census census. Reintegra $\frac{1}{2}$ octave census census ut sit census census, et illud est ut ducas eum in 16. Duc ergo

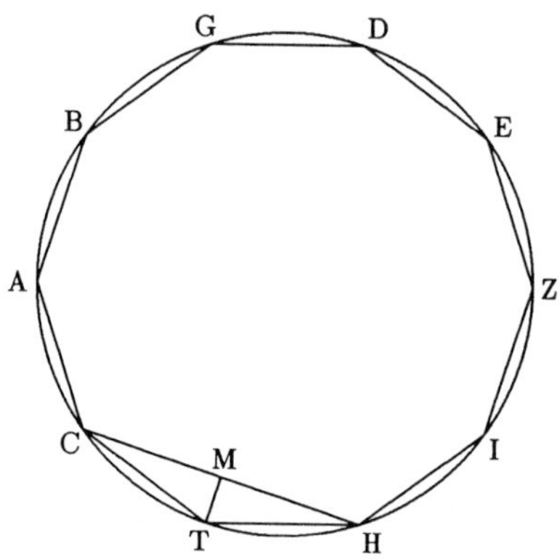

Prop. 20

totum quod habes in 16, et proven̄ census census dim̃ radice $\frac{4}{5}$ census census census <census> equalis 1600. Reintegra etiam illud ut sit census census integer, quod est ut ducas eum in 5 et radicem 20: et duc totum quod habes in 5 et radicem 20, et proven̄ census census equalis 8000 et radici 51200000 cuius radix est linea CH, que est corda $\frac{1}{5}$ circuli circumdantis decagonum. Et hoc est quod voluimus exponere.

---

437 CTE] *lege CTH* *438* in⟩ se *add. et del.*

X

PSEUDO-EUCLID ON THE POSITION OF THE IMAGE IN REFLECTION: INTERPRETATIONS BY AN ANONYMOUS COMMENTATOR, BY PENA, AND BY KEPLER

PSEUDO-EUCLID

In theorem 16 of the pseudo-Euclidean treatise *Catoptrica* the position of the image of a point reflected in a plane mirror is found as the intersection of the ray between eye and miror and the perpendicular from the object to the plane of the mirror.[1] In Fig. 1, which is taken from pseudo-Euclid, with Roman letters replacing Greek, *A* is the object, *B* the eye, *D* the point of reflection, and *E* the image. *E* is found as the intersection of *BD* and the perpendicular *AG*. In theorems 17 and 18 the same is demonstrated for convex and concave spherical mirrors, the perpendicular being replaced by the line joining the object and the centre of the sphere. In the three theorems Pseudo-Euclid gives the standard ancient and medieval procedure to find the position of the image in catoptrics: Ptolemy,[2] Alhazen,[3] Grosseteste[4] and Witelo[5] all give this procedure. It is also to be found in Risner's own work on optics[6] as well as in his editions of Alhazen and Witelo. Decades after the publication of Kepler's *Paralipomena in Vitellionem* (1604) the princple was discussed by Tacquet and Barrow.[7] It even survived into the eighteenth century.[8] In many of these authors the principle is discussed together with a similar principle for refraction, but only reflection will be considered here.

Pseudo-Euclid's proposition 16, whose enunciation, curiously enough, states only that the object is seen on the perpendicular and does not mention any intersection, says that the object will be seen at *E* because it will be along *AG* and along *BD*. The reason for its being seen along *AG* is "it is supposed

1. Pseudo-Euclid, 312–13.
2. Ptolemy, III, para. 3, p. 88.
3. Alhazen, V, paras. 2, 3, 4, 6, 9, 10, pp. 125–31.
4. In *De Iride*. See Turbayne, 471, footnote. Turbayne's conclusions are expressed tentatively, but this part, on reflection, is unambiguous.
5. Witelo, e.g. propositions 37, 38, pp. 176–82 (text), 300–07 (translation), 413–21 (commentary).
6. Risner, III, 8, p. 200.
7. Barrow's *Lectiones Opticae*, London, 1669. See Turbayne, 469–70.
8. Robert Smith's optics of 1738, para. 12. See Turbayne, 467.

J.D. North and J.J. Roche (eds.), The Light of Nature. ISBN 90-247-3165-8.

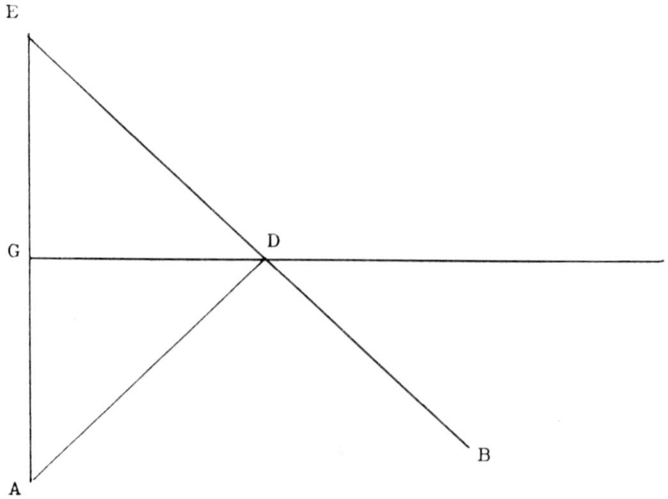

Figure 1.

in the phenomena that, the place *G* being occupied (καταληφθέντος), *A* is not seen''. The ''phenomenon'' referred to is the fourth postulate at the beginning of the book:[9]

> In plane mirrors, the place being occupied at which the perpendicular from the thing seen falls, the thing is seen is no longer seen.

It is scarcely surprising that, though the principle for finding the image was understood, the postulate and its application caused some perplexity. Theorems 17 and 18 are strictly analogous to theorem 16 and rely on postulate 5:[9]

> And in convex mirrors, the place being occupied through which [the straight line] from the thing seen to the centre of the sphere is drawn, the thing seen is no longer seen. The same also happens with concave [mirrors].

Various explanations have been given for these postulates. But since 1949, when Lejeune published his article demonstrating that the postulates as a whole

9. Literal translation of Pseudo-Euclid, 286.

include the equivalent of Ptolemy's laws of reflection, there has been little room for doubt that postulates 4 and 5 refer to the established procedure in Greek optics for finding the point of reflection by covering it with some substance (say, wax) and observing that the image disappears.[10] These postulates, then, locate the image on the perpendicular from the object to the mirror (or on the line joining the object and the centre of the spherical mirror) by the experimental fact, or phenomenon, that, when the eye is on this perpendicular, the image disappears when the foot of the perpendicular is covered.

To justify his conclusions, Lejeune points to Heron's catoptrics, known only in Latin translation as *De speculis*, theorem VI, and some theorems in Ptolemy's *Optics*, also known only in Latin translation – in all of which wax, or some other opaque object is used to find the point of reflection. For instance, Heron, *De speculis*, theorem VI reads:[11]

> In planis speculis est aliquis locus, quo apprehenso non amplius videtur idolum. . . . apprehenso ergo loco cera vel aliquo alio non amplius videtur *A*. si autem signum *D* excidat a speculo, apparebit idolum in speculo. . . .

Lejeune points out that *apprehenso loco* is probably equivalent to "τοῦ τόπου καταληφθέντος".

Two observations may be made. The first concerns the practicability of the postulate. Lejeune thought that postulates 3–5, though called "phenomena", were artifices made points of departure for logical or pedagogical reasons. In the fourth and fifth postulates, for instance, either the eye would get in the way of the ray from the object or the object would get in the way of the ray to the eye. But this does not square with experience. Postulate 4 may be confirmed by a very simple experiment with a pin and the bathroom looking-glass. The second observation is that the procedure to find the image implies that there is an image to find – one image, unique and unmoving, for all parts of the mirror. While this is acceptable – and correct – for plane mirrors, it is not so for spherical, let alone general, mirrors.

A COMMENTATOR

In one of the oldest manuscripts of the medieval Latin translation of the *Catoptrica*, unfortunately incomplete, there is a commentary worked into, and so

10. Lejeune, 608–10. One need not agree with all Lejeune's conclusions in detail.

11. Heron, 330. I have changed the letters (*D* to *A*, *H* to *D*) to make the text conform to Fig. 1. Heron's diagram does not show point *E*. For Ptolemy see II, para. 4, p. 89. There is an English translation in Cohen and Drabkin, 269.

contemporary with, the text, which is perhaps of the 12–13th century. The fourth postulate reads:[12]

> In planis speculis loco occupato in quem cathetus cadit a conspecto, iam non res conspecta videtur.

Over *occupato* there is the interlinear gloss "scilicet a radio". The commentator thus interprets the postulate as meaning that when the ray is perpendicular, the object is not seen. The commentary to the postulate confirms this:

> verbi gratia: res conspecta sit *B*, a qua *BO* cathetus cadit ad *O* terminum *OG* planitiei, et *D* sit oculus. Quandocumque radius cadit in punctum planitiei, ad quem punctum a conspecto (i.e. a *B*. que est res conspecta) non potest duci linea perpendicularis (sicut a *B* ad *Q* non potest

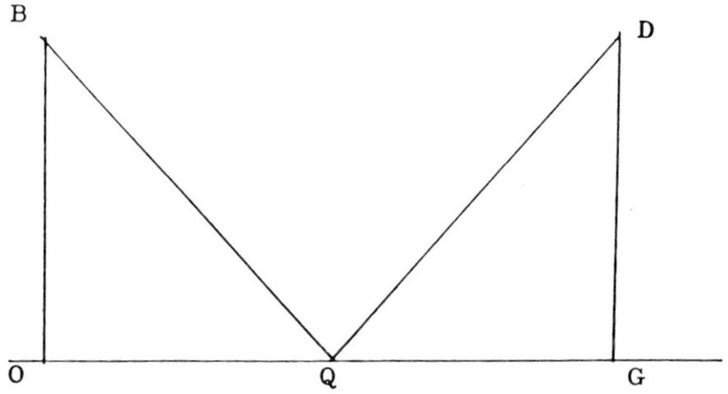

Figure 2

> duci linea perpendicularis) *B* videtur, quod est res conspecta. Si vero radius ad talem punctum planitiei cadit super quem linea ducta a *B* cadit perpendiculariter – ut si radius cadit ad *O*, ad quem ducitur *BO* perpendicularis – *B* prius conspectum non videtur.

> For example, let the thing seen be *B*, from which the perpendicular *BO* falls to *O*, the end of the plane *OG* [?], and let *D* be the eye. Whenever the ray falls on a point to which a perpendicular line from the thing seen (i.e. from *B*, which is the thing seen) cannot be drawn (as from *B* to *Q* a perpendicular line cannot be drawn), *B*, which is the thing seen, appears. But if the ray to such a point of the plane, on which the line drawn from *B* falls, falls perpendicularly – so if the ray falls to *O*, to which the perpendicular *BO* is drawn –, *B* seen before, does not appear.

12. MS British Library Add. 17368, fol. 60r. On the manuscripts of this work, see Lindberg, 47–50.

Perhaps the commentator thought, as Lejeune did, that seeing the image when the eye was on the perpendicular from object to mirror was a practical impossibility, but deduced from this that the image was on the same perpendicular.

JOHANNES PENA (?1528–1558)

Johannes Pena (Jean de la Pène), pupil and collaborator of Petrus Ramus (Pierre de la Ramée),[13] published the *editio princeps* of Euclid's *Optica* and the pseudo-Euclidean *Catoptrica* in Paris in 1557, together with a Latin translation. In the preface to the *Optica, De usu opticae*, he made a number of assertions about binocular vision, the motion of the Earth, the nature of comets, and the existence of the celestial spheres.[14] These ideas became better known after Kepler had subjected them to a detailed criticism in the preface to his *Dioptrice* (1611).[15]

Pena's edition and translation of the *Optica* and *Catoptrica* were influential and went into several editions,[16] and in particular were taken over by Gregory in his edition (1703) of Euclid's works with Latin translation. Gregory says in the preface that he emended the text as far as he could:

> Versione latina Joannis Penae sumus usi, Graecumque textum quam potuimus castigatum fecimus:

In fact he also altered the Latin, often by removing unnecessary phrases included by Pena.

Pena's translation was by no means the first. Apart from the medieval translations there was another made by Bartholomeo Zamberti, printed in 1505 and often afterwards, another by Conrad Dasypodius (1529–1600, originally Rauchfuss), this only of the *Catoptrica*, printed in the same year as Pena's, and yet another by Maurolico (1494–1579), never printed.[17] Pena says in the dedication of his translation that the old translation was perverse and quotes Maurolico's support to say that though the translator knew Greek, he made many mistakes; some of the proofs did not tally with the theorems, some of the theorems and many of the scholia were left out, and so on. The "old translation" is probably the much-printed Zamberti translation. Ironically, Pena's work is

13. Hooykaas, 45–46; Verdonk, 59–66. I am grateful to Prof. M. Folkerts for translating the latter.

14. Pena; Frisch III, 573–74; Thorndike, V, 304–6, VI, 19–20, 71–72; Hooykaas, 47–49. Pena's preface was reprinted with Risner's optics.

15. Kepler, IV, 334–41.

16. Verdonk, 65, 431–32; Ver Eecke, xxxviii–xli, xliv–xlvi.

17. Ver Eecke, xxxvii–xxxix.

often compared to its disadvantage with Zamberti's in Jean Buteo's diatribe on errors in the translators of Euclid,

> *Io. Buteonis annotationum liber in errores Campani, Zamberti, Orontii, Peletarii, Io Penae interpretum Euclidis.*

Buteo was a monk[18] who wrote on such matters as Noah's ark, duplicating the cube, the value of pearls, and various topics in civil law.[19] Among the last is a tract *Geometriae cognitionem Iureconsulto necessariam.*

In Buteo's criticism of Pena's translation of the *Catoptrica*, which he treats before the translation of the *Optica*, he begins with Pena's rendering (*radium*) of the very first word in the treatise, *ὄψιν*, which he finds inferior to *visus* or *visio*. He then turns to Pena's translation of the first postulate,

> Ponamus radium esse rectam lineam, cuius media omnia extremis officiunt [Pena had *officiant*].

Nothing could be more absurd than *officiunt* here; even Zamberti's *correspondent* is better; and still better would have been *obtendentur*. So Buteo proceeds through the postulates, his energy flagging only with the theorems. Now Pena had translated the fourth postulate as:

> In planis speculis oculo posito in eo speculi loco, in quem cadit perpendicularis ducta a re aspectabili ad speculum, rem aspectabilem non cerni.

thus putting the eye on the mirror at the foot of the perpendicular from the object. He translated the reference to this postulate in theorem 16 by

> Quia igitur in primo Phaenomeno positum et concessum est punctum α cerni non posse ab oculo posito in puncto γ, igitur punctum α videbitur in aliquo puncto lineae $\alpha\gamma$ productae in continuum et rectum.

The fifth postulate and theorems 17 and 18 are similarly treated. In all three theorems the reasoning is obscure and the translation inexact, so we should expect some vituperation from Buteo. We are not disappointed. After quoting the above rendering of the fourth postulate he says:[20]

18. See Joecher, I, 1278.
19. *Ioan. Buteonis Delphinatis Opera Geometrica*, Lyons, 1559. Copy used, MStB Math. P. 58m.
20. Buteo, 272–73.

> The translator is clearly proved not to have understood the whole *phenomenon*, two others, and everything that is demonstrated from them in the following [theorems] – he renders τόπου καταληφθέντος, i.e. *occupato loco*, by such an absurd paraphrase as *oculo posito in eo speculi loco* and then by *oculo occupante locum*. Who has ever seen an eye placed on a mirror? . . . Zamberti, not at all far from the truth, said *loco assumpto*. Such errors proceed from no further off than ignorance of the matter.

Not surprisingly, he returns to the attack[21] when considering theorem 16.

Buteo could have saved his ammunition if he had noticed the verso of the title-page of Pena's translation. Under the title *Lectori* there is, after an explanation of the use of asterisks, a series of *corrigenda*:

> Porro errata emendentur hoc modo, . . .
>
> Et in Catoptricis primo phaenomeno[22] lege: item in planis speculis occupato eo speculi loco in quem cadit etc. . . .
> Pag. 55 lin. 13 *α* cerni non posse occupato loco in quo est *γ*. . . .

Thus all the passages putting the eye on the mirror are removed. Gregory evidently saw these *corrigenda*, for he changed the text accordingly – but not in other passages found objectionable by Buteo. As an example of Gregory's trimming of Pena's text we may compare his rendering of the part of theorem 16 given above in Pena's original:

> Quia igitur in phaenomenis positum est, punctum *A* cerni non posse occupato loco in quo est *Γ*. igitur punctum *A* videbitur in aliquo puncto lineae *AΓ* productae in directum.

It is worth noting that postulates 4 and 5 form one of the reasons why Gregory considered the *Catoptrica* spurious:[23]

> If the place on the mirror (plane, convex or concave) is occupied in which the perpendicular from the object falls onto the mirror, the object is still seen. I wonder that this could be proposed as a phenomenon by any philosopher, let alone Euclid, when daily experience with plane mirrors teaches the contrary.

Other reasons for doubting the authenticity of the text are that the last *phaenomenon*, on refraction, is inappropriate in a work on catoptrics; that theorem 25 is false; that the author sometimes speaks in the first person, sometimes in the second, and sometimes in the third; and so on.

21. Buteo, 274–75.
22. I.e. the fourth postulate. Various authors gave various names to the postulates, but this is not discussed here.
23. Gregory, penultimate page of preface (unpaginated).

JOHANNES KEPLER (1571–1630)

Kepler begins cap. 3 of his *Paralipomena in Vitellionem* with a section entitled *Refutatio Euclidis, Vitellionis et Alhazeni.*[24] In theorems 16, 17 and 18, he says, Euclid made a false assumption (*assumit falsum*). His diagram is Fig. 1, only upside down and with *C* for *G*.

> That the place of the image of the thing *A* is on perpendicular *AC* he proves thus. *Nam*, he says, *assumpto loco speculi* C *in quem* AC *perpendicularis incidit, visibile* A *non amplius videtur*. If you are to understand *assumpto*, i.e. occupied or hidden *[Si intelligas, assumpto, id est occupato, aut tecto] loco* C. the axiom is false, though Euclid puts this among his postulates borrowed from experience (or, otherwise, what he in this book calls by the name of *phenomena*) at the beginning of the book. For, even if *C* is hidden or completely removed, so long as *D* remains, *A* will notwithstanding be seen at *E* by the eye *B*.

Now although Kepler gives the word *phenomena* (and other words later) in Greek, it is unlikely that he had seen the Greek text. For, if he had, he would surely have given the Greek for the critical word *assumpto*. It is therefore improbable that he had Pena's edition of 1557 to hand. Further, the translation he quotes is a modification of that of Zamberti, who translates the fourth postulate and its citation in theorem 16, in the Munich manuscript, as:[25]

> In speculis planis assumpto loco in quem ab inspecto perpendicularis cadit, inspectum non amplius videtur.
> . . . assumpto loco *C* ipsam *A* non videtur.

The editions printed at Basel in 1537 and 1546 agree in the citation and for the postulate have:

> In planis nanque speculis loco assumpto in quem ab inspecto perpendicularis cadit, non amplius spectatur visibile.

The important *assumpto* is always present, and this is not so for the other translations, including Pena's. The later reference to Pena in cap. 4, section 7, proposition xi[26] is avowedly taken from Tycho Brahe. We may assume that Kepler first saw Pena's translation between 1603, when he wrote the *Paralipomena*, and 1611, when he publishd the *Dioptrice*.

Kepler suggests[27] that the axiom would be true if the eye were at *A*, but that it would then not prove the conclusion, except for perpendicular vision. Al-

24. Kepler, II, 61–64.
25. Zamberti, CLM 6, 439r, 443v.
26. Kepler, II, 120–21.
27. Kepler, II, 62.

hazen and Witelo seem to have sensed this and left the axiom out as absurd.

It would be inappropriate here to go into Kepler's detailed treatment of the image in plane, spherical, and more exotic, mirrors. Perhaps we may leave the subject with two quotations, one from Kepler himself,[28] having shown that for a spherical mirror the image is *not* on the line joining object and centre.

> Ex his igitur apparet, non esse universaliter verum, locum imaginis esse in perpendiculari, nisi et haec limitatio accedat, ut visus respectu speculi collocetur ita, ut natura monstrat.

The other quotation is from Johann Brengger,[29] a physician from Kaufbeuren who corresponded with Kepler in 1604–05 on the subject:[30]

> Euclidem enim tam turpiter lapsum non credo.

BIBLIOGRAPHY

MStB = Munich, Staatsbibliothek; MDM = Munich, Deutsches Museum.

Alhazen, *Opticae Thesaurus. Alhazeni Arabis libri septem, nunc primum edit . . . a Federico Risnero*, Basel, 1572, MDM 1927 B4.

Buteo, *Io. Buteonis annotationum liber in errores Campani, Zamberti, Orontii, Peletarii, Io. Penae interpretum Euclidis*, attached to his *De quadratura circuli libri duo*, Venice, 1559, as pp. 207–283. MStB Math. P. 63.

Caspar, Max, *Kepler, 1571–1630*, English translation by C. Doris Hellman, New York, 1959.

Cohen, M.R. and Drabkin, I.E., *A Source Book in Greek Science*, Cambridge, Mass., 1958.

Frisch, C.H. (ed.), *Ioannis Kepleri Astronomi Opera Omnia*, Frankfurt am Main, 1858–71.

Gregory, *Euclidis quae supersunt omnia ex recensione Davidis Gregory*, Oxford, 1703.

Heron, *Herons von Alexandria Mechanik und Katoptrik*, herausg. u. übersetzt von L. Nix und W. Schmidt, Leipzig, 1900. Vol. II, Fas. I of the *Opera Omnia*.

Hooykaas, R., *Humanisme, science et réforme. Pierre de la Ramée (1515–1572)*, Leiden, 1958.

Jöcher, C.G., *Allgemeines Gelehrten-Lexicon*, Leipzig, 1750.

Kepler, Johannes, *Gesammelte Werke*, herausg. von Max Caspar [u.a.], Munich, 1938–

Lejeune, Albert, "Les 'Postulats' de la Catoptrique dite d'Euclide", *Archives Internationales d'Histoire des Sciences*, 1949, 7, pp. 598–613.

Lindberg, David C., *A Catalogue of Medieval and Renaissance Optical Manuscripts*, in *Subsidia Medievalia*, IV, Toronto, 1975.

Pena, *Euclidis Optica et Catoptrica e Graeco versa per Ioannem Penam Regium Mathematicum*, Paris, 1557. MDM 1927 A 183.

Pseudo-Euclid, *The* Catoptrica *in* Euclidis Optica, Opticorum recensio Theonis, Catoptrica, cum scholiis antiquis, ed. I.L. Heiberg, Leipzig, 1895. Vol. VII of *Opera Omnia*.

Ptolemy, *L'optique de Claude Ptolemée dans la version latine d'après l'arabe de l'émir Eugène de Sicile*, édition critique et exegétique par Albert Lejeune, Louvain, 1956.

28. Kepler, II, 75.
29. See Caspar, 177–78.
30. Kepler, XV, 83.

Risner, *Opticae libri quatuor ex voto Petri Rami novissimo per Fredericum Risnerum eiusdem in Mathematicis adiutorem*, Kassel, 1606. MStB 4° Math. A 245.

Thorndike, Lynn, *A History of Magic and Experimental Science*, New York, 1923–56.

Turbayne, Colin M., "Grosseteste and an Ancient Optical Principle", *Isis*, 1959, 50, pp. 467–72.

Ver Eecke, *Euclide, l'Optique et la Catoptrique*, oeuvres traduites . . . avec une introduction et des notes par Paul Ver Eecke, Paris, 1959.

Verdonk, J.J., *Petrus Ramus en de Wiskunde*, Assen, 1966.

Witelo: Smith, Albert Mark, "Witelo on the Principles of Reflection: a Critical Edition and English Translation, with Notes and Commentary", unpublished Ph.D. thesis, Wisconsin, 1956.

Zamberti, MStB MS CLM 6, 16c.

Zamberti, *Euclidis Megarensis . . . Elementorum Geometricorum . . . cum expositione Theonis in priores XIII a Bartholomeo Veneto latinitate donata, Campani in omnes, et Hypsiclis Alexandrini in duos postremos . . . Phaenomena, Catoptrica et Optica* . . . , Basel, 1537. MStB 2° A. gr. b. 552. Ditto, 1546, 2° A. gr. 553.

XI

Al-Khāzinī's "Sphere That Rotates by Itself"

Acknowledgments

I am most grateful to the Alexander von Humboldt-Stiftung for the fellowship that enabled me to carry out the research at the Institut für Semitistik der Universität München. I should also like to thank Professor Dr. Paul Kunitzsch for checking through the entire text and translation, for considerable help with its dedicatory preface, and otherwise for many helpful suggestions.

Summary

Al-Khāzinī wrote his description of the "sphere that rotates by itself" before his *Zīj* and the *Mīzān al-ḥikma*, and thus in the first decade or so of the twelfth century. The sphere, which is inscribed with the normal celestial circles, is a combination of an automatic demonstration-instrument and a *dhāt al-kursī*, which functions rather like an astrolabe, but in three dimensions. The drive, a weight resting on top of a leaking reservoir of sand, is traced through Muḥammad ibn Yūsuf al-Khwārizmī (10th century) to Heron of Alexandria. After comparison with earlier self-rotating spheres in China, an origin for the device in Hellenistic antiquity is suggested – a possibility corroborated by passages in Pappus (c. 300) and Proclus (5th century). Other descriptions of the celestial globe are mentioned to illustrate the device, parts of al-Marrākushī's (13th century) being given *in extenso*. After the text and English translation, there is a commentary, which *inter alia* treats the units and dimensions.

I INTRODUCTION

1. *The Author*

Al-Khāzinī's "sphere that rotates by itself" is a solid sphere – that is, a sphere marked with the stars and the usual celestial circles – half sunk in a box and propelled by a weight falling in a leaking reservoir of sand. In addition, the sphere is used to find directly several arcs of importance in spherical astronomy. Although the short treatise that describes the device is ascribed to "al-Khāzimī" in the text, it is clearly by ᶜAbd al-Raḥmān al-Khāzinī,[1] the author

1. For the variations in the name of al-Khāzinī and his patron, see Hall, *DSB*, at the beginning. Here, and in the notes which follow, references by short title are to the bibliography at the end of the paper.

of the *K. Mīzān al-ḥikma* (Balance of Wisdom). For not only is ᶜAbd al-Raḥmān's patron Abū'l-Husayn ᶜAlī ibn Muḥammad mentioned with fulsome flattery at the beginning (with the additional name "ibn ᶜĪsā"), but there is a striking parallel between passages in the present text and the *K. Mizan al-hikma.* At the same time the authorship of the *Mīzān* text is confirmed.[2] The details of the parallel passage and of other matters mentioned in this introduction are to be found in the commentary to the text.

It is likely that the present text (hereafter called the "sphere" text for short) was written before both *K. Mīzān al-ḥikma,* written in 1121-2,[3] and *al-Zīj al-Sanjarī,* for these are dedicated not to ᶜAlī b. Muḥammad but to Sanjar b. Malikshāh, Saljuq ruler of Eastern Persia 1097-1157. He was appointed governor of Khurāsān from 1097 at the age of ten and became supreme Sultan of the Saljuq family from 1118.[4] No argument can be attached to his minority or to his being only a governor for the first twenty odd years, for there is a coin of his from Marw "probably minted in 499/1105-6"[5] But it is likely that al-Khāzinī was directly employed by Sanjar later in life and that the "sphere" treatise, dedicated to the patron who had owned him as a slave and had given him a first-class education, was a product of earlier years. ᶜAlī b. Zaid al-Bayhaqī (d. 1174), to whom we are indebted for the few facts of al-Khāzinī's life that we have – including some entertaining details of his asceticism – mentions only the *Mīzān* treatise and the *zīj* (here called *al-Muᶜtabar al-Sanjarī,* "the estimable [book] of Sanjar").[6] For this passage more credence should be accorded to al-Bayhaqī than usual, since he was close enough to al-Khāzinī's circle to have a horoscope he had assembled presented to him for criticism. But the biographical notice is so sketchy that we cannot be sure that the author had included everything written to date.

A further reason for placing the "sphere" text early is his mentioning in the dedicatory preface the great disparity amongst the *zīj*es, their errors and bad method – something he would surely have phrased differently if he had written one himself. In fact the rotating sphere is presented as a preliminary to a series of observations to correct the old *zīj*es. The "sphere" text may thus be confidently placed prior to *al-Zīj al-Sanjarī.* True, al-Khāzinī may have calculated the *qibla*-table[7] before writing section 15 of the second part, but this table probably does not depend on new observations and anyway does not occur in the *zīj.*[8]

The star table in the *zīj* is calculated from Ptolemy's and corresponds to

2. It was doubted by Khanikoff, pp. 113-6, but, it is true, apparently by no one since.
3. Khanikoff, p. 16.
4. Bosworth, *Isl. Dyn.*, p. 115 and *CHI* V, p. 135.
5. *Ibid.*, p. 135.
6. Meyerhof, pp. 196-7.
7. Le Strange, pp. 27-31. The table is on pp. 30-1. See Lorch, *Qibla.*
8. Le Strange, p. 27 n.

the year 1115.[9] If this date is accepted for the whole *zīj*, the sequence of al-Khāzinī's writings is: the "sphere" text, the *zīj* (1115), the *Mīzān* (1121-2). The short treatise on observational instruments[10] would presumably come somewhere near 1115.

2. *The Mechanism*

i. The Sand-drive

As may be seen from the diagrams (figs. 4, 5) and the facsimile from the Damascus manuscript, the slowly falling weight pulls the string taut; the string, after passing over pulleys, turns an axle by being wrapped round a wheel or drum; this axle turns another by means of a pair of toothed wheels, and the second axle turns the sphere, which is half sunk in the top of a box representing the plane of the horizon.

A falling weight in a reservoir of finely divided solid matter was a source of power used by Heron of Alexandria for the moving figures in his automatic theatre.[11] In the reservoir he put millet (*kegkhros*) or mustard-seed (*napu*), because they are "light and slippery".[12] The string was led round pulleys and wound round a drum mounted on an axle that caused the figure to move. Apart from the substance in the reservoir, the mechanism is very similar to that of al-Khāzinī over a millenium later. Heron's reservoir even had a square (*tetragōnou*) cross-section.[13]

This type of mechanism was certainly known in the Arabic-speaking world a century or so before al-Khāzinī. For Abū ᶜAbdallah Muḥammad b. Yūsuf al-Khwārizmī, who flourished at the end of the tenth century, devotes a paragraph of his encyclopaedic dictionary *Mafātīḥ al-ᶜulūm*[14] to it. In the section on mechanics, the eighth *bāb* of the second *maqāla*, we find as the heading of the first *faṣl*[15]

9. Destombes, *Etoiles*, p. 343. Kunitzsch, *Chrysokokkes*, contains an edition of al-Khāzinī's star-catalogue. The date 1115 comes directly from the rubric of the catalogue, which adds 15° longitude to Ptolemy's.

10. Sayılı.

11. See diagram in Needham et al., *CCCW*, 207. I am grateful to Cdr. Waters of the National Maritime Museum, Greenwich, for pointing out this illustration, which is taken from Heron, *Automatentheater* (see next note).

12. Heron, *Op. Om.* I, p. 346 (German translation p. 347). Here and elsewhere Greek words are transcribed letter by letter. A similar passage occurs on p. 368 (German translation pp. 367–9). Other passages (pp. 396 and 402; German pp. 397 and 401) refer to millet alone, but no doubt any appropriate substance was intended.

13. *Ibid.*, p. 356. For Heron's automaton–theatre, see Beck.

14. The section on mechanics is translated by Wiedemann, *Aufsätze* I, pp. 173–228. For other parts translated by him, see Sezgin VI, pp. 239–40. The passages quoted have been translated afresh.

15. Al-Khwārizmī, p. 249; Wiedemann, *Op. Cit.*, p. 200. The extended quotation is from pp. 250 and 202 respectively.

> On the mechanics of the movement of water, making wonderful vessels., and related matters on making instruments that move by themselves [بذاتها].

The second section is

> As for the movements that arise from [means] other than water, from those that work by sand and those that work with mustard-seeds [خردل] or millet [جاورش]: an instrument is made in the elongated form of the tube [على هيئة البربخ طويلة]. Its lower part is pierced with a small hole, and its top is open. Then it is filled with sand or mustard-seed, or something similar. A piece of lead [رصاص] is placed on top of it. The lead draws a thread [خيط] or cord [حبل] tight; to the thread is attached what is necessary for the motion. Then the tube is placed in a vertical position, so that the sand or other [material] can come out of the hole at the bottom of it. As the sand gradually diminishes, the weight is moved downwards and it moves what is connected to it. In this way, wonderful motions of various types are set up.

It will be noticed that there are some differences of terminology between al-Khwārizmī and al-Khāzinī. Where al-Khwārizmī has رصاص and البربخ for "lead" and "tube" (reservoir) respectively, al-Khāzinī has أسرب (on the diagram) and خزانة. Wiedemann's interesting observation[16] that the tubes are shorter in Heron writings than amongst the Arabs certainly finds confirmation in the "sphere" text, for al-Khāzinī's reservoir is over six foot high, while its cross-section is a square of just over 6½ inches long. Either there was a large amount of empty space in the box (the biggest wheel inside was about ten inches in diameter) or the reservoir stood up like a factory chimney.

Apart from the possibility that water will freeze in a cold climate, two advantages of sand should be noted: it flows more or less uniformly without a constant head.[17] and a heavier weight can be used. Of course, sand is always prone to become clogged,[18] especially with slow rates of flow.

ii. Related Devices in Antiquity

Devices to mimic the heavens were known in antiquity. In the eighth book of his *Mathematical Collection*,[19] Pappus (fourth century A.D.) lists those who may be called mechanics (*mechanikoi*). In the list the makers of engines for war, self-moving devices like Heron's automata, and water-clocks are mentioned. Finally

> They [the ancients] also called mechanics those who were skilled in making globes [*sphairopoiias*] and constructed a likeness of the heavens by means of uniform circular motion of water.

It is probable that the globes described here are of the same basic type that

16. *Ibid.*, p. 202.

17. Needham, *SCC* IV ii, p. 509 note. For a discussion of how sand flows in a sand-clock of the hour-glass type, see Balmer pp. 624–32.

18. Needham, *SCC* IV ii, p. 510.

19. Pappus, *Collection* III, p. 10.

al-Khāzinī made, albeit driven by water. Hultsch[20] uses this passage and a reference by Proclus (410-84) to

> the globe [made] in imitation of the heavens, such as Archimedes concerned himself with

to argue that Archimedes' complicated device was water-driven. His conclusion is possible – though the latest discussion of Archimedes' globe[21] does not represent it as water-powered – and anyway does not exclude the possibility that the passage from the *Collection* refers also to simple globes. Anaphoric clocks, well known in antiquity, are probably not meant by Pappus, since in these clocks a disc, and not a sphere, was rotated. One end of a cord was attached to a float in a water-reservoir that was filled every 24 hours, and at the other end was a counterweight to provide power and to keep the cord tight; and in between the cord was wrapped round the axle of the anaphoric disc representing the heavens.[22]

A possible source for al-Khāzinī's globe is the lost book on waterclocks by Heron.[23] It was at least partly concerned with clocks designed for astronomical purposes.[24] Less likely is an Arabic treatise on "Wheels that rotate by themselves [من ذاتها]" ascribed to Philon of Byzantium, in which eight models are described,[25] since Carra de Vaux says it is on perpetual motion.[26]

iii. *Related devices in China*

In their various publications Prof. Needham and his associates have described for us a number of self-moving globes constructed in China. The early examples (up to 1100) are the following:

> 1. By *Chang Hêng* (AD. 78–142), about 132 A.D.; descriptions soon after 310, quoted in 635, and 656. A globe (armillary sphere?) was turned by a clepsydra, or "dripping water". The sphere was inscribed with various circles and turned "following the trip-lug [?] and the turning of the auspicious wheel".[27]
>
> 2. *Chhien Lo-Chih*, 436 A.D.; descriptions 500, 656 A.D. An armillary sphere, of diameter about six feet, with the Earth in the middle was rotated by a clepsydra. It was made after the recovery of the remains of Chang Hêng's instruments.[28]

20. Hultsch, *Himmelsglobus*. The Proclus quotation comes from his commentary on the first book of the *Elements*, Friedlein's edition, p. 41.
21. Zhitomirskii, p. 297, fig. 6. This is not the clock with jackwork ascribed to Archimedes in an Arabic treatise, for which see Jazari, p. 10.
22. See e.g. Drachmann, *Mech. Tech.*, p. 193.
23. This book is mentioned by Pappus (see Rome 87–9) and Proclus in *Hypotyposis*, pp. 120–3.
24. Drachmann, *Ktesibios, Philon and Heron*, pp. 98–9, sees no reason to suppose that the *application* to finding the apparent diameter of the Sun was treated by Heron.
25. Krause 443. MSS Aya Sofia 2755. 2°, ff. 61v–69v; Seray 3466, 2°, f. 7 ff.
26. Philon, *Appareils pneum.* pp. 5, 6. The text is also in Bodleian MS Marsh 669 (*Ibid.* 5).
27. Needham, *SCC* III, p. 359; IV ii, p. 485, *HC*, p. 100 et seq.
28. Needham, *SCC* IV ii, p. 483; *HC* pp. 95–9.

2′. The same, 440 A.D. A globe of diameter 2.2 feet, on which the Sun, Moon, and five planets were attached to the ecliptic.[28]

3. *Thao Hung-Ching* (452-536), about 520 A.D. description c. 670, copied 983. An armillary sphere over three foot high with the Earth in the middle, was mechanically rotated.[29]

4. *Kêng Hsün*, late sixth century; descriptions 636, and about 670 AD. His armillary sphere (or globe) was rotated "through the power of (falling) water".[30]

5. *I-Hsing* and *Liang Ling-Tsan*, eighth century; descriptions 945, 1061, and 1267 A.D. I-Hsing introduced here the "Ptolemaic ecliptically mounted sight-tube". The whole device was "made to turn automatically by the force of water acting on a wheel". Later descriptions say scoops are used. The globe was half sunk in wooden casing and rotated by a complicated mechanism, possibly with an escapement.[31]

6. *Su Sung*, end of eleventh century; description by Su Sung. Both a globe, half sunk into the casing, and an armillary sphere are rotated by a very elaborate mechanism powered by a water-wheel controlled by an escapement. The armillary sphere alone weighed fifteen tons. The whole clock was between 30 and 40 foot high.[32]

No more need be said here about the globes of I-Hsing and the famous Su Sung, since they are evidently more sophisticated than al-Khāzinī's primitive device. But the four earlier self-rotating spheres sound remarkable similar, except that they are water-powered. In some of them an armillary sphere is substituted for the globe, but the principle remains.

Needham maintains that the power of a sinking float would be inadequate to drive a sphere of any size and therefore supposes that such spheres were powered by scoop water-wheels.[33] This is gratuitous, because a sinking float can provide as great a force as a scoop of water of the same weight and can, besides, be made very large. The torque given to the axle depends on the size of the barrel round which the cord is wrapped. It is clear from the fragment of a Roman anaphoric clock found near Salzburg in 1897 that heavy objects could be rotated by a sinking (or rising) float. This fragment, rather less than a quarter of the entire disc, weighs 5.5 kg[34] – so that the disc weighed a few pounds short of half a hundredweight – and we have no reason to suppose that the float was an exceptionally large one (so that it could support an exceptionally large counter-weight). Of course, a solid globe (no mention seems to have been made of hollow globes) would be much heavier than an anaphoric dial of comparable

29. Needham, *SCC* IV ii, p. 482.
30. *Ibid.*, p. 482.
31. *Ibid.*, p. 471–4.
32. Needham, *SCC* IV ii, p. 498; *CCCW* p. 312; *HC passim* esp. pp. 48–59.
33. Needham, *SCC* IV ii, p. 481.
34. Ungerer, pp. 32–4.

size, but several reports concern armillary spheres, which are comparatively light. We may conclude that the early Chinese self-rotating spheres were rotated either by a water wheel, like Su Sung's or by a sinking float, like the Hellenistic and Byzantine clocks and (with the substitution of sand) al-Khāzinī's sphere.

Clocks worked by sand seem to have been of late introduction in China. Ricci, the sixteenth-century Jesuit missionary, says that the Chinese had a few sand-clocks.[35] The earliest seems to be of the fourteenth century.[36]

iv. The Question of Transmission

Clocks exhibiting jackwork and astronomical information were well established in the Islamic and Byzantine cultures and are clearly derived from the early Hellenistic period. Vitruvius attributes a clock with jackwork to Ctesibius. who lived in the early third century B.C.[37] Needham explains their later appearance in China as probably due to "stimulus diffusion".[38] That is, the Chinese craftsmen worked independently from a few ideas transmitted to them, and there was no wholesale transmission of Hellenistic mechanics to China. The reason given is that the Hellenistic and Byzantine clocks used only the sinking float principle, whereas the Chinese used (the more powerful) water wheels. Some doubt has already been cast on the latter assumption. Further, water-wheels of various kinds do occur in the Hellenistic-Arabic tradition. There is a water-wheel in the last last chapter of the Arabic version of Philon's *Pneumatics*,[39] though not of exactly the same type. According to Prof. Drachmann[40] the chapter is "a later interpolation of uncertain date", but even if this is so – and certainly the text is in pretty bad shape and the oldest manuscript is probably the thirteenth century[41] – there is no reason to suppose it came from China. Vitruvius describes a similar device. Drachmann considers the apparatus in the Philo text as "an ancient instrument, possibly partly improved by a later hand".[42] The water wheels in al-Jazarī are well known,[43] if late (early thirteenth century). Finally, Dr. D.R. Hill, the translator of several Arabic books on machines, considers that, "The use of both types [horizontal and vertical] of water wheel was widespread in Islam from the 3rd 9th century onwards".[43a] While independent invention or transmission from China to the West cannot be

35. "Pauca enim alia conficiunt rotulis ab arena velut aqua circumactis...". *De Chr. Exp.*, p. 22.
36. Needham, *SCC* IV ii, p. 510.
37. Drachmann, "Ctesibius", *DSB*.
38. Needham, *CCCW*, p. 228. See also table 59 in *SCC* IV ii, p. 553.
39. Philon, *Appareils pneum.*, pp. 94–6. There is a picture, a modern reconstruction, on p. 186.
40. Drachmann, review of *HC*, p. 202.
41. Krause, p. 443. See the introductory chapters to Prager, *Pneum.*
42. Drachmann, *Ktesibius, Philon and Heron*, p. 66.
43. E.g. Jazarī p. 103.
43a. Hill, *Ibn Muᶜādh*, p. 39.

ruled out, it is likely that one or more types of water-wheel used in clocks were transmitted to both the Chinese and Islamic cultures from Hellenistic antiquity, along with striking mechanisms, jackwork and mechanical toys. From the previous section it appears that the same could be said of self-rotating globes; and I have elsewhere[44] argued a similar case for the steelyard clepsydra, or balance-water-clock.

Certainly none of these devices were beyond the ingenuity of the people who constructed the Antikythera mechanism or Archimedes' sphere showing the motions of the Sun and Moon.[45] Whether the Alexandrians invented these things themselves is another matter. The means of transmission to China is a matter for speculation, though it might be pointed out that several waves of western influence in astronomy reached India in antiquity[46] and a self-rotating globe is reported there of about the year 500 A.D.[47] Again, there are instances recorded of clocks being sent by Muslim rulers to western kings as presents: Harūn al-Rashīd's present to Charlemagne and Saladin's to Frederick II of Sicily.[48] Direct borrowing is possible.

As Needham points out,[49] "It almost seems as if some mad dictator had contrived to expunge all details of Hellenistic astronomical models from the records". The self-rotating sphere of al-Khāzinī makes a small contribution to recapturing these details, especially when compared with the Chinese constructions. Perhaps we may hope to recover, some time, Heron's lost book on water-clocks.

3. *The Solid Sphere*

So far, the "sphere that rotates by itself" can be seen as a demonstration instrument, or possibly a clock. But the sphere also serves as an instrument known in the Islamic world as ذات الكرسي and in the Latin Middle Ages as *sphera solida*.[50] This instrument is to be distinguished from the spherical astrolabe, in which the horizontal coordinates are marked on a sphere and a close-fitting rete, pivoted at the poles, carries the celestial circles and representations of a few stars. Different again are the armillary sphere, a demonstration instrument showing only the principle celestial circles, and the ذات الحلق (*habens armillas*), described by Ptolemy in *Almagest* V, 1 for direct observations, and its modifications.

44. Lorch, *Balance-Clock*.

45. See note 21. For translations of the classical citations of Archimedes' model, see Price, *Origin of Clockwork*, pp. 89-90.

46. Pingree, *DSB* XIV, p. 533.

47. Needham, *SCC* IV ii, p. 539.

48. Beckmann, pp. 343, 349.

49. Needham, *HC*, p. 185.

50. Bibliographical details of solid spheres etc. in Lorch, *Sphera Solida*.

Qusṭā b. Lūqā's ذات الكرسي consists of two fixed rings, representing the horizon and meridian and attached to a stand, and a solid sphere, marked with the celestial equator, ecliptic, etc., and pivoted to the meridian ring at points corresponding to the North and South poles. All the circles are graduated. Most of the problems for which the plane astrolabe is normally used may be solved with this instrument. In al-Battānī's بيضة (egg) we have the same instrument but with an extra colural armilla carrying movable sights and another meridional armilla to allow for continuous variation in geographical latitude.

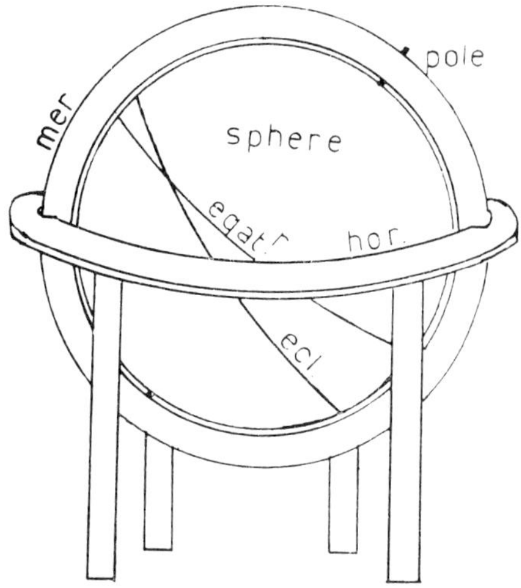

Fig. 1
Qusṭā b. Lūqā's *dhāt al-kursī*

Al-Khāzinī's solid sphere is similar to Qusṭā b. Lūqā's, but is half sunk in the top of a box representing the horizontal plane – in this detail resembling the Chinese examples. In addition to the graduated horizon and equator, on box and sphere respectively, there is a loose quadrant, also graduated. Details of its use are given in the commentary to the text. An apparently similar quadrant is mentioned in a short appendix to the Spanish translation of Qusta b. Luqa's treatise on the solid sphere. This appendix, included to make the work on the sphere more complete, is entitled,[51] "On knowing how the rings [circles?] of the *tasyir* are made on the sphere and [how] to 'equalize the houses' according to the opinion of Hermes, and how to operate with them":

> De saber cuemo se fazen las armillas dell atacyr en la espera. et egualar las casas segund la opinion de Hermes. et cuemo obren con ellas.

In the first paragraph one end of the quadrant is described as being fixed to the pole of the ecliptic with a pin (*clavo delgado*).

The solid sphere described by al-Marrākushī (13th century) has a loose quadrant. According to the description in his جامع المبادئ و الغايات[52] the sphere is made, and then its diameter is found by a geometrical construction. With this diameter the horizon and meridian are made, together with a three-legged support. Holes are made in the meridian at intervals of a degree and a device (ماسكة) is provided, pivoted to the horizon, to engage any one of these holes, so

51. *Libros del Saber*, I, p. 206. The chapter continues to p. 209. For Hermes, see Sezgin IV, pp. 41–3. For *tasyir*, see Kunitzsch, *Glossare*, pp. 44–5.

52. MS Paris BN 2508 (old no. 1148) ff. 13r-16r (old pagination, pp. 25–32). See Sédillot, *Mémoire* pp. 110.4.

that the sphere is adjustable for geographical latitude. A graduated quadrant of the same diameter is made;[53]

ثم نعمل ربع دائرة من نحاس مساوياً لربع حلقة
نصف النهار وليكن عليه ا ب ج د ونقسم قوس ج د منه
بتسعين قسماً مساوية ونكتب عدد هذه الأجزاء على ما تراه
في الصورة ونجعل لهذا الربع موضعاً بكفة[54] اما في حلقة
نصف النهار واما في حلقة الأفق

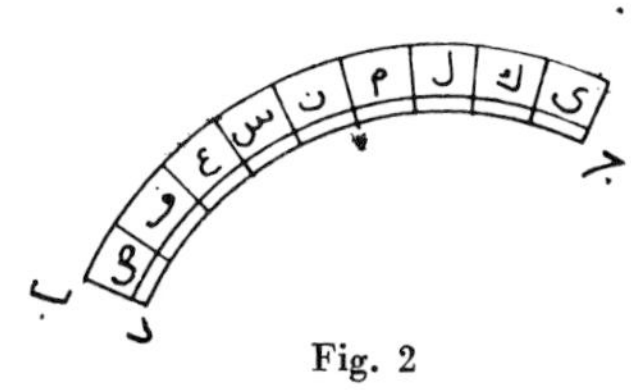

Fig. 2

Then we make a quadrant from brass equal to a quarter of the meridian ring. Let there be [p(ints] *ABGD* on it. We divide its arc *GD* into ninety equal divisions and write the number[s] of the degrees, as you see in the diagram. We make a place for this quadrant at the edge, either in the meridian ring or in the horizon ring.

Al-Marrākushī now turns his attention to the sphere. A point is chosen [O in diagram][55] as the North pole of the equator and with this as centre a (small) circle is drawn with compasses (بالبركار) and divided into four equal parts [at points *UXYZ* in diagram]. A great circle – the circle through the four poles, i. e. of the equator and the ecliptic – is now drawn through the pole and two of the points of intersection [*U*, *Y*, say]. For this one should use an آلة الخرط, which is surely not a lathe, but needs both poles to draw a circle:[56]

والأصلح في تخطيط هذه الدائرة ان تخط بآلة الخرط وان يطلب قطباها بالربع المجزى وهو ربع ا ب ج د أو بالاستقراء وذلك ان الربع المجزأ اذا وضع احد طرفيه على القطب الشمالي وجعل حرفه المجزى على احدى النقط الأربع كان طرفه الآخر واقعاً على أحد قطبي الدائرة المطلوبة واذا وضع أحد طرفيه على القطب الشمالي وجعل حرفه المجزى على النقطة المقابلة للنقطة التي كان عليها في الوضع الأول من النقط الأربع كان طرفه الآخر واقعاً على القطب الآخر من قطبي الدائرة المطلوبة .

In the delineation of this circle it is best that it is inscribed with the turning-instrument and that its poles are sought with the graduated quadrant, the quadrant *ABGD* (or by investigation): if one of the ends of the graduated quadrant is placed on the North pole and its graduated edge is laid over one of the four points, its other end will lie on one of the poles of the circle sought. When one of its ends is placed on the North pole and its graduated edge is laid over the point opposite [the one] that it was over in the first case of the four points, its other end will lie on the other of the two poles of the circle sought.

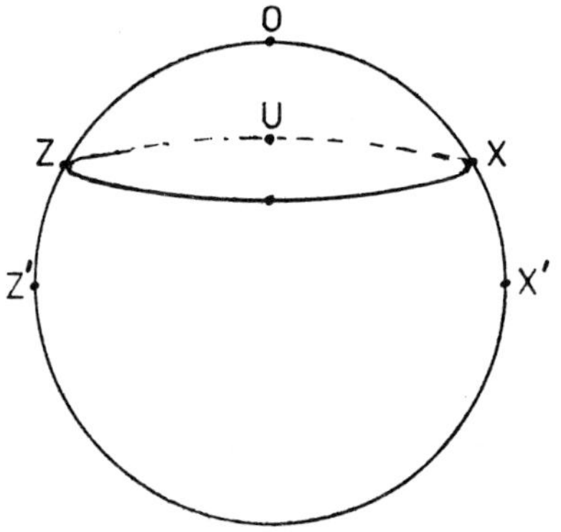

Fig. 3

53. F. 15r, lines 13–6.
54. Sédillot, *Mémoire* p. 114, has يكفيه, [a place] that suffices.
55. Fig. 3. There is no diagram in the MS.
56. BN 2508 f. 15 lines 5–11. The preparation of the sphere is not given by Sédillot.

Thus, in the diagram, poles *X'* and *Z'* of circle *UOY* are found. This circle is now drawn and graduated. After this a similar treatment is accorded to the equator and the ecliptic, the poles being first found on circle *UOY*. Unfortunately no value for the obliquity of the ecliptic (الميل الأعظم) is given. Holes are made at the poles not only of the equator but also of the ecliptic; and the sphere is now mounted, by the ecliptic's poles, in the meridian ring, so that its graduations may be used to put the fixed stars on the sphere. A fourteenth-century celestial globe in the Oxford Museum of the History of Science is equipped with a hole at the South pole of the ecliptic, though its frame has only the lower half of the meridian.[57]

Since, by contrast, al-Khāzinī's account of making the sphere is very short, we may, perhaps, use the above description to fill it out. Caution must be used, however, since al-Khāzinī apparently uses compasses to draw great circles. Al-Battānī says much less of the practical details of inscribing the circles, and Qusṭā b. Lūqā gives so little on constructional matters that four prefatory chapters had to be supplied in the Spanish translation.[85]

Of the uninvestigated texts on the globe the most promising are an extended treatment by al-Ṣūfī (903-86), author of enormous treatises on the astrolabe and on the fixed stars, in 157 chapters[59] and an anonymous treatise in 43 folios, both in Istanbul.[60] The latter quotes five works on the sphere: by Autolycus, Qusṭa b. Lūqā, Heṛon, Philon and Theon of Alexandria. There are also several tracts in Latin, one of which, "De horologio secundum alkoram id est speram rotundam", is evidently a translation from Arabic.[61] For further treatments the various commentaries on the *Almagest* might well repay investigation, since the solid sphere is probably to be regarded as a development of *Almagest* VIII, 3.

II THE TEXT

1. *General*

Al-Khāzinī's description appears in two manuscripts: Ẓāhirīya 4871, ff. 73r-74r, and Oxford Thurston 3, ff. 118r–119r, denoted by sigla *D* and *T* respectively. The part of Thurston 3 that contains our text is a treasure-trove of astronomical treatises and scraps written in a single hand, which is not easy to read and often confused by an imprint of the opposite page. The scribe and date may be identified by reference to the colophon of Quṭb al-Dīn al-Shīrāzī's

57. See plate XI of Mayer *Islamic Astrolabists*. A third hole is also visible. Nasr, *Islamic Science*, p. 123, shows the globe mounted on the ecliptic's poles. For a list of surviving Islamic globes up to the time of Ulugh Beg, see Destombes, *Globes*.

58. *Libros del Saber* I, p. 151 et seq..

59. Istanbul MS Seray 3505 (item 1, 60ff). Krause, pp. 463–4.

60. Istanbul MS Aya Sofia 2673 ff, 41v–84v, 864H. Krause, pp. 525–6.

61. Millás Vallicrosa, *Est. un. Cat.* 288–90.

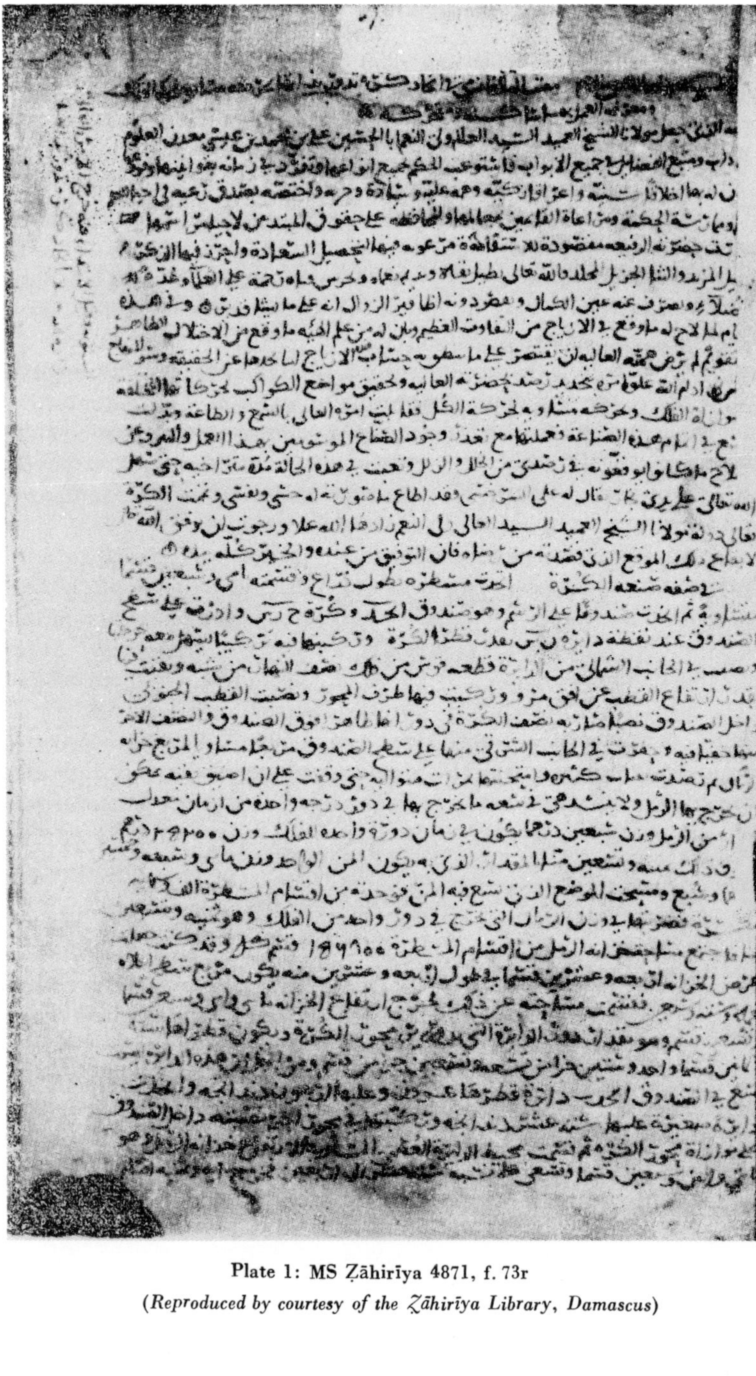

Plate 1: MS Ẓāhirīya 4871, f. 73r

(*Reproduced by courtesy of the Ẓāhirīya Library, Damascus*)

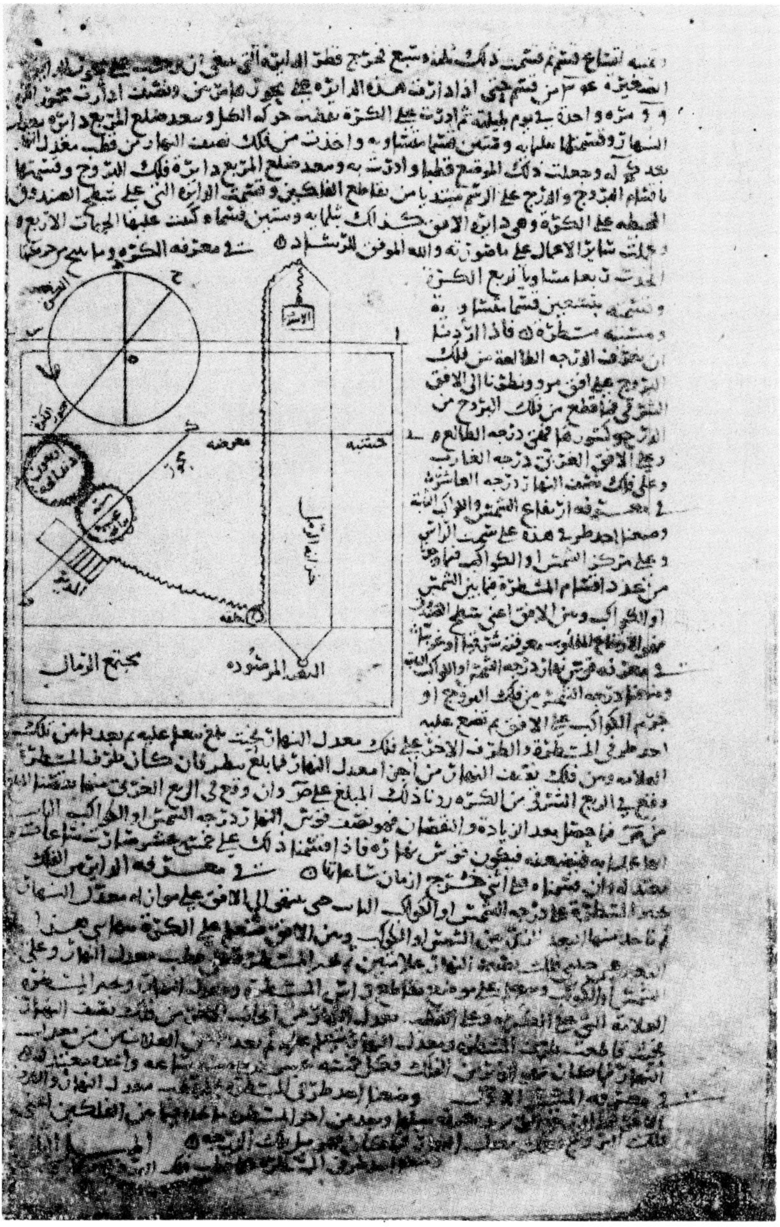

Plate 2: MS Ẓāhirīya 4871, f.73v

(Reproduced by courtesy of the Ẓāhirīya Library, Damascus)

Plate 3: MS Ẓāhirīya 4871, f. 74r

(Reproduced by courtesy of the Ẓāhirīya Library, Damascus)

abbreviation of Jābir b. Aflaḥ's commentary on the *Almagest*, ff. 73v–92v: Suhrāb b. Amīr al-Ḥājj b. Muḥammad b. al-Ḥasan and 675H (1276 A.D.)[62] The Damascus manuscript, which also contains a number of interesting items, is described elsewhere.[63] It appears to be of later date, but sometimes has better readings. In addition, it carries the laudatory introduction, which Thurston 3 omits.

There are indications that the text is in an imperfect state. For instance, a whole paragraph on the ascendant and similar points obtrudes itself between the introduction of the rule (73v 9–11) and the first instruction (73v18 et seq.) on the use of "this [rule]". Again, the number $34\frac{64}{99}$, the diameter of the wheel labelled *i* in the diagram (fig. 5), had to be restored from marginal annotations and calculation – in the Damascus manuscript part of the wreckage of the fraction appears in the next line. But only the most certain and obvious corrections or emendations have been made and the translation has been kept very literal.

The numbers in the text refer to the lines in *D*. Punctuation and hamzas have been silently added and nothing appears in the critical apparatus when the reading is clear in *T* but is obscured in *D* by a covering in the right-hand margin of f. 73r and the top of f. 74r. There are a few partly illegible calculations in the margins of *T* f. 118r, but they appear to add nothing to the text and have not been transcribed. Otherwise everything has been recorded.

[ ] at the beginning of a line in *D* means that the MS is obscured and that what is inside, if anything, is a guess; elsewhere it indicates an editorial addition.

< > means an addition from the margin.

[. . .] indicates an omission in the text.

2. *The Principal Diagram*

Since the diagram in *D* can be seen in the facsimile, it will be of more use here to reconstruct it and give it in plan (fig. 4) and elevation (fig. 5). The latter carries all the words in the manuscript diagrams. They are indicated by lowercase Roman letters. Arabic letters are transcribed from the manuscripts, and the capital Roman letters are added for ease of reference.

The following differences in the MSS

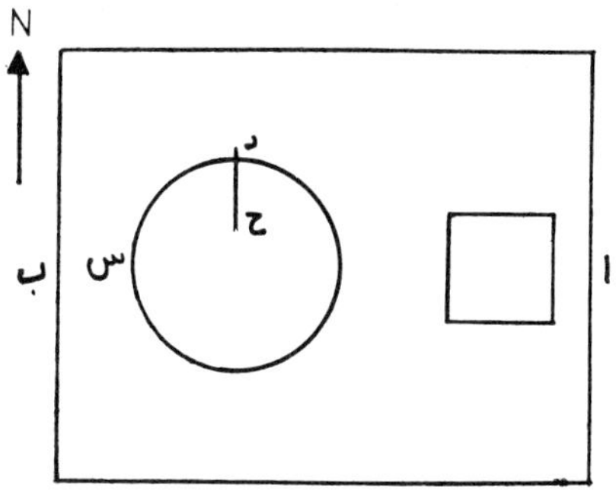

Fig. 4

62. For the scribe, see Kunitzsch, *Ibn aṣ-Ṣalāh*, pp. 29-30.
63. See Kennedy and Rageb.

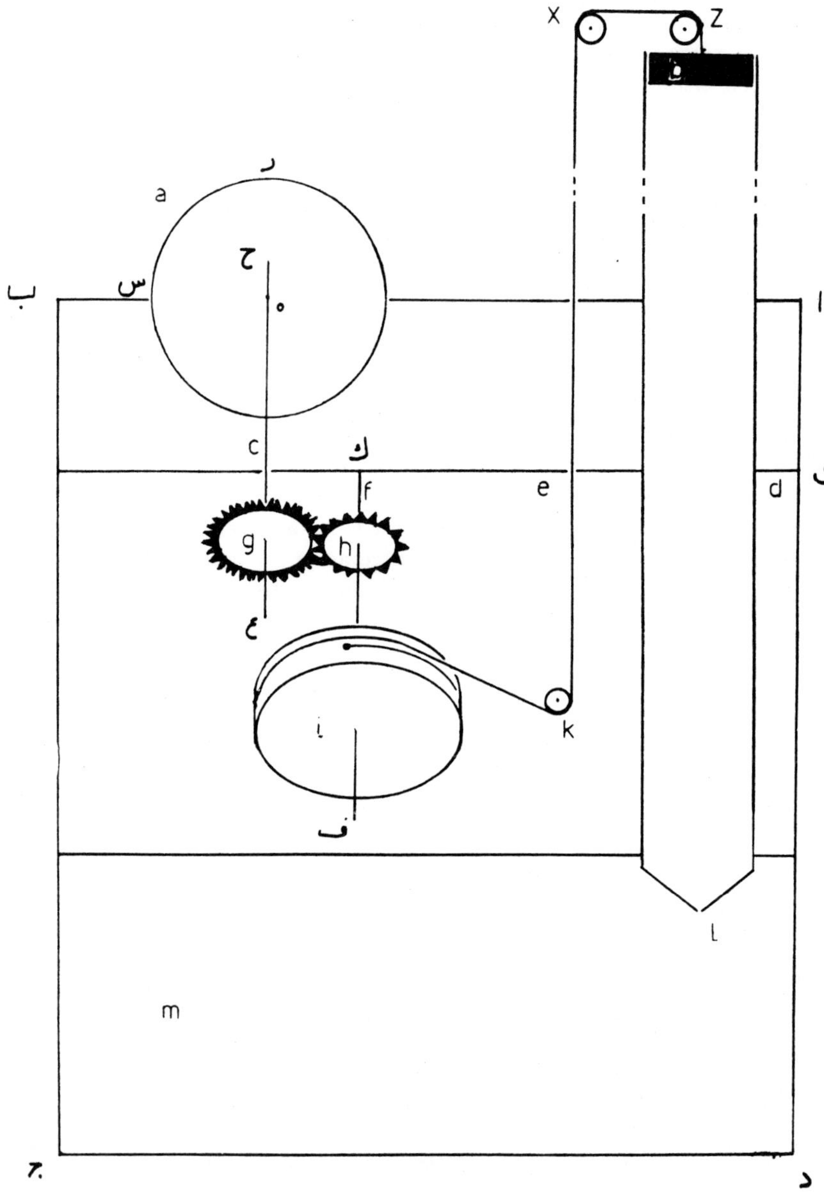

Fig. 5

(*D* f.73v, *T* f. 118r) may be noted. (1) In *T* the reservoir is placed entirely in the box, but the text says that a hole 24 units square must be cut in the top of the box for it. (2) In *T* the sphere is put in the middle of the top of the box; in *D* it is on the left, as here. (3) In *T* the cord is shown like a chain; *D* is not clear. It is also not clear – *T* is vaguer still – what the lines on the drum *i* are meant to be. (4) *D* omits pulley *X*. (5) ج and د are omitted in *D*.

On the whole the diagram in *T* is more neatly executed. For instance, the toothed wheels are shown with just 16 and 40 teeth.

To correct the perspective, the points ع and ف have been moved from the left-hand edge and are now supposed to be in the front of the box. The pulley *Z* has been added. For convenience, the reservoir, which is over eleven times as tall as it is broad, has been foreshortened.

Legend for Fig. 5

a الكرة sphere
b الأسرب lead
c محور الكرة axle of sphere
d-e خشبة معرضة (خشبة معارضة *T*) crossbeam (?)
f محور axle
g اربعون دندانجة 40 teeth
h ستة عشر دندانجة 16 teeth
i المدير turner [i.e. wheel, drum]
j خزانة الرمل sand-reservoir
k حلقة pulley
l الثقبة المرصودة observed hole
m مجتمع الرمال *D* (مجمع الرمال *T*) sand-container.

3. *Arabic text and Translation*

In the name of God, the Merciful, the Compassionate.

THE CHAPTER OF AL-KHĀZIMĪ ON SETTING UP A SPHERE THAT ROTATES BY ITSELF WITH A MOTION EQUAL TO THE MOTION OF THE CELESTIAL SPHERE AND KNOWLEDGE OF ITS USE, STATIONARY OR IN MOTION.

[——] who made our master, the shaykh, the great man, the wise lord, the great benefactor, Abū'l-Husayn ᶜAlī ibn Muḥammad ibn ᶜĪsā, the mine of the sciences and letters and the fountainhead of merits in all fields. For he comprehended sciences of every kind and was unique in his time in [understanding] their principles and branches – because he had splendid and extravagantly pure qualities of character, high aspiration and qualities of leadership. He [God] tested him and distinguished him in the candour of his protection in revitalizing science, applying wisdom, care for those who practice its achievements, and guarding the rights of those who undertake the revitalization of its principles. May his sublime excellency [always be] sought after, in order to draw profit from his help in them [the sciences] to reach a happy state; and may he reach in them a long and rich reputation and great and everlasting praise. May God (be He exalted!) prolong his being, make his grace continue, and prevent his decease, in mercy for the scholars and help for the excellent; and may He turn from him the evil eye, and clip before him the claws of Death, for He is capable of everything He wants.

In these days, when it was apparent to him what a great disparity there was among the *zījes*, and when it was clear to him from astronomy what discernible deficiencies there were in the correct setting up [of the *zījes*], his high endeavour did not rest content with confining himself to what the calculators of these *zījes* had pronounced, because of their [the *zījes*'] deviation from the truth,, and bad method. Thus he ordered (God perpetuate his exalted orders!) me to make a new observation in the presence of his Highness and to ascertain the positions of the stars in their various uniform motions. Before it began, the exalted order (God increase his excellence!) was issued to set up a sphere that turns by itself in parallel with the celestial sphere with a motion equal to the motion of the universe.

I received his exalted order with attention and obedience, and did my best to complete this design. I made it although it was difficult to find workers specialized in this work; and declaring myself innocent of [all] that they had

بسم الله الرحمن الرحيم 73r

مقالة للخازمي في اتخاذ كرة تدور بذاتها بحركة مساوية لحركة الفلك ومعرفة العمل بها ساكنة ومتحركة

[ ]ه الذي جعل مولانا الشيخ العميد السيد العالم ولي النعم ابا الحسين علي بن محمد بن عيسى معدن العلوم [والا]داب ومنبع الفضائل في جميع الابواب ، فاستوعب الحكم بجميع انواعها وتفرد في زمانه بقوانينها وفروعها، [فا]ن له بها اخلاقا سنية واعراقا زكية وهمة علية وسيادة ، وجربه واختصه بصدق رعية في احياء العلم [ ] وممارسة الحكمة ومراعاة القائمين بمعالمها والمحافظة على حقوق المبتدئين لاحياء مراسمها، [ ]ت حضرته الرفيعة مقصودة للاستفادة من عونه فيها لتحصيل السعادة ، واحرز فيها الذكر [الطو]يل المزيد والثناء الجزيل المخلد، فالله تعالى يطيل بقاءه ويديم نعماءه ويحرس فناءه رحمة على العلماء وعدة[للف]ضلاء ويصرف عنه عين الكمال ويقص دونه اظافير الزوال ، انه على ما يشاء قدير .

وفي هذه [للا]يام لما لاح له ما وقع في الازياج من التفاوت العظيم وبان له من علم الهيئة ما وقع من الاختلال الظاهر [في التـ]ـقويم لم يرض همته العالية ان يقتصر على ما ينطق به حساب هذه الازياج لتباعدها عن الحقيقة وسوء المنهاج [ ] نى ادام الله علو امره بتجديد رصد بحضرته العالية وتحقيق مواضع الكواكب بحركاتها المختلفة <المستوية وقبل الابتداء فيه خرج الامر العالي زاده الله علاء باتخاذ كرة تدور بذاتها> [٠]وازاة الفلك وبحركة مساوية لحركة الكل . فقابلت امره العالي بالسمع والطاعة وبذلت [ ]ـع في اتمام هذه الصناعة فعملتها مع تعذر وجود الصناع الموسومين بهذا العمل والتبرؤ عن [ ]لاح ما كانوا يوقعونه في رصدي من الخلل والزلل وتعبت في هذه الحالة مدة متراخية حتى سهل الله تعالى على يدي نجار يقال له علي السرخسي فقد اطاع

1. بسم ... الرحيم om *T*.

3–18. به ... بيده [ ] om *T*.

11. لتباعدها : لساعدها *D*.

12. < المستوية ... بذاتها > from margin in *D*.

brought into my observation[s] of corruption and mistakes, I became weary in this matter for a [long] tedious time – until God (be He exalted!) made it easy through the hands of a carpenter, whose name was ᶜAlī al-Sarakhsī. He followed what my senses and mind portrayed to him; and so I completed the sphere.

May the high rank of our master, the Shaykh, the great man, the wise lord, the great benefactor, be exalted and may God increase it in excellence! I asked that God (be He exalted!) let me reach that place of His satisfaction that I aspire to, because success comes from Him and the Good is entirely in His hands.

Description of the construction of the sphere

I took a rule a cubit long and divided it into 72 equal parts. Then I took a box as in the diagram – it is box *ABGD* – and a sphere *HRS*. On the surface of the box I drew a circle *RS* about point *E* [whose diameter is] of the amount of the diameter of the sphere. I mounted it in it so that its motion was easy against it. On the northern side of the circle I erected a portion of arc of the meridian [made] of brass. I made a hole in it at the amount of the altitude of the pole from the horizon of Marw. I mounted the end of the axle in it, and set up the South pole inside the box so that in its rotation half of the sphere became visible above the box and the other half was hidden in it. On the eastern side of it [i.e. the circle], on the surface of the box, I carved out a square equal to the square [cross-section] of the sand reservoir.

Then I prepared many holes and tested them repeatedly, until I discovered the narrowest hole through which sand comes out without getting clogged. It was of such a width that in the rotation of one degree of the equator a weight of 70 *dirhams* of sand comes out, and, in the time of one revolution of the celestial sphere, a weight of 25200. This is 98 *manns*, by the quantity in which one *mann* is the weight of $257\frac{1}{7}$ of them [*dirhams*]. I measured the volume [*mawḍi*ᶜ] in which the *mann* is contained and found it to be 1600 cubic [units] in divisions of the rule. I multiplied it by the weight of sand that comes out in one rotation of the celestial sphere, which is 98 *manns*, and the volume of the sand reservoir came to 156800 in divisions of the rule. I had made the breadth of the reservoir 24 divisions and the length 24, [so that] the square of its upper surface is 576. I divided its volume by that, and the height of the reservoir is produced – $272\frac{2}{9}$ divisons. It is the quantity of a rotation [i. e. circumference] of the wheel that turns the axle of the sphere. Its diameter would be $86\frac{61}{99}$ divisions. Because this circle could not be accommodated in the box, I took a wheel whose diameter was 20 and on which there were 40 teeth. Further, I took a small wheel on which there were 16 teeth and mounted it on another axle, which I erected inside the box parallel to the axle of the sphere. Then I divided

ما صورته له حس ونفس وتمت الكرة[17] تعالى دولة مولانا الشيخ العميد السيد العالي ولي النعم زادها الله علاء، ورجوت ان يوفق الله تعالى[18] لايقاع ذلك الموقع الذي قصدته من رضاه فان التوفيق من عنده والخير كله بيده .

[19] في صفة صنعة الكرة

اتخذت مسطرة بطول ذراع وقسمته باثنين وسبعين قسما [20] متساوية، ثم اتخذت صندوقا على الرسم وهو صندوق ا ب ج د وكرة ح ر س ، وادرت على سطح[21] الصندوق عند نقطة ه دائرة ر س بقدر قطر الكرة وركبتها فيه تركيبا يسهل معه حركتها[22] ونصبت في الجانب الشمالي من الدائرة قطعة قوس من فلك نصف النهار من شبه وثقبت فيها[23] بمقدار ارتفاع القطب عن افق مرو، وركبت فيها طرف المحور ، ونصبت القطب الجنوبي[24] داخل الصندوق نصبا صار به نصف الكرة في دورانها ظاهرا فوق الصندوق والنصف الآخر[25] منها خفيا فيه ، وحفرت في الجانب الشرقي منها على سطح الصندوق مربعا مساويا لمربع خزانة[26] الرمال، ثم رصدت ثقبات كثيرة وامتحنتها بمرات متوالية حتى وقفت على ان اضيق ثقبة يمكن[27] ان يخرج بها الرمل ولا ينسد هي في سعة ما يخرج بها في دور درجة واحدة من ازمان معدل [28] النهار من الرمل وزن سبعين درهما، يكون في زمان دورة واحدة للفلك وزن ٢٥٢٠٠ درهم[29] ويكون ذلك ثمنية وتسعين منا بالمقدار الذي به يكون المن الواحد وزن مائتي وسبعة وخمسين[30] منها وسبع ومسحت الموضع الذي يسع فيه المن فوجدته من اقسام المسطرة الف وستمائة[31] مكسرة، فضربتها في وزن الرمال التي تخرج في دور واحد من الفلك وهو ثمنية ونسعين[32] منا، فاجتمع مساحة خزانة الرمل من اقسام المسطرة ١٥٦٨٠٠ قسم، وقد كنت جعلت[33] عرض الخزانة اربعة وعشرين قسما في طول اربعة وعشرين منه يكون مربع سطح اعلاه[34] خمسمائة وستة وسبعين، فقسمت مساحته على ذلك فخرج ارتفاع الخزانة مائتين واثنين وسبعين قسما[35] وتسعي قسم، وهو مقدار دور الدائرة التي تدير محور الكرة ويكون قطرها ستة[36] وثمنين قسما وأحدا وستين جزءاً من تسعة وتسعين جزءاً من قسم، ومن اجل ان هذه الدائرة ليست [37] تسع في الصندوق اتخذت دائرة قطرها عشرون وعليها اربعون دندانجة واتخذت دائرة صغيرة[38] عليها ستة عشر دندانجة وركبتها في محور آخر نصبته داخل الصندوق [39] على موازاة محور

20. صندوق ا ب ج د : ا ب ج د T. 21. نقطة ه : نقطة D.
26. الرمال : الرمل T. ان om T. 27. ان يخرج بها : ان يخرج منها T. من دور : في دور T.
30. منها : [ ]ها (?) D. 34. على : عن D. مائتين : مائتى MSS.
35. قطرها : قطراها D. 36. وأحدا : واحد MSS. 36. ثمنين : ثمانين D.

the circumference of the greater wheel equal to the height of the reservoir, which is $272\frac{2}{9}$ divisions, in the ratio of 16 to 40, and it came out to $108\frac{8}{9}$. Then I divided that by $3\frac{1}{7}$ to produce the diameter, $34\frac{64}{99}$ divisions, of the wheel, which must be mounted on the axle of the small wheel. Thus, if this wheel turns on its axle $2\frac{1}{2}$ times, it turns the axle of the sphere once [and this happens] in a day and a night.

Then, with the pole of the motion of the universe [as pole] and with a distance [i.e. opening of compasses] of the side of the square [inscribable in a great circle of the sphere] I drew the circle of the equator on the sphere. I divided it into 360 equal parts. Along the meridian I took a distance of 23 [degrees] 35 [minutes] from the pole of the equator and made that place a pole; and with it [as pole] and with a distance of the side of the square, I drew the circle of the ecliptic. I divided it into the signs of the zodiac and [into] degrees, according to the drawing, starting from the intersection of the two circles [equator and ecliptic]. I divided the circle that is on the surface of the box and surrounding the sphere, the circle of the horizon, likewise into 360 equal parts. I inscribed the four directions on it, and made everything according to what I have sketched. God helps us to find the right way.

[THE SECOND PART: ON THE USES OF THE INSTRUMENT]

[1] *On knowing the sphere and what ensues from its motion*

I took a quadrant equal to a quadrant of [a great circle of] the sphere and divided it into 90 equal parts. I called it a "rule".

If we want to know the ascendent degree of the ecliptic at the horizon of Marw, we look at the eastern horizon; and what it cuts off from the ecliptic in degrees and their fractions is the degree of the ascendant. On the western horizon it is the degree of the descendant; and on the meridian it is the degree of the tenth [house] [i.e. the culminating degree].

[2] *On knowing the altitude of the Sun and fixed stars*

We place one of the ends of this [rule] on the zenith and [let the rule pass] through the centre of the Sun or star. What number of divisions of the rule we find between the Sun or star and the horizon – that is, the plane of the box – is the altitude required to be known, be it eastern or western.

الكرة، ثم قسمت محيط الدائرة العظمى المساوية لارتفاع خزانة الرمل وهو[40] مائتين واثنين وسبعين قسما وتسعين على نسبة ستة عشر الى اربعين فخرج مائة وثمنية اقسام [73v] وثمنية اتساع قسم، ثم قسمت ذلك على ثلثة وسبع فخرج قطر الدائرة التي ينبغي ان تركب على محور الدائرة [2] الصغيرة [وهو]٣٤ [قسما] و٦٤ [جزءا] من ٩٩ جزءا من قسم حتى اذا دارت هذه الدائرة على محورها مرتين ونصف ادارت محور الكرة [3] مرة واحدة في يوم بليلته، ثم ادرت على الكرة بقطب حركة الكل وببعد ضلع المربع دائرة معدل[4] النهار وقسمتها بثلثمائة وستين قسما متساوية، واخذت من فلك نصف النهار من قطب معدل النهار [5] بعد كجه له وجعلت ذلك الموضع قطبا وادرت به وببعد ضلع المربع دائرة فلك البروج وقسمتها [6] باقسام البروج والدرج على الرسم مبتدئا من تقاطع الفلكين، وقسمت الدائرة التي على سطح الصندوق [7] المحيطة على الكرة وهي دائرة الافق كذلك بثلثمائة وستين قسما متساوية وكتبت عليها الجهات الاربع[8] وعملت سائر الاعمال على ما صورته. والله الموفق للرشاد.

في معرفة الكرة وما ينتج من حركتها

[9] اتخذت ربعا مساويا لربع الكرة [10] وقسمته بتسعين قسما متساوية[11] وسميته مسطرة، فاذا اردنا[12] ان نعرف الدرجة الطالعة من فلك[13] البروج على افق مرو ونظرنا الى الافق[14] الشرقي فما قطع من فلك البروج من[15] الدرج وكسورها فهي درجة الطالع، [16] وعلى الافق الغربي فهي درجة الغارب [17] وعلى فلك نصف النهار [فهي] درجة العاشر.

[18]في معرفة ارتفاع الشمس والكواكب الثابتة

[19]وضعنا احد طرفي هذه [المسطرة] على سمت الرأس[20] وعلى مركز الشمس أو الكوكب فما وجدنا[21] من عدد اقسام المسطرة فيما بين الشمس[22] أو الكوكب وبين الافق اعني سطح الصندوق[23] فهو الارتفاع المطلوب معرفته شرقياً أو غربياً.

40. مائتين : مائتى MSS. وتسعين : وتسعى MSS.

73v 1. على ذلك : ذلك *D*. 2. و٦٤ ... ٣٤ : added in margin in *T*; ٣٤ *D*.

من ٩٩ om D, but ٩٩ appears as first word in next line!

6. على سطح : في سطح *T*. 7. متساوية om *D*. 8. والله الموفق للرشاد om *T*.

11. وسميته : ومسيته *D*., ? *T*; corrected in *T* in margin. 16. فهي om *D*.

20,22. الكوكب : الكواكب *D*.

[3] *On knowing the arc of day of the degree of the Sun or fixed stars*

We place the degree of the Sun in the ecliptic, or the body of the star, on the horizon. Then we place one of the ends of the rule on it [i.e. the horizon, where the degree of the Sun or star now lies] and the other end on the equator. Where it reaches on it [the equator] we put a mark. Then we count what [the number of degrees] is between this mark and the meridian in graduations of the equator. We examine what [place] it reaches: if the end of the rule lies in the eastern quarter of the sphere, we add that amount to 90; and, if it lies in the western quarter of it, we subtract the amount from 90. What arises after the addition or subtraction is half the arc of day of the degree of the Sun or fixed star, however we operate with it. We double it, and it becomes its arc of day. If we divide this by 15, [the number of] equal hours results; and if we divide it by 12, [the length of the] seasonal hours is produced.

[4] *On knowing the rotation of the sphere*

We pass the rule over the degree of the Sun or fixed star so that it reaches to the horizon [and lies] parallel to the equator. Then we take from it the distance that is between the Sun or star and the horizon and put two marks on the sphere at the extremities of this distance on either side of the meridian. Then we pass the rule over the pole of the equator and over the Sun or star, and we put a mark at the place of intersection of the head of the rule and the equator. We pass the rule over the mark on the sphere and over the pole of the equator, [the mark chosen being] on the other side of the meridian; and where the end of the rule and the equator intersect we make a mark on it [equator]. Then we count [the graduations] between the two marks along the equator. What [we] have is the rotation of the celestial sphere. Every 15 degrees of it is one equinoctial hour.

[5] *On knowing the first inclination* [*declination*]

We place one of the ends of the rule on the pole of the equator and the other end [!] on the degree whose inclination we want to know. We count from the foot of the rule [the graduations] that we find between the two circles, i.e. the ecliptic and equator. What [we] have is the inclination of that degree.

[6] *The second inclination*

We place one of the ends of the rule on the pole of the ecliptic and [let the rule pass] over the degree of the Sun or degree whose inclination we are looking

[24]في معرفة قوس نهار درجة الشمس أو الكواكب الثابتة

[25] وضعنا درجة الشمس من فلك البروج او [26] جرم الكوكب على الافق ثم نضع عليه[27] احد طرفي المسطرة والطرف الآخر على فلك معدل النهار فحيث بلغ نتعلم عليه ثم نعد ما بين تلك[28] العلامة وبين فلك نصف النهار من اجزاء معدل النهار فما بلغ ننظر فان كان طرف المسطرة[29] وقع في الربع الشرقي من الكرة زدنا ذلك المبلغ على ص وان وقع في الربع الغربي منها نقصنا المبلغ[30] من ص فما حصل بعد الزيادة والنقصان فهو نصف قوس نهار درجة الشمس او الكوكب الثابت[31] ايها عملنا به فنضعفه فيكون قوس نهاره ، فاذا قسمنا ذلك على خمسة عشر صارت ساعات[32] معتدلة وان قسمناه على اثني عشر خرج ازمان ساعاتها .

في معرفة الدائر من الفلك

[33] نجيز المسطرة على درجة الشمس او الكوكب الثابت حتى تنتهي الى الافق على موازاة معدل النهار[34] ثم نأخذ منها البعد الذي بين الشمس او الكوكب وبين الافق فنتعلم على الكرة بنهايتي هذا[35] البعد عن جنبتي فلك نصف النهار علامتين ، ثم نجيز المسطرة على قطب معدل النهار وعلى[36] الشمس او الكوكب ونتعلم على موضع تقاطع رأس المسطرة ومعدل النهار ، ونجيز المسطرة[37] العلامة التي على الكرة وعلى قطب معدل النهار من الجانب الآخر من فلك نصف النهار[38] فحيث قاطعت طرف المسطرة ومعدل النهار نتعلم عليه ، ثم نعد ما بين العلامتين من معدل[39] النهار فما كان فهو الدائر من الفلك ، فكل خمسة عشر درجة منه ساعة واحدة معتدلة .

[40] في معرفة الميل الأول

وضعنا احد طرفي المسطرة على قطب معدل النهار والطرف [41] الآخر على الدرجة التي نريد معرفة ميلها ونعد من آخر المسطرة ما نجده فيما بين الفلكين اعني[42] فلك البروج وفلك معدل النهار فما كان فهو ميل تلك الدرجة .

الميل الثـــاني

[43] وضعنا احد طرفي المسطرة على قطب فلك البروج وعلى درجة **[f. 74r]** الشمس

24. الكواكب : الكوكب *T*. 26. الكوكب : الكواكب *D*. 27. معدل : نصف *D*.
31. عملنا : كلنا ؟ *D*. 32. عشر om *D*. 35,37. فلك om *T*.
41. ونعد : وبعد *D*. 42. النهار ... فلك : البروج والمعدل *T*. تلك om *T*.
43. فلك om. *T*. وعلى درجة : ودرجة *T*.

for. We count off from the rule what we find on it between the two circles. What [we] have is the second inclination of the degree.

[7] *On knowing the right ascension of the signs*

We place the rule on the pole of the equator pass it over the degree whose right ascension we want to know. We see how many [graduations lie] between one of the equinoxes and the point of its [the rule's] intersection with the equator. What [we] have is the right ascension of that degree.

[8] *On knowing the ascension of the signs for the locality*

We place the degree whose ascension we want to know on the horizon of that locality. We count off the times from the nearer of the equinoxes to the ascending[-point] of the equator. What [we] have is what is required.

[9] *On knowing the nature of the day*

We have already explained the nature of the determination of half the arc of day. We add it to 90 or subtract it from it. It is the equation of the day for that degree.

[10] *On knowing the ortive amplitude*

We have already explained the determination of the ascendant. We count off the distance between the rising[-point] of the degree of the ascendant and the rising[-point] of the equator in degrees of the horizon. What [we] have is the ortive amplitude.

[11] *On knowing the latitude of the climate of observation*

Which is the altitude of the pole of the ecliptic. We place the rule on the zenith and on the pole of the ecliptic. We count off the graduations of the rule from the zenith to where it intersects the ecliptic. What [we] have is the latitude of the climate of observation.

The argument of the latitude of the region of observation: we count off the graduations of the rule between the zenith and the equator. What [we] have is what is required. The equation of the ascendant and of the tenth [house]: we count off the graduations of the ecliptic between the rule (on the zenith and the pole of the ecliptic) and the meridian. What [we] have is what is required.

والدرجة التي نطلب ميلها وعددنا من المسطرة ما نجد منها فيما بين الفلكين فما كان فهو الميل الثاني لتلك الدرجة .

[2] في معرفة مطالع البروج في الفلك المستقيم

وضعنا المسطرة على قطب معدل النهار ونجيزها على الدرجة[3] التي نريد معرفة مطالعها في الفلك المستقيم وننظر كم بين احد الاعتدالين الى نقطة تقاطعها مع دائرة معدل النهار[4] فما كان فهو مطالع تلك الدرجة في الفلك المستقيم .

في معرفة مطالع البروج في البلد

وضعنا الدرجة[5] التي نريد معرفة مطالعها على افق ذلك البلد ونعد الازمان من اقرب الاعتدالين الى مطالع معدل النهار فما كان[6] فهو المطلوب .

في معرفة كيفية النهار

قد بينا كيفية استخراج نصف قوس النهار فزدناه على ص̄ او [7] نقصناه منه فهو تعديل النهار لتلك الدرجة .

في معرفة سعة المشرق

قد بينا استخراج الطالع فنعد البعد[8] بين مطلع درجة الطالع وبين مطلع معدل النهار من اجزاء دائرة الافق فما كان فهو سعة المشرق .

في معرفة عرض[9] اقليم الرؤية

وهو ارتفاع قطب فلك البروج . نضع المسطرة على سمت الرأس وعلى قطب فلك البروج ونعد من[10] سمت الرأس الى حيث قاطعت فلك البروج من اجزاء المسطرة فما كان فهو عرض اقليم الرؤية .

حصة عرض اقليم الرؤية :[11] عددنا من اجزاء المسطرة فيما بين سمت الرأس ودائرة معدل النهار فما كان فهو المطلوب .

تعديل الطالع والعاشر : [12] نعد من اجزاء فلك البروج فيما بين المسطرة على سمت الرأس وقطب فلك البروج وبين فلك نصف النهار فما كان فهو [13] المطلوب .

2. معدل النهار : المعدل *T*. 4. في البلد : بالبلد *T*. 5. ونعد : وبعد *D*. التي : الى *D*.
7. فنعد : فبعد *D*. 8. بين مطلع : بين مطالع *T*.
9. الراس : illeg. in *T*. وعلى قطب فلك : وقطب *T*. 9. حيث om *T*. 12. فلك om *T*, 3 times.

[12] *On knowing the azimuths of [points having] altitudes*

We place the rule on the zenith and the degree of the Sun or star, and count off the degrees of the horizon between the intersection of the rule with the horizon and the rising[-point] of the equinox. What [we] have is the desired azimuth for that altitude. If we place the rule on the rising[-point] of the equator and the zenith, the point [s] under the rule in the turning of the sphere are [points having] altitudes, but no azimuth.

[13] *On knowing the ascension of the "direction" [of the Sun or star]*

We pass the rule over the zenith and the centre of the Sun or star having latitude, so that what is between the intersection of the rule and the equator [and the intersection of the horizon and the equator] is the mean ascension of the "direction". The equation of the ascension of the "direction": we place the rule similarly, letting it pass through the pole of the equator and on the star as an arc. What [we] have between these circles in graduations of the equator is the equation of the ascension of the "direction". The sum of the two, [i.e.] with the equation of the day, is the rotation of the celestial sphere.

[14] *On knowing the distance between the two places*

From the West of the equinox we count off the graduations of the equator in the quantity of the longitude of that place. We pass the rule over it and over the pole of the equator. Then we count off the graduations of the rule from the equator in the direction of its pole, so that the latitude of that place is counted off. Where it reaches, there is the zenith of that place. We make a mark on it and let the rule pass along the [great arc] between the zeniths of the two places. What [we] have between the two zeniths is the distance between them in degrees. For every degree we calculate $66\frac{2}{3}$ miles.

[15] *On knowing the direction of the qibla*

We mark the zenith of Mecca on the sphere. Then we place the rule so that it goes through the zeniths of it [Mecca] and of Marw. The distance between its intersection with the horizon and the meridian line in graduations of the horizon is the direction of the *qibla*. If we fix the rule at this place, and then the sphere rotates, the shadows of the altitudes of the Sun for all the degrees of the ecliptic always fall on the

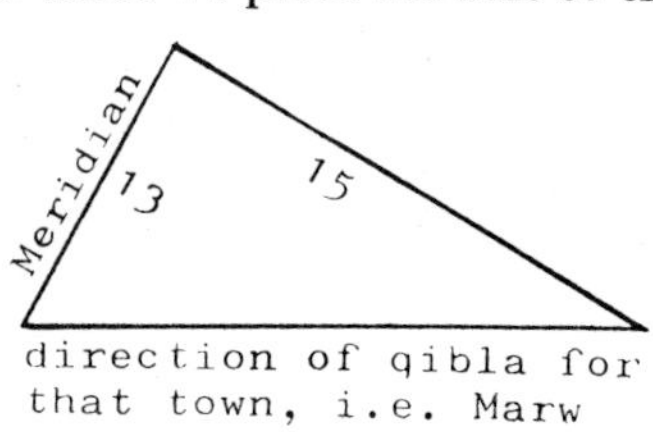

Fig. 5A

في معرفة سمت الإرتفاع

نضع المسطرة على سمت الرأس ودرجة الشمس او الكوكب ونعد من اجزاء[14] الافق فيما بين تقاطع المسطرة والافق وبين مطالع الاعتدال فما كان فهو السمت المطلوب لذلك الارتفاع واذا[15] وضعنا المسطرة على مطالع معدل النهار وسمت الرأس كانت النقطة التي تحت المسطرة في دوران الكرة ارتفاعات[16] لا سمت لها .

في معرفة مطالع السمت

نجيز المسطرة على سمت الرأس ومركز الشمس او الكوكب ذي العرض فان[17] ما بين تقاطع المسطرة ودائرة معدل النهار [.....] فهو مطالع السمت الوسطى .

تعديل مطالع السمت : نضع المسطرة[18] كذلك ونجيزها على قطب معدل النهار وعلى الكوكب قوسا فما كان فيما بين هاتين الدائرتين من اجزاء معدل النهار فهو[19] تعديل مطالع السمت ومجموعها مع تعديل النهار هو الدائر من الفلك .

في معرفة بعد ما بين البلدين

[20]نعد من مغرب الاعتدال من اجزاء معدل النهار بمقدار طول ذلك البلد ونجيز المسطرة عليه وعلى قطب معدل النهار[21] ثم نعد من اجزاء المسطرة من معدل النهار الى ناحية قطبه بقدر عرض ذلك البلد، فحيث بلغ فثمة سمت رؤوس اهل ذلك[22] البلد، فنتعلم عليه ونجيز المسطرة فيما بين سمتي رؤوس اهل البلدين فما كان فيما بين سمت رؤوسهما فهو البعد فيما بينهما[23] بالدرج، ونحسب لكل درجة ستة وستين ميلا وثلثي ميل .

في معرفة سمت القبلة

نعين سمت رؤوس اهل[24] مكة على الكرة ثم نضع المسطرة مارة على سمت رؤوس اهلها واهل مرو وكان بعد ما بين تقاطعهما والافق[25] وبين خط نصف النهار من اجزاء الافق هو سمت القبلة، فاذا اثبتنا المسطرة على هذا الموضع ثم دارت الكرة[26] فان اظلال

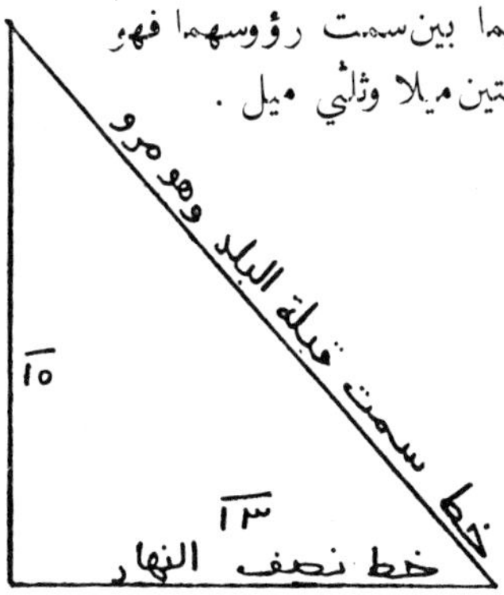

[شكل ٥ أ]

17. فهو : وهو *D*. 18. نحيز على : illeg. in *T*. 18–20. فهو ... اجزاء معدل النهار om *T*.
20. البلد : البعد ?*T*. 21,22. رؤوس : روس MSS. 22. رؤوسهما : روسهما MSS.
23. رووس : روس *T*, رووش *D*. 24. مكة : اهل مكة *D*.
24. رووس : روس MSS. The diagram is back to front in *D*.

direction of the *qibla* at Marw. He asked me – God extend his exalted enquiry – about the ratio of the obliquity of the direction of the *qibla* at Marw to its meridian line. I have calculated it: the ratio is 13:15, as in this diagram [see fig. 5A].

[16] *On knowing the operations for the fixed stars*

We pass the rule over the pole of the equator and the star and we mark where it cuts the equator and the ecliptic. The distance between the mark on the equator and the star in graduations of the rule is the distance of the star from the equator. The mark on the ecliptic is the degree of the transit of the star in the meridian line [i.e. it is the degree of the ecliptic that culminates at the same time as the star]. Then we pass the rule over the pole of the ecliptic and the star and put a mark on each of the two circles of the equator and ecliptic. The [distance] between the position of the star on the sphere and the ecliptic in graduations of the rule is the latitude of the star. The [distance] between it and the equator is the argument of its latitude. The distance between the two circles is the second inclination of its [the star's] degree. The mark that is on the ecliptic is the degree of the star. The [distance] between the two marks in graduations of the equator is the variation of the distance of the star. The [distance] between the two marks in graduations of the ecliptic is the variation of the degree of the transit.

[17] *On knowing the parallax of the Moon in longitude and latitude*

We pass the rule over the zenith and the centre of the Moon. Its [true] altitude is known and its apparent altitude is known by calculation. We put a mark on the sphere at each of its position[s]. Then we pass the rule over the pole of the ecliptic and on the mark of its true position, and the rule cuts the ecliptic is its true position [in longitude]. What is between it and the ecliptic is its true latitude. Then we pass the rule over the pole of the ecliptic and over the second mark. Where it cuts the ecliptic is the apparent position [in longitude]. The surplus between the two intersections with the ecliptic is the parallax [in longitude. The surplus between the two latitudes is the parallax] in latitude. The difference between it and its [the Moon's] latitude (or their sum) is the apparent latitude of the Moon and is called the established latitude of the Moon. From it are calculated the solar eclipses, the sighting of the new moons, and what is done by calculations of the quarters of the Moon.

This is the total of the canonical operations that are known with such a sphere. We have reported them with the utmost brevity. We shall report later what results from them in every question that might arise, God willing (be He exalted!).

ارتفاعات الشمس في كل درجة من درج فلك البروج تقع ابدا على سمت قبلة مرو . وقد كان[27] سألني ادام الله علو سؤاله عن نسبة انحراف سمت قبلة مرو عن خط نصف نهارها[28] فحسبتها فكانت نسبة ثلاثة عشر الى خمسة عشر كما في هذه الصورة .

[29]في معرفة أعمال الكواكب الثابتة

نجيز المسطرة على قطب معدل النهار والكوكب[30] ونتعلم حيث قطعت معدل النهار وفلك البروج ، فالبعد الذي بين العلامة على فلك معدل النهار وبين الكوكب[31] من اجزاء المسطرة هو بعد الكوكب من معدل النهار ، والعلامة التي على فلك البروج هي درجة ممر الكوكب في خط[32] وسط السماء . ثم نجيز المسطرة على قطب فلك البروج والكوكب ونتعلم على كل واحد من فلكي معدل النهار والبروج[33] علامة فيكون ما بين موضع الكوكب من الكرة وبين فلك البروج من اجزاء المسطرة هو عرض الكوكب[34] ، وما بينه وبين معدل النهار هو حصة عرضه ، وما بين الفلكين ميل درجته الثاني ، والعلامة التي على فلك البروج[35] هي درجة الكوكب ، وما بين العلامتين من اجزاء معدل النهار هو اختلاف بعد الكوكب ، وما بين العلامتين من اجزاء فلك[36] البروج هو اختلاف درجة الممر .

في معرفة اختلاف منظر القمر في الطول والعرض

نجيز المسطرة على سمت[37] الرأس ومركز القمر فيكون ارتفاعه معلوماً ويكون ارتفاعه المرئي بالحساب معلوماً ، فنتعلم على كل واحد من[38] موضعيه علامة من الكرة ثم نجيز على قطب فلك البروج وعلى علامة موضعه المقوم المسطرة فحيث قاطعت المسطرة[39] فلك البروج فهو موضعه المقوم ، وما بينه وبين فلك البروج عرضه المقوم ، ثم نجيز المسطرة على قطب فلك[40] البروج وعلى العلامة الثانية فحيث قاطعت فلك البروج فهو موضعه المرئي ، وفضل ما بين التقاطعين[41] من فلك البروج هو اختلاف منظره في [...] العرض ، وفضل ما بينه وبين عرضه أو مجموعهما هو عرض القمر[42] المرئي ويسمى عرض القمر المحكم ومنه تحسب الكسوفات الشمسية ورؤية الأهلة وما يعمل بحساب ارباع[43] القمر

فهذه جملة الأعمال المشهورة التي تعرف بمثل هذه الكرة قد ذكرناها على نهاية الاختصار و[44] سنذكر ما يتولد منها في كل سؤال يجري فيما بعد إن شاء الله تعالى .

26–7. وقد ... سؤاله : فقد سالني (?) *T.* — 30. الذى : التى *T.*

31,35 هى درجة : هو درجة MSS. — 38. موصعيه : موصعه MSS.

38. قاطعت : قاطع MSS. — 39. عرضه : عرض MSS.

4. *Commentary*

The Introduction

In 73r9 عين الكال , "the evil eye", "is applied to an eye believed to have the power of killing by its glance", according to Lane's *Arabic-English Lexicon*, pp. 2216 col. 1 and 2423 cols. 1–2. الزوال, "Death", should really be "vanishing" or "extinction".

The most interesting part of the introduction is the mention (73r16) of the carpenter. Clearly al-Khāzinī was – at any rate at this stage – not competent to do the practical work himself, despite his using the first person singular in his account of the construction. The failures before the arrival of ᶜAlī al-Sarakhsī may perhaps account for the curious order in which the construction was carried out – and in particular for al-Khāzinī's apparently having a supply of square piping ($6\frac{1}{2}'' \times 6\frac{1}{2}''$) both when he was making the present instrument and when writing the *K. Mīzān al-ḥikma* (see below).

On Making the Sphere

A rod one cubit, or *dhirā'*, long and divided into 72 parts is taken as the standard for measuring lengths. The procedure is as follows, square brackets being used for operations not mentioned explicitly:

[1. The reservoir, sphere and box are made].

2. A circular hole is cut in the box for the sphere; and the sphere is mounted so that its axle points to the poles of the world and just half the sphere is visible above the box.

3. A square hole 24×24 (in divisions of the rule) is cut beside the sphere, to the East, to fit the reservoir.

4. The height of the reservoir is calculated on the assumption that the sand inside lasts just 24 hours.

5. The wheel whose circumference is equal to this height is calculated to be too big for the box; so a pair of toothed wheels are introduced so that a smaller one can be used. The diameter is calculated [and the reservoir, string, pulleys, wheel to wrap the string around, and the two toothed wheels are mounted as in the diagram].

6. The equator and ecliptic circles are drawn on the sphere and divided into 360°. The signs of the zodiac are marked. The horizon-circle, the circle on the box that surrounds the sphere, is also graduated.

It is curious that al-Khāzinī makes the box first and then investigates the size of the components inside. He also seems to have the cross-section, but not the height, of the reservoir before he starts.

To calculate the height of the reservoir, al-Khāzinī finds the amount of sand required, on the understanding that the reservoir must hold just enough to last 24 hours and that the sand flows at the least possible rate without getting clogged. Having found the volume of sand, and hence of the reservoir, he divides it by the area of the cross-section to find the height. The calculations are as follows:

By experiment, in one degree of the rotation of the heavens, 70 *dirhams* of sand flow out.

$\therefore$ in 360° [i.e. 24 hours], $360 \times 70 = 25200$ flow out.

$\because 257\frac{1}{7}$ *dirhams* = 1 *mann*, this makes 98 *manns*.

By experiment, 1 *mann* occupies 1600 cubic units (divisions of the rule):

$\therefore$ volume of reservoir $= 98 \times 1600 = 156800$ cu. units.

$\therefore$, its cross-sectional area being 576 ($= 24 \times 24$), the height is $272\frac{2}{9}$.

Note. The Hyderabad edition of the *K. Mīzān al-Hikma* has 257 *dirhams* per *mann*, the $\frac{1}{7}$ having been lost. A text with a similar error misled Wiedemann into thinking that subsequent figures were wrong.[64]

So far the calculation finds an exact parallel in the *K. Mīzān al-ḥikma*, 8.1.2.1–3,[65] where al-Khāzinī gives the details of the water– or sand-reservoir in a balance-clock, ميزان الساعات, used in astronomical observations. In this instrument the reservoir is attached to one end of a balance-beam. The positions of compensating weights – on special scales reading directly in units of time or equivalent angle of motion of the celestial sphere – are noted at the beginning and end of the interval to be measured.[66] The procedure to find the height of the reservoir, which here contains either sand or water, is the same as in the present text. Even the cross-section of the reservoir is the same, 24 × 24, the units of length being likewise $\frac{1}{72}$ of a *dhirā'*. True, the weight of water or sand that flows out in one degree of celestial motion is given as 60 *dirhams*, i.e. 21600 *dirhams* or 84 *manns* per revolution. But al-Khāzinī says the observation is repeated for many revolutions continuously, the sand or water being weighed and returned to the vessel, and goes through the calculation without specifying the result, saying only that it is preserved (in *manns*) as the "first preserved [quantity]", المحفوظ الأول. He points out that from this the corresponding quantity for an hour, or other interval, may be calculated. The "second preserved [quantity]" is the volume of sand or water to be accommodated in the reservoir, and this is to be divided by 576, as in the present text, to find the height (here طول) of the reservoir.

The cases of sand and water are considered separately, in chapters 8.1.2.2 and 8.1.2.3 respectively. In the latter we find that one cubic *dhirā'*, or 373248 cubic units, contains 157 *manns* of water, a figure he took from al-Bīrūnī's

64. Wiedemann, *Stundenwage*, p. 61.

65. *Mīzān*, p. 153 et seq..

66. Abstract of Rozhanskaya's article in *Referativnyi Zhurnal*. I am most grateful to Dr. V. Bialas for translating this summary.

calculation, which is quoted in a passage of the third *maqāla* of the *Mīzān*, by leaving out the fractions.[67] Al-Bīrūnī arrived at this figure, however, by taking 1 *mann* = 182 *mithqāls* or 260 *dirhams* (the normal ratio[68] of 7:10 being maintained), instead of 180 and $257\frac{1}{7}$, as al-Khāzinī for some reason assumes here. In the following calculations, results for both 182 and 180 *mithqāls* per *mann* will be given (implying that 1 cubic *dhirā'* of water weighs about 157.17 or 158.92 *manns* respectively). But first we estimate the *dirham* and the *dhirā'*.

All the values given by Hinz for the *dirham* lie between 3 and 3.3 grams. According to the relation given by al-Bīrūnī before introducing the *mann*, the equivalent of 1 cubic *dhirā'* = 28605.647 *mithqāls*, the *dhirā'* will be between 49.68 and 51.28 cm. The only suitable values in Hinz's table are 49.875 cm, which, as the *dhar^c-e shar^{c}i*, was the canonical *dhirā'* in Iran, and the less likely *dhirā^c al-dūr* and the Egyptian *dhirā' al-yad* (according to one calculation), both given at 50.3 cm. These yield c. 3.04 and 3.11 gm, respectively, for the *dirham*. As an average value in Persia Hinz suggests 3.2 gm, but one of his figures (1 *mithqāl* = 4.3 gm) yields 3.01 gm. If the *dirham* is c. 3.04 gm, the *mann* is about 789.3 or 780.7 gm. Perhaps such exact calculations should be treated with some reserve. Certainly the rough figures given by both Wiedemann and Hill,[69] 3 gm for the *dirham* and ½ metre for the *dhirā'* fit al-Bīrūnī's relation well enough – they imply a density of water of about 1.02 gm/cc.

The figure 1600 given in both texts for the volume in cubic units of 1 *mann* of sand is curious. Comparison with the relation that 157.17 or 158.92 *manns* of water occupy 373248 cubic units yields a specific gravity of the sand of about 1.48 or 1.47. According to Brady's *Materials Handbook*,[70] "the weight of sand varies from 2,600 to 3,100 lb per cubic yard, depending on the composition and size of grain"; so that the specific gravity of sand lies between 1.54 and 1.84. Therefore the sand (رمل) used by al-Khāzinī was either especially light or else not pure sand.[71] Wiedemann's discussion,[72] in which the specific gravity of sand is taken as about 2, is vitiated by his tacit assumption that the unit of weight used in this passage for water is the *mann*, but for sand is the *mana*. Our text shows that they are the same.

Furthermore, 1600 is a suspiciously round number, which al-Khāzinī says he found by measurement. Again, 70 *dirhams* is a curiously exact amount of sand to escape during the revolution of one degree of the celestial sphere. Were

67. Khanikoff, pp. 75-7 and 121-2. Al-Bīrūnī gives "157 manns, and 6 ístârs and $\frac{1}{2}$ and $\frac{1}{4}$ and $\frac{1}{5}$". In the course of the calculation he approximation $\frac{2}{45}$ by $\frac{3}{60}$, but this does not materially affect the result.

68. Hinz, p. 1.

69. Wiedemann and Hauser, p. 47, and Jazarī, p. 238 respectively.

70. Brady, *Materials Handbook*, p. 155.

71. For the various types of sand etc. used in simple sand-clocks later in the West, see Balmer, p. 623.

72. Wiedemann, *Stundenwage*, p. 61.

there not some very awkward fractions in the subsequent calculations, al-Khāzinī might be thought to have taken a round number with a seven in it to make multiplying by $257\frac{1}{7}$ easier. The 60 *dirhams* in the equivalent passage in the *Mīzān* is meant, no doubt, as an estimate or example.[73] After all, it is unlikely to serve for both water and sand and the figure is anyway soon dropped from the discussion. It should be noted that the figure 1600 is used only to calculate the height of the reservoir and will not affect the accuracy of the clock. But in the "sphere" text the figures 70 and 1600 are subsequently used in minute calculation of the size of the components. Only two conclusions seem likely: 1. that the calculations were imposed after the instrument was made, if indeed it was made, and do not reflect the actual results and reasoning of the artificer; or 2. that the figures are indeed approximate and the accuracy of the subsequent calculations spurious.

The rest of al-Khāzinī's calculations are relatively easy. If the string were wrapped round a wheel attached to the axle of the sphere, the circumference of this wheel would have to be equal to the height of the reservoir, for this empties itself once in 24 hours. But this would make the diameter of the wheel $86\frac{61}{99}$ $(=272\frac{2}{9}/3\frac{1}{7})$ units, which would not fit in the box. Instead, the string is wrapped round a smaller wheel whose axle is connected to the axle of the sphere by gears in the ratio 16:40. This smaller wheel must have a circumference $\frac{16}{40}$ of the circumference of the above hypothetical wheel, i.e. $108\frac{8}{9}$ units. Hence the diameter is $34\frac{64}{99}(=108\frac{8}{9}\ /\ 3\frac{1}{7})$ units.

The numbers of teeth, 16 and 40, on the toothed wheels are of some interest, because the examples of wheels and moulds for wheels found in China, apparently from the fourth century B.C., also have round numbers of teeth, e.g. 16, 24 and 40.[74] Perhaps this argues less for influence fron China than for the simplicity of making such wheels. Al-Bīrūnī described a geared astrolabe and there is an example of a similar instrument from fourteenth–century Islam with simplified gearing in the Museum of History of Science at Oxford. Both instruments involve gear-ratios less simple than 16:40.[75]

The figure for the obliquity of the ecliptic, $23^{\circ}35'$, is the same as that used in *al-Zīj al-Sanjarī*[75a] According to a passage from this *zīj* translated by Nallino,[75b] al-Khāzinī quotes various decreasing values of the obliquity, and describes the astronomical model that accommodates it, but for his own pur-

73. *Ibid*, p. 60.
74. Price, *Origins of Clockwork*, pp. 83-4.
75. *Ibid.*, pp. 98-100.
75a. Kennedy, *Survey of Tables*, p. 159.
75b. *Battānī*, I, p. 161. The passage occurs in MS Vatican Arabic 761, f. 10v.

poses keeps to the relatively recent value of 23°35′ given by al-Battānī and others. This value was used by many astronomers, including Ḥabash, Abū'l-Wafā', and al-Bīrūnī.[75c]

Finally, the approximate dimensions should be mentioned. The reservoir is about 188.6 cm or 6′2″ tall and 16.6 cm or 6½″ across. The box is not wide enough to accommodate a wheel 60 cm, or nearly 2', diameter, but may have been quite tall. The wheel round which the string was wrapped is 25.4 cm or 10″ in diameter and the two gears are 13.9 cm (or 5½″) and 5.5 cm (or 2″) respectively. Wiedemann[76] notes that, if the sand escapes at a rate of 60 *dirhams* per degree, the weight of the reservoir would be about 65 kg. At a rate of 70, given in our text as an actual measurement, the reservoir must weigh 76608 gm, a little over 1½ cwt.

The Uses of the Instrument

Section 1. The rule, here introduced, is used as a straightedge (or rather as a rigid arc of a great circle) in sections 4, 7, 12, 13, 15, 16, 17, and as a graduated straightedge in sections 2, 5, 6, 11, 14, 16, 17. In sections 1, 8, 10, the rule is not used. In section 3 (and 9), geometrically the most interesting application, the rule is used as an exact quadrant of a great circle.

Section 3. The procedure may be justified as follows. If (in fig. 6) ABC is the the equator and N its northern pole, and if the quadrant is PQ, the equation of the day, which must be added to 90° to make half the arc of day, is BA. To justify al-Khāzinī's taking CP instead, we need only prove that AP is a quadrant. But since $PN = PQ =$ quadrant, P is the pole of circle AQN; and so PA is a quadrant. A similar demonstration could be provided for the case where Q is on the other side of the equator and P is perforce in the western quarter of the sphere. As al-Khāzinī says, PC must be subtracted from 90° in this case.

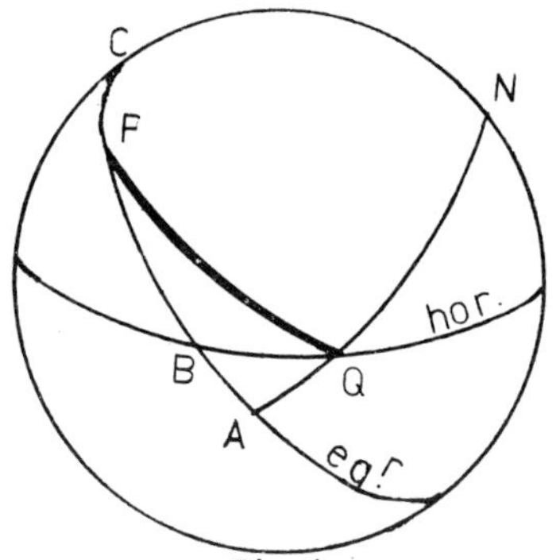

Fig. 6

Section 4. The obvious way to measure the rotation, i. e. of the Sun, S, in fig. 7, since dawn, on a sphere with a quadrant is to put the quadrant on N, the North pole, and S and to see where it cuts the equator (at X, say), and then to do the same with N and H, where H is on the horizon and $NH = NS$, to find Y: XY is the arc required. But such a procedure is impossible when the sphere is half sunk in a box. No doubt the confused instructions in this section are an

75c. *Battānī* I, pp. 159–60; Kennedy, *Bīrūnī's Taḥdīd*, p. 50.
76. Wiedemann, *Stundenwage*, p. 60.

attempt to obviate this difficulty. What al-Khāzinī apparently does is to copy the small arc SH by means of the quadrant at a more accessible place. Presumably K is marked, so that $KS = SH$, and then U is found by laying the quadrant on N and K and seeing where it cuts the equator. Finally UX is measured in graduations of the equator. What is not clear from the text is finding K. The difficulty comes from the obtrusion of the meridian into the procedure at 73v35 and 73v37, the first case having the additional problem of "*two* marks". It is to be assumed that the text is corrupt and that the two marks refer to the final measurement along the equator or to the points S and K. Perhaps the author assumed the Sun was high enough in the sky for the point K to be on the other side of the meridian.

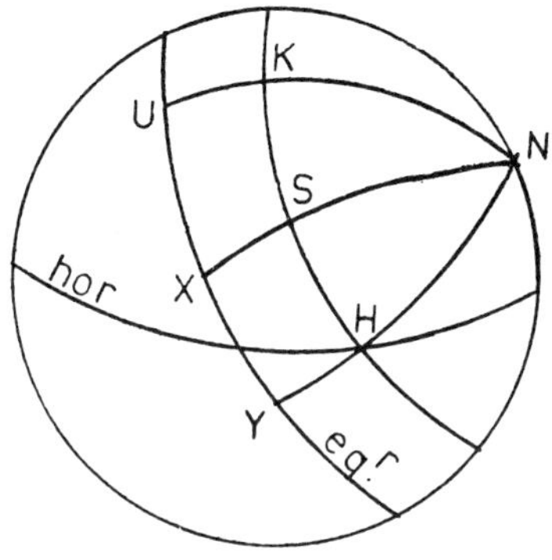

Fig. 7

Sections 5 and 6. The first inclination is the declination SU and the second inclination is the arc SQ in fig. 8, in which S is the point whose inclination is to be found and P is the pole of the ecliptic SE, PQ the rule, and EQ the equator. For neither inclination does al-Khāzinī consider the case when the rule would have to be longer than a quadrant, as in the present diagram. It looks as if he has half-converted the description of an ordinary globe – one, that is, which is not self-moving – for if the South pole were available this problem would be solved.

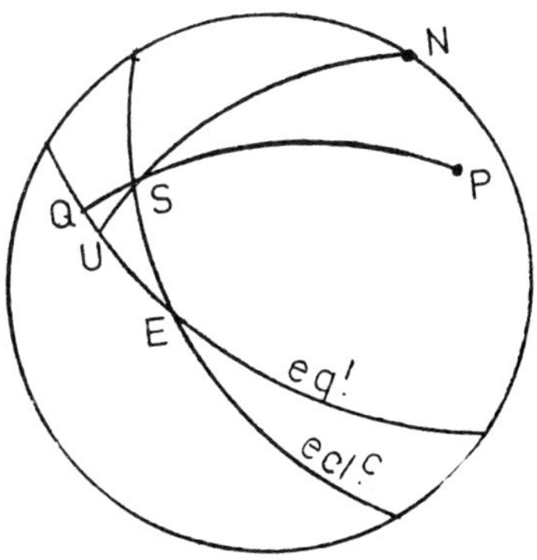

Fig. 8

Section 7. Here, again, only half the cases are considered.

Section 9. This section seems pointless.

Section 11. The "latitude of the climate of observation" (عرض إقليم الرؤية), here defined as the altitude of the pole of the ecliptic, is the shortest distance from the zenith to the ecliptic (ZQ in fig. 9). Prof. Kennedy translates a similar phrase in Persian as the "latitude of visible climate". It was used in the calculation of eclipses and apparently came from Indian astronomy.[77] The "ar-

77. Kennedy, *Parallax*, pp. 37–8.

gument" (حصة)[78] of this quantity is ZX. The "equation"(تعديل) of the ascendant and of the tenth house is the arc QA. Again, the rule will often be too short.

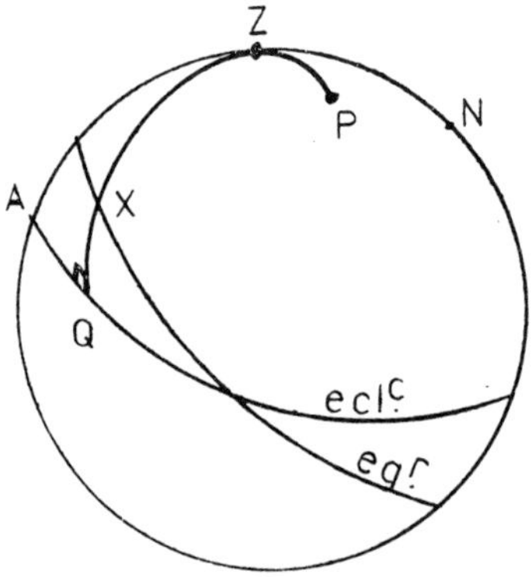

Fig. 9

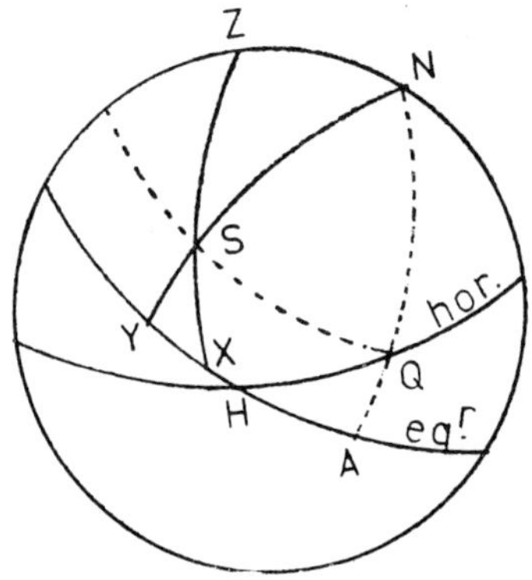

Fig. 10

Section 13. The quantity found here is YH in fig. 10, in which N is the North pole of the equator HXY, Z the zenith, and S the Sun or star. It is divided into "mean ascension", XH, and the "equation", XY. If the equation of the day, HA, is added, the rotation YA is obtained.

Section 14. The title is translated "... between *the* two places" because probably Marw and Mecca are meant, rather than any two places. To support this conjecture it could be adduced that only one zenith is found on the sphere (see discussion of section 15). Further, the next section, on the *qibla*, appears as a continuation.

The value $66\frac{2}{3}$ miles for one degree is common and was used by al-Bīrūnī.[79]

Section 15. In describing how the *qibla* is found, al-Khāzinī omits the inportant condition that the measurement must be taken when the sphere is aligned with the terrestrial coordinates, that is, when the equinox corresponds with the zero of longitude. If the sphere is rotating in concord with the heavens, the measurement must be taken at the time when the celestial equinox is directly above the zero of longitude. This condition must also hold for finding the distance between Marw and Mecca if – as it appears – the position of Marw is not found on the sphere in the same way as the position of Mecca (or whatever the second town is) but is taken to be the zenith on the stationary meridian-circle.

The sentence beginning, "If we fix the rule at this place" (74r25-6) should probably be interpreted in this way: the rule is held fixed to the stationary zenith and horizon, and the sphere rotates; if the Sun is at the point of the

78. See Kunitzsch, *Glossare*, p. 33; also King, *Khalīlī* p. 103.

79. Kennedy, *Bīrūnī's Taḥdīd*, p. 131.

ecliptic intersected by the rule, it will have the same azimuth as Mecca; and therefore all shadows will be aligned with Mecca. Of course, they will point in exactly the opposite direction, since both Mecca and the Sun are to the South of Marw.

No doubt it was his patron ᶜAlī b. Muḥammad who asked him (74r26-7) to find the *qibla*. It is of interest to note that he calculates it rather than using his new instrument.

At first sight, the figure given for the *qibla*, equivalent to cot $q = \frac{13}{15}$ (which yields q, the westward inclination of the *qibla* to due South, $= 49°5'$) looks like a rough approximation. But the approximations given by al-Bīrūnī,[80] on whom al-Khāzinī relies for other numerical information,[81] for the *qibla* of Ghazna are fairly accurate. For $q = 70°47'6''$ he gives the equivalent of sin $q = \frac{17}{18}$; and 17 cosec q is 18.003. Even the inferior approximation he gives, which is equivalent to $\frac{1}{3}$ for cos q means taking 3.036 (= cosec q) as 3. Prima facie, therefore, we should take the $\frac{13}{15}$ fairly seriously. A value remarkably close to this may be found by interpolation in a *qibla* table attributed to al-Khāzinī by al-Qazwīnī in 1340.[82] If the values in al-Bīrūnī's Qānūn[83] are accepted for the longitude and latitude of Marw (86°30′, 37°40′) and Mecca (67°, 21°20′), q comes out as 49°6′ and 13 tan q = 15.01. 67° was the most popular medieval longitude for Mecca,[84] and al-Khāzinī himself takes the latitude of Marw to be 37°40′ in his zij.[85] 21°20′ appears to be the value of φ_M (the latitude of Mecca) underlying the table. It must be admitted that one of the values on which the interpolation is based may be an error,[86] but the table may not be al-Khāzinī's own and, if it is, he may have made the mistake himself.

Other means of calculating q yield results not so close to that implied by cot $q = \frac{13}{15}$. For instance, interpolation in the table with $\varphi_M = 21°40'$, as given by al-Qazwīnī, produces $q = 49°47'$ and 13 tan $q = 15.37$. If the same values are assumed, the figures for q given by any of the correct methods of calculation – and several were known to the medieval Muslim astronomers[87] – are 51°13′ and 51°55′ (13 tan q = 16.26 and 16.59) for $\varphi_M = 21°20'$ and 21°40′ respectively.

80. *Ibid.* pp. 214–5.
81. Hall, *DSB*, pp. 341–3.
82. See note 7.
83. Vol. II, pp. 571 and 551.
84. See King, *Khalīlī*, p. 84; Le Strange, p. 28; handlist drawn up by E. S. Kennedy and F.I. Haddad – for a description of which see Haddad and Kennedy.
85. Kennedy, *Survey of Tables*, p. 159.
86. $(\Delta\varphi, \Delta L) = (17,20)$. See Lorch, *Qibla*, pp. 262-3.
87. See King, "Kibla" *EI*.

Similar figures for the approximate fornulae given by King[88] are $q = 49°55'$ and $50°27'$ ($13 \tan q = 15.43$ and 15.74) for the simpler formula, and $q = 47°52'$ and $48°22'$ ($13 \tan q = 14.37$ and 14.63) for the more complicated one.

It is therefore tempting to speculate that the value 15/13 for $\tan q$ came from the table reproduced by al-Qazwīnī. Even the most likely formula to give rise to it[89] yields $q = 49°19'$ and $13 \tan q = 15.12$ – a result not so close.

Section 16. This section is largely a repetition of what has already been said for points on the ecliptic. In the diagram N is the pole of the equator EX and P is the pole of the ecliptic EY. S represents the star. The points X, Y and U, V are found with the quadrant – which again would not always suffice for the purpose. SU is the "argument' (حصة) of its latitude; UV is the second inclination; V is the degree (i. e. longitude) of the star. XU is the "variation of the distance" of the star (اختلاف بعد الكوكب) and VY is the "variation of the degree of the transit" (اختلاف درجة الممر).

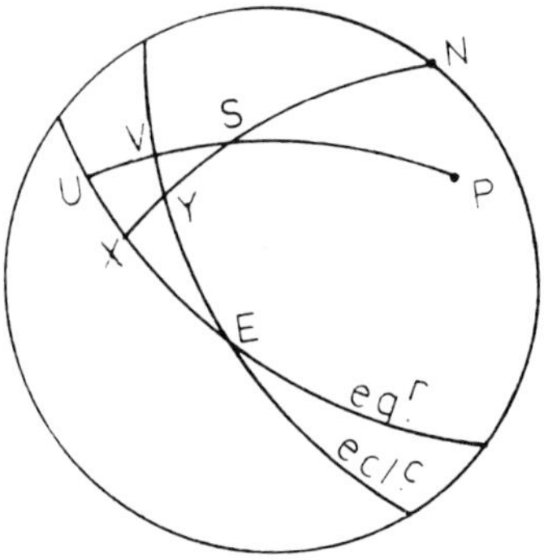

Fig. 11

When speaking of the meridian line in connection with the transit, al-Khāzinī uses the expression خط وسط السماء instead of the usual فلك (or خط) نصف النهار.

Section 17. The omission in 74r41 by homoioteleuton shows the poor state of the text. But the idea is clear: the true position of the Moon is marked on the sphere; its altitude is measured with the rule; its parallax in altitude is calculated to give its apparent altitude, which is then marked on the sphere; and the true and apparent longitudes and latitudes are found directly by using the rule.

The term عرض محكم (established latitude) in 74r42 is also found in tables attributed to Yaḥya b. abī Manṣūr (c. 890).[90]

The last paragraph seems to promise a more extended treatment of the sphere. Certainly there is nothing on the subject immediately following in the Oxford MS, which continues with a short chapter on compound proportion.

88. *Ibid.* p. 84, col. 2.

89. Lorch, *Qibla*. Formula (4).

90. Kennedy, *Parallax* p. 44. The MS is Escurial 927. See Kennedy, *Survey of Tables* pp. 132 and 145 et seq..

Bibliography

R. T. Balmer, "The Operation of Sand–clocks and Their Medieval Development", *Technology and Culture*, 19 (1978) 615–32.

Battānī. Al–Battani sive Albatenii Opus Astronomicum, ed. C. A. Nallino, (Milan, 1899–1907).

Tho. Beck, "Herons des Älteren Automatentheater", *Beiträge zur Geschichte der Technik und Industrie*, I (1909), 182–99.

J. Beckmann, *A History of Inventions, Discoveries and Origins* (translated W. Johnston), vol. I, 1846.

Bīrūnī, *Qānūn*. Abū Rayhān Muḥammad b. Aḥmad al–Bīrūnī, *Al–Qānūnu'l–Mas^cūdi* (*Canon Masudicus*), 3 vol., (Hyderabad, 1954–6).

C. E. Bosworth, *The Islamic Dynasties*, Islamic Surveys 5, (Edinburgh, 1967).

Bosworth, *CHI*. C. E. Bosworth, "The Political and Dynastic History of the Iranian World (A.D. 1000–1217)", *Cambridge History of Iran* V (1968), 1–202.

Geo. Brady, *Materials Handbook. An Encyclopaedia for Purchasing Agents, Engineers and Foremen* (New York, 1963).

M. Destombes, "L'Orient et les catalogues d'étoiles au Moyen Age", *Archives Internationales des Sciences*, 9 (1956), 339–44.

M. Destombes, "Globes celestes et catalogues d'étoiles orientaux du moyen–âge", *Actes du VIIIe Congrès Internationale d'Histoire des Sciences, Florence–Milan* 3–9 *Septembre* 1956, (Florence, 1958), vol. I, pp. 313–24.

A. G. Drachmann, "Ctesibius", *Dictionary of Scientific Biography* III (New York: Scribners, 1971), pp. 491–2.

A. G. Drachmann, "Ktesibius, Philon and Heron. A Study in Ancient Pneumatics", *Acta Historica Scientiarum Naturalium et Medicinalium* (Copenhagen), IV (1948).

A. G. Drachmann, Review of *Heavenly Clockwork* by J. Needham et al., *Centaurus* X (1964), 201–3.

A. G. Drachmann, "The Mechanical Technology of Greek and Roman Antiquity", *Acta Historica Scientiarum Naturalium et Medicinalium* (edidit Bibliotheca Universitatis Hauniensis) 17, (Mnnksgaard, 1963).

F. I. Haddad and E. S. Kennedy "Geographical Tables of Medieval Islam", *Al–Abhath* 24 (1971), pp. 87–102.

Robt. E. Hall, "Al–Khāzinī", *Dictionary of Scientific Biography*, VII (1973), pp. 335–51.

Heron, *Op. Om.* I. *Heronis Alexandrini Opera quae supersunt Omnia*, vol. 1 Pneumatica et Automata, recensuit Guilelmus Schmidt, (Leipzig, 1899).

Hill, *Ibn Mu^cādh*. D. R. Hill, "A treatise on Mechanics by Ibn Mu^cadh Abū ^cAbd Allāh al–Jayyānī", *Journal for History of Arabic Science*. 1 (1977), 34–46.

Walther Hinz, *Islamische Masse und Gewichte, umgerechnet ins Metrische System*, Handbuch der Orientalistik, Erganzungsband 1, Heft 1, (Leiden, 1955).

F. Hultsch, "Ueber den Himmelsglobus des Archimedes", *Zeitschrift fur Mathematik und Physik* (Leipzig), 22 (1877), 106–7.

Jazarī. Ibn al–Razzāz al–Jazarī, *The Book of Knowledge of Ingenious Mechanical Devices*, translated and annotated by Donald R. Hill, (Dordrecht, 1974).

Kennedy, *Parallax*. E. S. Kennedy, "Parallax Theory in Islamic Astronomy", *Isis*, 47 (1956), 33–53.

Kennedy, *Survey of Tables*. E. S. Kennedy, "A Survey of Islamic Astronomical Tables", *Transactions of the American Philosophical Society*, New Series 46, part 2 (1956), 121–77.

Kennedy, *Bīrūnī's Taḥdīd*. E. S. Kennedy, *A Commentary upon Bīrūnī's Kitāb Taḥdīd al-Amākin, an Eleventh Century Treatise on Mathematical Geography* (Beirut, 1973).

E. S. Kennedy and Jimal Rajeb, "A Description of the Contents of Ẓāhiriya MS 4871", to appear.

N. Khanikoff, "Analysis and extracts of *Kitāb mīzān al–ḥikma* (Book of balance of wisdom), an Arabic work on the water–balance, written by al–Khāzinī in the XII century", *Journal of the American Oriental Society*, 6 (1859), 1–128.

Al–Khwārizmī. *Liber Mafâtîh al–Olûm explicans vocabula technica scientiarum tam arabum quam peregrinorum auctore Abû Abdallah Mohammed ibn Ahmed ibn Jûsof al-Kâtib al-Khowarezmi*, edidit G. Van Vloten, (Leiden, 1895).

David King, "Al-Khalīlī's *qibla* Table", *Journal of Near Eastern Studies* 34 (1975), 81–122.

David King, "Ḳibla", *Encyclopaèdia of Islam*, 2nd edition, (Leiden: E. J. Brill, 1960 to present).

Max Krause' "Stambuler Handschriften islamischer Mathematiker", *Quellen und Studien zur Geschichte der Mathematik. Astronomie und Physik*, Abt. B, Bd. 3 (1936) Heft 4, pp. 437–532.

Kunitzsch, *Glossare*. Paul Kunitzsch, "Mittelalterliche astronomisch–astrologische Glossare mit arabischen Fachausdrücken", *Bayerische Akademie der Wissenschaften. Philosophisch–historische Klasse, Sitzungsberichte*, Heft 5 (1977), 1–59.

Kunitzsch, *Ibn aṣ–Ṣalāḥ*. Paul Kunitzsch, *Ibn aṣ–Ṣalāḥ: Zur Kritik der Koordinatenüberlieferung im Sterkatalog des Almagest* (Göttingen, 1975). [Also in *Abh. d. Ak. d. Wiss. in Gottingen* III 94 (1975)].

Kunitzsch, *Chrysokokkes*. Paul Kunitzsch, "Das Fixsternverzeichnis in der 'Persischen Syntaxis' des Georgios Chrysokokkes", *Byzantinische Zeitschrift*, 57 (1964), 382–411.

Le Strange. *The Geographical Part of the Nuzhat–al–Qulūb composed by Hamd–allāh Mustawfī of Qazwīn in* 740 (1340), edited and translated by G. Le Strange, vol. II (translation), (Leiden, 1919).

Libros del Saber de Astronomía del Rey D. Alfonso X de Castilla, compilados, anotados y comentados por Don Manuel Rico y Sinobas, (Madrid, 1863–7).

R. P. Lorch, *Sphera Solida*. "The *sphera solida* and Related Instruments", *Centaurus*, 24 (1980), 153–61.

R. P. Lorch, *Qibla* "The *Qibla*–Table Attributed to al–Khāzinī", *Journal for the History of Arabic Science*, 4 (1980), 259-64.

R. P. Lorch. *Balance–Clock*. "On al–Khāzinī's Balance–Clock and the Chinese Steelyard Clepsydra", to appear in *Archives Internationales d'Histoire des Sciences*.

C. A. Mayer, *Islamic Astrolabists and Their Works* (Geneva, 1956).

Max Meyerhof, "Alī al-Bayhaqī's Tatimmat Ṣiwān al-Ḥikma, A Biographical Work on Learned Men of the Islam", *Osiris*, 6 (1948), 122–217.

Millás Vallicrosa, *Estudios universitaris Catalans, serie monografica*. 1, vol. I, (Barcelona, 1931).

Mīzān. ᶜAbd al-Raḥmān al-Khāzinī, *Kitāb mīzān al-ḥikma* (Hyderabad, 1941).

S. H. Nasr, *Islamic Science*; *An Illustrated Study* (London, 1976).

Needham, *CCCW*. Joseph Needham, *Clerks and Craftsmen in China and the West* (Cambridge, 1970).

Needham, *SCC*. Joseph Needham and Wang Ling, *Science and Civilization in China* (Cambridge, vol. 1 1954, vol. III 1959, vol. IV part ii, 1965).

Needham, *HC*. Joseph Needham, Wang Ling, D. de S. Price, *Heavenly Clockwork. The Great Astronomical Clocks of Medieval China* (Cambridge, 1960).

Pappus, *Collection. Pappi Alexandrini Collectionis Quae Supersunt*, ed. F. Hultsch, (Berlin, 1876–8).

Philo, *Appareils pneum.* Baron Carra de Vaux, *Le livre des appareils pneumatiques et des machines hydrauliques par Philon de Byzance* (Paris, 1902).

David Pingree, "History of Mathematical Astronomy in India", *Dictionary of Scientific Biography*, XV (1978), 533–633.

F. D. Prager, *Philo Byzantius. Philo of Byzantium, Pneumatica. The Latin treatise on experimental Physics: Western Version and Eastern Version* (Wiesbaden, 1974).

D. de S. Price, "On the Origins of Clockwork, Perpetual Motion Devices and the Compass", Constributions from the Museum of History of Technology, U. S. National Museum [Smithsonian] *Bulletin* 218, paper 6 (1959), 82–111.

Proclus, *Hypotyposis. Procli Diadochi Hypotyposis Astronomicarum Positionum*, edidit C. Manitius, (Leipzig, 1909).

Referativnyi Zhurnal, *Astronomiya*, Akademiya Nauk SSSR, Volume for 1979, item 5.51.15.

Ricci, *De Chr. Exp. De Christiana Expeditione apud Sinas suscepta ab Societate Jesu. Ex P. Matthaei Ricii eiusdem Societatis Commentariis, libri V*, auctore P. Nicolao Trigantio, (Augsburg, 1615).

A. Rome, "Commentaires de Pappus et Théon d'Alexandrie sur l'Almageste", *Studi e Testi*, 54 (1931).

M. M. Rozhanskaya, "The Astronomical Clock of al-Khāzinī" [in Russian], *Istoriko-matematicheskie Issledovaniya*, 14 (1978), 189–98.

Aydin Sayılı, "Al-Khâzinî's Treatise on Astronomical Instruments", *Ankara Üniversitesi Dil ve Tarih-Cografya Facültesi Dergisi*, 14 (1956), 18–9 [English. Turkish orig. pp. 15–7].

L. Am. Sédillot, "Mémoire sur les instruments astronomiques des Arabes", *Mém. prés. par divers savants a l'Acad. Roy. des Inscriptions et Belles Lettres*, (Paris, 1845).

Fuat Sezgin, *Geschichte des arabischen Schrifttums* (Leiden, 1974-).

A. Ungerer, *Les Horloges Astronomiques et Monumentales les plus Remarquables de l'Antiquité* (Strasbourg, 1931).

Eilhard Wiedemann, "Zur Mechanik und Technik bei den Arabern", Beiträge zur Geschichte der Naturwissenschaft, VI, *Sitzungsberichte der Physikalisch-Medizinischen Sozietät zu Erlangen*, 38 (1906), 1–56. Reprinted in *Aufsätze zur Arabischen Wissenschaftsgeschichte* (Hildesheim, 1970), I, 173–228.

Eilhard Wiedemann, "Ueber die Stundenwage", Beiträge [see previous item] 37, *Sitzungsberichte*, 46 (1914) 27–38; = Aufsätze II, 57–68.

Eilhard Wiedemann and F. Hauser, "Ueber die Uhren in Bereich der Islamischen Kultur", *Nova Acta. Abhandlungen der Kais. Leop. Carol. Deutschen Akademie der Naturforscher* 100 (Halle, 1915), 1–272.

S. V. Zhitomirskiĭ, "Archimedes' 'Celestial Globe'" [in Russian], *Istoriko-matematicheskie Issledovaniya*, 14 (1978), 271–302.

XII

The *sphera solida* and Related Instruments

1. *Introduction*

The "solid sphere" is essentially a globe, on which the stars and principal celestial circles are depicted, and a frame of horizon and meridian circles. Astronomical problems are solved with the help of graduations on the circles. An elaborate version of the instrument is described in a Latin text, at least as old as the fourteenth century, beginning "Totius astrologie speculationis radix . . . ex hiis que per visum percepta sunt non sine congruis instrumentis sumpsit exordium". For convenience, the term *sphera solida* will be reserved for this text or this instrument. It carried a further quadrant, of which one end was pivoted at the zenith and the other could move along the horizon. A hemispherical bowl was added to the lower half of the apparatus. Four vertical cords served to support the instrument and were also adapted for taking altitudes.

The uses of the instrument were much the same as those of the plane astrolabe. There are several other instruments similar in purpose and design. In the spherical astrolabe the inner sphere carries the horizontal coordinates and the rete carries the stars and celestial coordinates. The only complete example is in the Oxford Museum of the History of Science.[1] A fragment of another was exhibited at the Science Museum, London, in 1976 in the exhibition associated with the World of Islam Festival.[2] Of the Arabic texts on the spherical astrolabe, the earliest are those by Qusṭa b. Lūqā (d. *c.* 912), though

I am most grateful to the Alexander von Humboldt-Stiftung for the fellowship that enabled me to carry out the research for this paper. I should also like to thank Professor Dr. Paul Kunitzsch for many helpful suggestions.

the attribution is uncertain, and al-Nairīzī (d. *c.* 922).[3] The instrument was mentioned by al-Khwārizmī (d. *c.* 850) and it is probable that it was known in Hellenistic times or earlier.[4] Al-Bīrūnī also included the spherical astrolabe in his book on "all possible methods of constructing the astrolabe"[5] and al-Marrākushī placed it with the other spherical instruments in his encyclopaedic work on astronomical instruments.[6] However, no description of the instrument could be found when the *Libros del Saber* were being drawn up for Alfonso X of Castille in the thirteenth century, and the work had to be done afresh[7] by Isaac b. Sid. Of course, Alfonso's astronomers had at least heard of the instrument and must have procured one for Isaac b. Sid to describe. A late treatise on the spherical astrolabe was written by the seventeenth-century ascetic al-Rūdānī, who flourished in Marrakesh and died in Damascus in 1683.[8]

Not to be confused with the spherical astrolabe is the armillary instrument described by Ptolemy in *Almagest* V, 1. It was a system of rings, or armillas, pivoted so as to rotate inside each other, and used for direct observation of the Moon and other celestial bodies.[9] But there were variants of the instrument that were applied to the same purposes as the plane astrolabe. Theon's "instrument with the rings", perhaps the same as the "meteoroscope" described by Proclus[10] is of this type. An elaborate treatise on a similar instrument was included in the *Libros del Saber,* having no fewer than seventy chapters on its uses. The armillary instruments of al-Kindī (d. 874),[11] Dunas ibn Tamīm[12] and Ḥabash al-Ḥāsib (9th century)[13] need to be further investigated. There is a Greek text on an instrument with rings, but no sphere, in Florence, but it has not yet been published.[14]

Of other types of three-dimensional universal instruments, al-Marrakushi mentions only the "shāmila" instrument. In essence this was a hemispherical bowl carrying an ecliptic-plate and half-equator-plate that could be revolved together about the pole of the universe. It was invented by al-Khujandī (d. *c.* 1000) and improved by Hibatallāh al-Asṭurlābī (d. 1140).[15]

2. *The* sphera solida

The text exists in numerous MSS, some of them of the fourteenth century,[16] and was printed in Venice twice in 1518[17] (ff. 139vb–143ra in both), with Sacrobosco's *Sphere* and other texts, and once in 1531 (ff. 202v–206r) with Campanus' *Sphere.* In MS British Library Arundel 268 it is described as "compositio a magistro Johanne de Harlebeke medico anno Domini 1303 Parisius"; and in MS Cambrai it is "compilatus a magistro Johanne de Halebeke Flandrensi experto".[17] This John of Harlebeke was perhaps the astrologer known to Gilles le Muisit (1272–1353) and a monk of the Benedictine abbey of St Martin, Tournai.[18] Other manuscripts are dated 1312 and most are anonymous.[19] Codex 46 Plut. 29 in the Laurentian library, Florence, after giving the name of John of Harlebeke and the date 1303, has the addition, "Dominus Accursius de Parma fuit principium huius operis".[20] Perhaps John was the author and Accursius the scribe of the Florence Manuscript.

After the prologue, the treatise is in two parts, nine chapters on making the instrument and thirteen on its uses.[21] The following is a translation of the prologue, based on the printed text of 1518 and Munich MSS CLM 27[22] and 10662.

The root and basis of all astronomical theory, and also its immense prolixity and inexhaustible depth of ingenuity, take their beginning from things that are observed with appropriate instruments. It is agreed amongst all the authorities on this subject that without instruments there would be no way of discovering the motions of the celestial bodies.

There is a great number of these instruments, but in this there generally seems to be agreement: that they all imitate the motion of the Heavens and are made in a likeness of it as if from the original, which not Man but God (may His name be blessed) has made. In the judgement of all those who philosophize aright [omnium recte philosophantium – the peripathetics], the Heavens are spherical. Hence, if it is also agreed that a copy is formed in a likeness to the original, it is plain that a spherical instrument is more similar to the spherical Heavens than all other instruments. In accordance with this, Ptolemy – being cognizant of the matter we have spoken of – made an armillary instrument with wonderful ingenuity [and described it] in the *Almagest* [V, 1]. The figure of the sphere is reflected in the form of this [instrument], but, contained only by the appropriate arrangement of circles, it had none of the stars in its composition – among the figures of the Heavens there is an empty space to mark the stars. By being so awkward, it is found unsuited [minus aptum sua tediositate] to determine the daily configurations of the Heavens.

Moreover, Ptolemy's astrolabe is elaborated in the likeness of that sphere [sc. of the Heavens], for it should be conceived as a sphere spread out in a plane. The geometrical proofs in this are ingenious, but in conceiving it the mind is affected by tedium, because things that are seen directly in the sphere we must there [in the astrolabe] conceive obliquely. Nor are the figures [imagines] in the sky [accurately] portrayed. [The astrolabe] is also partly truncated: it does not represent stars that are south of the tropic of Capricorn – not from necessity, but from custom and convenience. To place the latitudes of the planets in it is also difficult, if not impossible. It sometimes happens that the less experienced are misled in this matter and think that a planet in the East is above the Earth when it is below the Earth, and vice versa – and similarly for the other angle.[23] In the judgement of the experienced the astrolabe is deservedly held, by virtue of its ingenuity, numerous uses, simple operation, and portability, as nobler and better than the other instruments of the venerable Ancients.

The same Ptolemy mentioned a spherical instrument in the *Almagest* [VIII, 3], and indicated that the stars are inscribed in it according to longitude and latitude and that the celestial figures and constellations are drawn on it. But he does not intimate how this instrument might be brought to perfection so that it could be put to everyday uses, i.e. [finding] ascendants, equations of the houses, and other things necessary in this application [i.e. astrology]. We do not here set down the construction and uses of the astrolabe, as being sufficiently treated by the ancients. But what, by God's mercy, has come to us of the construction and uses of the instrument called "sphera solida" or "spherical astrolabe" we treat with the Lord's help – at the instance of some of my friends and for the use of those who might wish to look at this treatise.

We treat first, the construction of this instrument with all its parts, and secondly its uses. The instrument will be easy enough to construct, handsome and delightful in appearance, and because of its many uses, much desired and sought-after by the experienced. Of all other instruments it is the root and exemplar, except the two instruments that are called "perfect" – that is, they are universal – and other, imperfect, or particular, [instruments], such as the quadrant, cylinder, triquetrum, and shadow-instruments. Of these [latter] the everyday uses are few and they are, as it were, obtainable as parts of the perfect instruments. The Ancients have treated these carefully enough.

But enough of these things. We come now, by invoking divine help, to the book. The prologue ends.

It will be convenient to paraphrase the rest of the book. In the first part, the wooden[24] sphere is made (chap. 1) by turning it in a lathe about one axis and then turning it again about a perpendicular axis. On its surface are drawn (chap. 3) the equator, ecliptic, tropics, arctic and antarctic circles, the colure through the equinoxes, and the colure through the tropics. The fixed stars are indicated (chap. 4) according to their magnitudes by silver or tin wire of various thicknesses. This is pushed into the wood, flush with the surface, and the centre of the star is indicated by an indentation. The positions of the stars are found with the help of the meridian armilla, which is made (chap. 2) of brass (aurichalcum) and furnished with graduations and holes every degree or every two degrees (so that the instrument can be adjusted for any latitude). The support pieces (addidamenta) are placed symmetrically at the poles. In a similar, but slightly bigger, armilla, the horizon (chap. 5), slots are made to accommodate the meridian. Among the markings on the horizon are the signs of the zodiac and the months, but these have no other function than to be read off, one against the other, by means of a small cursor (*regula currens*).

A quadrant (chap. 6) is made and pivoted at the zenith with a device to allow for differences in geographical latitude.[25] The horizon can rest on the rim of a hemispherical bowl (chap. 7), which is supplied with a fork to hold the meridian. Finally the instrument – presumably *without* the bowl – may be suspended (chaps. 8, 9) by four vertical cords attached symmetrically to the horizon and fixed at the top to a "lily", a brass plate of four leaves, which may in turn be suspended from a ring at its centre. By means of a geometrical construction drawn separately, one of the cords is marked so that, when seen from a fixed point on the opposite cord, altitudes of a bead (almuri) aligned with some celestial object may be read off directly. Another of the cords is marked for shadows.

The first use of the *sphera solida* (part II) is as a demonstration instrument to show the various parts of the Heavens. To find the geographical latitude, one adds the latitude of a known star (or subtracts it, as appropriate) to its meridian altitude found with the cords. Alternatively, the mid-position between the greatest and least altitudes of a non-setting star may be taken.

To align the sphere with the Heavens in the daytime, and so obtain the *configuratio celi,* a pin is stuck into the degree of the Sun in the ecliptic and the sphere is turned until the pin has no shadow. At night the same can be achieved by the less spectacular method of taking the altitude of a known star and shifting the sphere till the representation of the star has the same altitude – just as in a plane astrolabe.

Apart from such easy exercises as finding the ortive amplitude, which is read off by putting the degree of the Sun on the horizon, the other uses of the instrument involve making marks on the equator-circle and finding the distance between them. For instance, to find the beginning of the eleventh house, a mark is put on the equator at its intersection with the horizon; and the process is repeated after the ascendant degree has been put under the meridian. If the arc between the two marks is trisected, and the first point of trisection is placed under the meridian, the degree of the ecliptic under the meridian is the beginning of the eleventh house. Finding the arc of day, the time, and the ascensions in the right and oblique spheres, are rather easier. The treatise finishes with a purely geometrical chapter on finding the height of a terrestrial object and like considerations.

3. Other Solid Spheres

Ptolemy's solid sphere in *Almagest* VIII 3 was probably taken as a starting-point for other instrument-makers. In this instrument, however, not even the equator was marked, because it would keep shifting with precession. Demonstration globes would seem to have been quite popular, even in the West,[26] and these could be supplemented by written descriptions such as that of Thābit b. Qurra (d. 901) in his *De imaginatione spere.*[27] But the earliest extant texts describing a solid sphere used like an astrolabe are by Qusṭā b. Lūqā (d. *c.* 912) and al-Battānī (d. 929).

The text of Qusṭā b. Lūqā's treatise on the solid sphere is very complicated.[28] There are at least two recensions in Arabic, dedicated to different people, and there were translations into Latin and Hebrew. It was also translated into Spanish for the *Libros del Saber,* prefaced by four extra chapters on constructing and mounting the sphere. At the end there is an extra chapter, taken from one Hermes, on finding the "atacyr"[29] and "egualar las casas", an astrological problem sometimes found in astrolabe-texts.[30] The Spanish version was subsequently translated into Italian.

However, the German translation of Schnell[31] based on MS Berlin 5836, the partial English translation of Worrell,[32] the Latin translation in MS Munich CLM 10662, ff. 118r–127v, and the Spanish seem to agree well enough, though the prologues are all different. The instrument is like the *sphera solida* without the quadrant, hemisphere, and cords. In the 65 chapters the author expatiates on many more uses than are to be found in the *sphera solida.*

Al-Battānī's instrument of this catagory,[33] the "bayda" (egg), is much more sophisticated, carrying another colural armilla with movable sights on one of its upper quadrants. It also had a movable meridian inside a fixed meridian, so that continuous adjustment could be made for geographical latitude.

There are two descriptions of solid spheres with separate quadrants.

One is by al-Khāzinī (fl. 1115–30) and is a complicated device mechanically rotated to match the Heavens.[34] Here the quadrant is used in most of the applications. The most interesting case is finding the arc of day, which is done by putting one end of the quadrant on the rising-point of the Sun. The distance along the equator between the place where the other end meets the equator and the meridian is the equation of the day. Al-Marrākushī also describes an instrument with a loose quadrant, but he gives its use only in a trivial application concerned with the construction of the instrument.[35] He gave no indication how the instrument was used. In the extra chapter, from "Hermes", to the *Libros del Saber,* a quadrant is mentioned as having one end pivoted to the pole of the ecliptic. Perhaps the original Hermes text belonged to the same tradition as those of al-Khāzinī and al-Marrākushī.

4. *Conclusion*

There are several unexamined Latin manuscripts and no doubt much has been lost. But no western treatise on the solid sphere considered so far is similar enough to the *Sphera solida* to suggest direct influence. It is therefore possible that it is a translation from the Arabic – a supposition confirmed in some places by its Latinity. Further, there are Islamic globes with a device like the quadrant on the *Sphera solida.* The famous globe in Dresden, made in 1279, has two movable semicircles pivoted at the zenith,[36] no doubt for the same purpose. It is also possible that the "perfect" instruments mentioned in the prologue are the shāmila instruments. But unfortunately there is no clear Arabic exemplar for the text of the *Sphera solida. Finit tractatus. Deo gratias.*

NOTES

1. See F. R. Maddison, "A fifteenth century Islamic Astrolabe", *Physis* 4 (1962) 101–9.
2. No doubt E. Cannobio, "An important fragment of a west Islamic spherical astrolabe", *Annali Ist. Mus. Stor. Sci. Firenze* 1 (1976) 37–41, is about this fragment, but I have unfortunately not seen it.
3. See H. Seemann and Th. Mittelberger, "Das kugelformige Astrolab nach den Mitteilungen von Alfonso X. von Kastilien und den vorhandenen arabischen Quellen", *Abhandlungen zur Geschichte der Naturwissenschaften und der Medizin* 8 (1925) Erlangen.

4. Maddison, *Op. Cit.*, p. 102.
5. *Kitāb al-Isti^cāb.* See Seemann and Mittelberger, *Op. Cit.*, p. 48.
6. *Compendium of beginnings (principles) and ends (results). Ibid.*, p. 44. The second part of al-Marrākushī's work (part relevant here) was given in French, abbreviated but with commentary, by L. Am. Sédillot in "Mémoire sur les instruments astronomiques des Arabes", *Mém. prés. par divers savants à l'Acad. roy. des Inscriptions et belles Lettres,* Paris 1845.
7. *Libros del Saber de Astronomía del Rey D. Alfonso X de Castilla,* compilados, anotados y comentados por Don Manuel Rico y Sinobas, Madrid 1863–7, vol. II, 103–223. See Maddison, *Op. cit.*, p. 102–3.
8. C. Pellat, "L'astrolabe sphérique d'Ar-Rūdānī", *Bulletin d'études orientales,* xxvi (1973) 7–82 and xxviii (1975) 83–166, Damascus. Also H. P. J. Renaud, "Additions et corrections à Suter . . .", *Isis* 18 (1932) 166–83.
9. See F. Nolte, "Die Armillarsphäre", *Abh. d. Gesch. d. Naturw. u. d. Medizin,* 2 (1922), Erlangen.
10. *Ibid.*, pp. 16–8 and 13. For two Arabic texts apparently by Theon, see F. Sezgin, *Geschichte des arabischen Schrifttums,* VI, 101–2.
11. *Ibid.* VI 153. An edition of this text is being undertaken.
12. S. M. Stern, "A treatise on the armillary sphere by Dunas ibn Tamīm", *Hommaje a Millás Vallicrosa,* II 373–80. Dunas lived in the tenth century.
13. *Ibid.* p. 377. The instrument may, however, be a solid sphere.
14. Florence, Laur. 89 sup. 48. I am indebted to Herr E. Reich (München) for lending me a copy of his transcription. There are other uninvestigated Greek texts.
15. J. Frank, "Ueber zwei astronomische arabische Instrumente", *Zeitschrift für Instrumentenkunde,* 40. Jarg., 7. Heft (Juli 1921) 193–200; and F. Rosenthal, *Osiris* 9 (1950) 555–64. The texts are being edited.
16. L. Thorndike and P. Kibre, *Catalog of Incipits of Medieval Scientific Writings in Latin,* 2nd ed., 1963. The incipit has only a superficial similarity to "Astrologie speculationis exercitium habere volentibus . . .", for which see Millás Vallicrosa, *Traducciones,* p. 321.
17. L. Thorndike, *Isis* 47 (1956) 393.
18. G. Sarton, *Introduction to the History of Science,* III 1746–7.
19. L. Thorndike, *Isis* 50 (1959) 37.
20. M. Steinschneider, *Zeitschrift für Mathematik und Physik* 16 (1871). Steinschneider suggests reading "fecit" for "fuit".
21. Steinschneider, *Op. cit.*, says that Oxford MS Ashmole 1796 has 14 chapters in the second part. The *explicit* he gives comes at the end of the 12th chapter of the 1518 edition. He also cites the last chapter of the first part in Cod. 16 Plut. 29 as the tenth.
22. CLM 27 often gives the best readings of the three, but there is a small lacuna at the end of part II chap. 11.
23. ". . . in reliquis angulis", i.e. plural, but there is only one angle in which the confusion could occur – in the West.
24. Metal is mentioned as a possibility, if an inferior one, but this does not square with the rest of the description.
25. The passage is not very clear.
26. E. Stevenson, *Terrestrial and Celestial Globes. Their History and Construction,* New Haven 1921, chapters 2 and 3; and Matteo Fiorini, *Erd- und Himmelsgloben, ihre Geschichte und Konstruktion,* nach dem italienischen frei bearbeitet von S. Günter, Leipzig 1895, ch. 3.

27. See the Latin translation edited by F. J. Carmody in *The Works of Thâbit b. Qurra,* Berkeley and L. A., pp 140–4.
28. See Sezgin, *Op. cit.,* VI 180–1; M. Steinschneider, *Hebräische Uebersetzungen,* 1893, 552–3. Steinschneider quotes his article in *Serapeum,* but I have not been able to see it.
29. See P. Kunitzsch, "Mittelalterliche astronomisch-astrologische Glossare mit arabischen Fachausdrücken", *Bayerische Akademie der Wissenschaft, Philosophisch-Historische Klasse, Sitzungsberichte,* Jahrg. 1977, Heft 5, 1–59, this on pp. 44–5 (article 22). Chapters 50–51 of the armillary instrument in the *Libros del Saber* are on the same question.
30. Verbal communication from Prof. P. Kunitzsch.
31. Hans Schnell, "Die Kugel mit dem Schemel", Inaug. Diss., Erlangen 1924.
32. W. H. Worrell, "Qusṭa ibn Lūqā on the Use of the Celestial Globe", *Isis* 35 (1944) 285–93.
33. *Al-Battani sive Albatenii Opus Astronomium* ed. C. A. Nallino, Rome 1899–1907: text III 210–4, translation I 139–42, commentary I 319–21 (partly supplied by Schiaparelli).
34. MSS Bodleian Thurston 3, ff. 118v–119v and Damascus 4871, ff. 73r–74r. An edition and translation have been prepared for publication.
35. Paris BN ar. 2508, f. 16r (new foliation). This part is not included by Sédillot, *Op. cit.,* p. 110–14.
36. Stevenson, *Op. cit.,* pp. 29–30. There is a fine picture in Helmut Grötzsch, *Dresden, Mathematisch-Physikalische Salon,* Leipzig 1978, plates 108, 109.

NOTE ADDED IN PROOF

Istanbul MS Aya Sofya 2624 (item 2) is perhaps a further Arabic text on the spherical astrolabe. See G. ᶜAwād, "Al-Asṭurlāb wa-mā ullifa fīhi ...", *Sumer* 13 (1957) 154–78, no. 71. His no. 141 is the treatise on the spherical astrolabe attributed to Qusṭā b. Lūqā.

Of further Arabic treatises on the solid sphere, that of ᶜAbd al-Raḥmān al-Sūfī (903–86) in 157 chapters in Istanbul MS Seray 3505 (item 1, 60ff) and the anonymous text in Aya Sofya 2673, ff 41v–84v, should be recorded. In Latin there is a fragment, from a very early translation from Arabic, on the use of the solid sphere, "De horologio secundum alkoram id est speram rotundam", published by Millás Vallicrosa in *Estudios universitaris Catalans, sèrie monogràfica, 1,* vol. I, Barcelona 1931, pp. 288–90. It bears little resemblance to the Qusṭā b. Lūqā text or the *Sphera solida.* There are MSS of both the latter in codex 224 of the State Library of Victoria, Australia. Further Latin texts are Jean Fusoris on the "Compositio revolutionum spere solide", inc. "Pro motu octave spere ..." in Paris BN 7295, ff 48–52v (see E. Poulle, *Un constructeur d'instruments ... Jean Fusoris,* Paris 1963, pp. 86 and 202–3) and a *De sphaera solida,* inc. "A phisicis spere gestibilis ...", in Florence bib. naz. conv. sopp. J. II 10, ff. 57r–64r.

XIII

ḤABASH AL-ḤĀSIB'S BOOK ON THE SPHERE AND ITS USE

1. Introduction

The instrument described in the text given below is a sphere set in a graduated vertical armilla (or ring) that is set in a graduated horizontal armilla. The sphere, on which the stars and various celestial circles are drawn, represents the heavens and the two armillas the meridian- and horizon-circles. In the form described by Ḥabash the instrument was used both as a demonstration instrument, for displaying the position and motions of the heavens, and as a model for solving problems in spherical astronomy in a direct way.

Celestial globes were well known in antiquity[1]. In the passage in Cicero's *Republic* in which Archimedes' globe was described[2] the original invention of the sphere (*sphaerae . . . solidae atque plenae*) is attributed to Thales of Miletus (6c. BC) and the marking of the stars to Eudoxus of Cnidus (4c. BC). Further, Aratus (early 3c. BC) used this instrument when writing his poem *Phaenomena*. Hipparchus may have used such a globe when compiling his star-coordinates[3]. It is not known when problems in spherical astronomy were first solved with the globe – though some, such as reading off the rising-places of the Sun or the stars were obvious enough. There is some evidence to suggest that mechanically propelled spheres with celestial markings were used in antiquity as demonstration devices, like the anaphoric clock[4].

[1] Schnell, 7–10; Schlachter.

[2] *De re publica*, I xiv (21, 22), pp. 13–14. Also described here is Archimedes' more complicated device showing the motions of the Sun, Moon and planets.

[3] Schnell, 9; Neugebauer, 931. For Aratus cf. Böker, 101–13. Cf. also Ptolemy, *Almag.* VIII 3 (see Neugebauer, 891 f.).

[4] Khāzinī, 290–91.

At least three types of celestial globe are described in Arabic texts. The simplest is that described by Ḥabash (fl. mid–9c.) and (some justification will be given later) by Qusṭā b. Lūqā (fl. late 9c.). In another type a further armilla is pivoted about the zenith and diametrically opposite point, so that altitudes and azimuths may be measured when not on the meridian. Al-Battānī (d. c. 930 AD) describes an elaborate instrument of this kind[5]. He called it *bayḍa*, a term used – apparently of an ordinary globe – also by the historian Abū ʿUmar Muḥammad b. Yūsuf al-Kindī al-Tujībī (896–971 AD)[6] to describe Aratus' globe and by Muḥammad b. Aḥmad al-Khwārizmī (late 10c. AD) in his *Mafātīḥ al-ʿulūm*[7]. Al-Battānī's *bayḍa* had an alidade that could be moved up and down the movable altitude-armilla, to sight the Sun directly. It seems that the Latin text edited by Millás, *Incipit de horologio secundum alkoram* . . ., was similarly equipped, for in the first paragraph it reads, "... *postea alzabis ipsam alkoram quousque sol inest per foramina super gradum solis* . . ."[8]; but the text has been so mangled in transmission that one cannot be sure. The *sphaera solida*, described in a text ascribable to John of Harlebeke in the early fourteenth century, who probably compiled it from texts deriving ultimately from Arabic sources[9], despite its name[10], had no alidade, but otherwise belonged to this class. Finally, the famous Dresden globe, made by al-ʿUrḍī in 1279 AD, also has a movable altitude semi-armilla[11].

The third type of globe is like the first, simple, type, but has a loose graduated quadrant. Such instruments are described by ʿAbd al-Raḥmān al-Ṣūfī (903–986 AD)[12], ʿAbd al-Raḥmān al-Khāzinī

[5] Battānī, cap. 57, I 139–42, 319–21; III 210–14. For the instrument in the mediaeval Latin translation see I lii–lv.

[6] Wiedemann, I 68; Sezgin, I 358.

[7] Wiedemann, I 80.

[8] Millás, 288. From two other places where *alzare* is used, it seems that its meaning is "to raise" – cf. Spanish "alzar", from Latin *altare*, from *altus*. See Bossong, 122, s. v. ALÇAR.

[9] Lorch, 155, 159.

[10] Geminus, e. g., speaks of making an armillary sphere and a "solid sphere", "καὶ αἱ κρικωταὶ σφαῖραι . . . καὶ στερεαί", V 65, p. 66 lines 9–10. The term *sphaera solida* could also be extracted from the passage in Cicero's *Republic* given above.

[11] Drechsler.

[12] Sezgin, VI 212–15.

(early 12c. AD), whose sphere is mechanically rotated[13], and Abū 'l-Ḥasan al-Marrākushī (13c. AD)[14]. Al-Khāzinī uses the quadrant for a variety of purposes, not only for measuring altitude and azimuth. Al-Marrākushī says that many globes lack the quadrant (and also a projection at the north pole), so that many of the necessary operations cannot be done[15].

Not until the requisite texts have been edited and studied will one be in a position to write a proper history of the instrument. There are not only the texts on the globe mentioned above – and there are others, too[16] –, but there are several other instruments with which comparison should be made, such as the astrolabe, plane and spherical, and the armillary sphere. In addition, there are numerous general descriptions of the heavenly sphere – by Autolycus, Euclid, Aratus, Theon, Thābit b. Qurra, and many others. Al-Marrākushī said he had seen five books on the sphere, the best being by Qusṭā b. Lūqā. The other four authors were Autolycus, Philon, Heron and Theon[17]. But we shall here largely confine ourselves to Ḥabash's description and a cautious comparison with Qusṭā b. Lūqā.

2. Ḥabash's globe text

Ḥabash's treatise on the globe is in two parts (*maqāla*), the first on the construction and the second on applications. For convenience these will be designated here by I and II respectively, the Arabic numeral following being the number of the chapter (*bāb*). At the beginning of the treatise the six chapters of the first *maqāla* are announced, but the text itself is not divided accordingly into numbered chapters. At the head of the second *maqāla* fourteen chapters are announced and the text is so divided. But in no extant manuscript does the numbering of the chapters run smoothly – two have an extra title "*al-bāb al-sādis*" in the middle of II 5 and the third is out of phase for several chapters. It is reasonable to suppose that

[13] Khāzinī. The astronomical applications are on pp. 308–17, 322–26.

[14] For the preparation of the quadrant see BN 2508, f 15r.

[15] BN 2508, f 13r lines 8–9.

[16] See SCHNELL's introduction. BN 2544, ff 79v–83v is a short variant o Qusṭā.

[17] BN 2508, ff 112r lines 5–8.

the chapter numbers and headings were added after the text was written.

There is some variation in the terminology. This may partly represent a difference between the two sections. Thus the term for armilla, *ḥalqa*, occurs twelve times in part I, but not once in part II. Again, the term *niṣf al-nahār*, for meridian occurs three times: twice in I 3, when the meridian-armilla is first introduced and in the phrase *irtifāʿ niṣf al-nahār*, and once in II 2 in the phrase *ẓill niṣf al-nahār*. In the last two cases it may be meant in the temporal sense of "midday". Elsewhere in part II (chapters 2, 8 [twice], 11 [twice] and 14) the term for meridian, referring to a place, is *wasaṭ al-samāʾ*, sometimes with a preceding *khaṭṭ*. There is a curious parallel in Ḥabash's treatise on the armillary instrument. In the first part of this text, describing the Ptolemaic instrument, he keeps to *niṣf al-nahār*, but when rings are added, so that the instrument may be used to find further arcs, Ḥabash uses – though with one exception – the term *wasaṭ al-samāʾ*[18].

The terminology for equator is more complicated, there being no fewer than four different terms. First, *muʿaddil al-nahār* is used once, in I 2. The regular term appears to be *al-falak al-mustaqīm* (here translated as *sphaera recta*). It is used when the graduations on the equator are being counted or marked (II 5, 6, 12), on its first introduction in I 1 – where it appears twice, in the curious compound *niṭāq al-falak al-mustaqīm*, perhaps to match the term for the ecliptic –, once in II 1, here apparently meaning the diurnal motion or its cause and once in the list of chapters in part II. Although Qusṭā b. Lūqā says that *khaṭṭ al-istiwāʾ* refers to the terrestrial equator corresponding to *muʿaddil al-nahār* in the heavens[19],

فالموضع من الأرض الذي يسامت خط معدل النهار من السماء يسمى خط الاستواء

and although this may explain the term *istiwāʾ* in the title of II 2 (which we may compare with the title of Qusṭā's fifth chapter – see

[18] For the manuscripts see SEZGIN, VI 173–75. See CELENTANO, 31–33 (app. crit.).

[19] In Q 4: M f 77r lines 3–4; A f 16r lines 7–9. The corresponding passages in SCHNELL and WORRELL agree.

below), Ḥabash uses it in different ways. In II 2 it means the east-west line, along which the shadow of a gnomon passes at the equinoxes on the terrestrial equator. In the other cases (I 4, II 3, 4) *khaṭṭ al-istiwā'* refers to some line – presumably the equator-circle or the east-west line – crossing the horizon at east- and west-points and making the zero for the rising-amplitude. Finally, there is yet another expression for equator in the last chapter, on declination. Here the equator is only considered in respect of its intersection with the meridian, and this intersection is called "the altitude of the beginning of Aries". This usage is to be found in other texts on instruments, such as the treatise on the armillary instrument attributed to Theon[20].

It is possible to see a common order, of a general kind, in the applications of the plane astrolabe in two early descriptions of its use, that of Philoponus and the first nineteen chapters of al-Khwārizmī's:

1. "static" problems, in which the instrument is set to correspond to the sky for one time and the desired information is read off;
2. problems involving two readings of the equator-circle;
3. finding various solar, lunar, planetary and stellar coordinates, beginning with the longitude of the Sun.

Both Philoponus and al-Khwārizmī have other chapters interpolated in this scheme, which is only intended to give a general view of the common parts of the two texts. No doubt refinements could be made to accommodate other astrolabe texts. The last twelve chapters of Ḥabash's globe text can be made to fit, more or less, into the scheme, as can be seen in the following table. Under the columns headed "Khw." and "Phil." are given the numbers of the corresponding chapters in Khwārizmī[21] and Philoponus. The column "Q", for Qusṭā, has been added for convenience in the discussion below.

[20] E.g., combined with *majrā* in the fourth *bāb*, MS Ahmet III 3505, f 119r last line (i. e., *majrā ra's al-ḥamal wa 'l-mīzān*).

[21] It should be remembered that there is as yet no critical text of Khwārizmī's treatise on the astrolabe, which appears to have suffered in transmission.

chap.	Ḥabash, *Sphere*	Q	Khw.	Phil.
3,4	rising amplitudes for the signs	31		
5	arc of day	(10)	5	(11)
5	length of day in equal hours	10	7	
5	length of temporal hour in degrees	12		11
5	time of day in temporal hours	14	2ii	5
6	oblique ascensions of points on ecliptic for a given star	33	9	10i
7	the co-rising point of the ecliptic	29	16	
8	the co-culminating point of the ecliptic	29	14	
8	and the co-setting point of the ecliptic	29	17	
9	place where year is a day and a night	45		
10	place where Taurus rises before Aries	48		
11	meridian altitude of Sun or star	42, 43	18i	12ii
12	oblique ascensions, given the longitude			
13	opposite of 12			
14	declination of points on the ecliptic	30	13	15

Several things stand out. First, there are no "static" problems in Ḥabash, though he speaks in II 5 of finding the temporal hours of daylight "if . . . you have already determined the ascendant". We shall return to this question later. Secondly, the treatment of special latitudes mentioned in II 9 and 10 is particularly suitable for demonstration in the globe. Thirdly, II 12 and 13 are odd in that chapter 12 repeats what has been done in chapter 6 and they otherwise have no equivalent in the other texts. Their oddness is confirmed by their terminology (see commentary). Of course, finding the time of day in temporal hours is a "static" problem for the astrolabe, with its temporal-hour markings, but not for the globe. If we remove this application and also chapters 9, 10, 12, 13, and compress the rest, we get the following, rather tidier, chart (p.o.e. means "point[s] on ecliptic"):

chap.	Ḥabash, *Sphere*	Q	Khw.	Phil.
3,4	rising amplitudes of p.o.e.	31		
5	arc-of-day problems	10, 12	5,7	11
6	oblique ascensions of p.o.e.	33	9	10
7,8	co-rising etc. p.o.e.	29	14, 16, 17	
11	meridian altitudes	42, 43	18i	12ii
14	declination of p.o.e.	30	13	15

The similarity of the order of these basic problems in Ḥabash, Khwārizmī 1–19 and Philoponus is immediate. At this stage we cannot tell if the curious, early, position of the rising-amplitude question is peculiar to Ḥabash or peculiar to an early form of the rehearsal of the instrument's uses.

3. The globe texts of Ḥabash and Qusṭā b. Lūqā

Both Ḥabash (II 5 iv) and Qusṭā (Q 13, 14)[22] describe how to tell the time with the globe if the ascendant is known; and Qusṭā also gives (Q 15, 16) the reverse procedure, to find the ascendant if the time of day is known. Now at night the ascendant can be found by observation, but for the daytime the *impasse* remains until chapter 54 in Qusṭā and is never resolved by Ḥabash. Perhaps it is for this reason that al-Marrākushī says that Qusṭā, though the best text on the instrument available to him, treated it inadequately[23]. For Marrākushī himself brings forward the problem of setting the globe (*waḍ' al-kura*) to match the sky to make his chapter 7. For this he uses essentially the same procedure as Qusṭā's for finding the altitude of the Sun (Q 53), namely making the horizon-armilla horizontal and turning it and the sphere until a peg put in the sphere at the degree of the Sun is in its own shadow. In chapter 11, on finding the altitude of the Sun "by the way that Qusṭā mentioned"[24], he refers back to this chapter 7.

[22] This convention will be used hereafter for convenience. Q 13 means Qusṭā's treatise, chapter 13.

[23] BN 2508, f 112r lines 11–13.

[24] BN 2508, f 116r line 5. Chapter 7 is on ff 114v–115r.

Now Qusṭā's text exists in at least two recensions by himself, dedicated to different people[25], and there is additional material in the Spanish translation made for Alfonso X in the thirteenth century[26]. In one of the additions, to Q 21, a quadrant is introduced so that a superior alternative method could be used for finding latitudes[27]. It is possible that the passage was already in the Arabic original followed by the translators. By analogy one could suppose that the quadrant introduced in Q 53 (and used in subsequent chapters) was an afterthought. If this is the case, it is likely that Ḥabash and Qusṭā's first 52 chapters belong to a tradition different from – and presumably older than – that of the globe with quadrant.

There are similarities between Ḥabash's and Qusṭā's treatment of the instrument – for instance in the description of the markings on the sphere and the example (II 10; Q 48) of latitude 78° as a place where Taurus rises before Aries[28]. The differences are also obvious: Qusṭā gives many more applications; with the exception of the description of the fixed stars he is always fuller and also more wordy; and his terminology is often different. Both similarities and differences are strikingly illustrated by the following extracts from Qusṭā[29]:

١ – وأما قطبا فلك وسط البروج فهما النقطتان اللتان تتقاطع عليهما الاثنتا عشرة دائرة الفاصلة للبروج، ...

٢ – الباب الخامس في معرفة علة استواء الليل والنهار في خط الاستواء، إذا أردت ذلك فضع قطب الشمال على الأفق من جهة الشمال فإن القطب الجنوبى لا محالة يصير على الأفق من جهة الجنوب، ...

[25] See e. g. WORRELL, 285.

[26] Alfonso, I 149–208. The first four chapters, giving details of construction, were added.

[27] Alfonso, I 181.

[28] Herr E. REICH (Munich) suggests that 78 was chosen as being midway between 66 (complement of the obliquity of the ecliptic) and 90. The special effects on latitude 66° were also treated by Qusṭā and others.

[29] 1: from Q 1, M f 76r lines 1–2, A f 12v lines 8–11. 2: from Q 5, M 77v, A f 18r–v. 3: from Q 48, M f 90v, A 43r.

٣ - الباب الثامن والأربعون في معرفة الموضع الذي يطلع فيه الثور قبل الحمل، إذا أردت ذلك فآرفع القطب الشمالى عن الأفق بقدر ثمانية وسبعين جزءًا وأدر الكرة فإنك ترى الثور يتقدم الحمل في دوران الكرة

A comparison with Ḥabash's I 2, II 2 and II 10 reveals great similarity, sometimes identity, in thought, but great diversity in expression. Sometimes every term is different. Other passages could have been taken to show the same thing. Some common source seems to be indicated, though how it came to be so different in the hands of the two authors remains to be explained.

4. The manuscripts[30] and editing conventions

E: Istanbul, Esat 2015(2), ff 28v–33v, 1181 H. It has a note of ownership on the flyleaf.

L: London University Library (kept at the School of Oriental and African Studies) 45806(3), ff 21v–26r. On f 21r, in another hand, is the following:

هذه رسالة معتبرة على الكرة وهو عبارة عن السماء والأرض عبارة من الجنوب والشمال [sic MS]

"This is a highly regarded treatise about the sphere, being an explanation of Heaven and Earth from the south and the north." This is followed by a note of sale.

B: Baghdad University, Philosophical Faculty, Maʿhad al-Dirāsāt al-Islāmīya, 171, ff 2v–4r, 1130 H. On f 1r, in a modern hand is the following:

كتاب معرفة الكرة والعمل بها، تأليف حبش الحاسب.

At the end of the treatise is a colophon giving the date of copying:

سوده لنفسه الحقير محمد على النظامى في ٢٨ رجب الفرد سنة ١١٣٠

[30] Many of the details are taken from Sezgin VI 173–75.

In the text all *hamza*s have been printed, but *shadda*s have been inserted only where necessary – in both cases irrespective of whether they appear in the manuscripts. Trivial differences in diacritical points (e. g. *yakūn* for *takūn*) have not been noticed in the apparatus; and the orthography of the *hamza* in the plural of *ra'ṣ* has been silently corrected. Otherwise it is intended that the reading of each manuscript can be deduced from the text and apparatus. Angle brackets, ⟨ ⟩, enclose words added by the editors.

In the translation care has been taken to reproduce technical terms as consistently as possible. Thus *niṭāq al-burūj* is rendered as "the belt of the signs", but *falak al-burūj* is "zodiac". On the terms for "meridian" and "equator", see above. The fifth form of *WSṬ* has been translated as "mediate", i. e. "come to the middle of the heavens". On the other hand *maʿrifa* is sometimes translated as "knowing" and sometimes as "determining". Square brackets, [], indicate material added by the translators.

بسم الله الرحمن الرحيم

هذا الكتاب فى معرفة الكرة والعمل بها من قول حبش الحاسب وهو مقالتان

المقالة الأولى فى معرفة الكرة تكون ستة أبواب، الباب الأول في معرفة ما شكل الكرة وما الخطوط المرسومة عليها وكم خط هي وما اسم كل واحد منها، الباب الثانى فى معرفة محاورها، الباب الثالث فى معرفة الحلقة التى تركب فيها هذه الكرة، الباب الرابع فى معرفة الحلقة التى تقطع تلك الحلقة على زوايا قائمة، الباب الخامس فى معرفة الكواكب المرسومة على هذه الكرة، الباب السادس فى معرفة نصبها ⋄

أما شكلها فنقول: إن شكل الكرة هو الشكل المجسّم المستدير المبيّن صنعته بالضرب والبرد والخرط والجرد والخطوط المرسومة عليها هي خطوط البروج الاثنى عشر ونطاق البروج وخط نطاق الفلك المستقيم، عدد الخطوط التى عليها أربعة عشر خطاً وأسماء الخطوط هى خطوط البروج الاثنى عشر تفصل ما بين كل برج وصاحبه وكل خط منها يقطع الكرة بنصفين ويمر على محورى فلك البروج الشمالى والجنوبى وفى وسط هذين المحورين خط يقطع هذه الخطوط الاثنى عشر كلها بنصفين وهو خط نطاق البروج وهو مقسوم بثلاثمائة وستين درجة ليقسم كل برج بثلاثين درجة، وعلى الكرة خط آخر زائل

1 Post الرحيم add B: وهو حسبى ونعم الوكيل.
4 منها B: منهما EL.
6 الحلقة B: علقة EL – تقطع تلك الحلقة EL: in marg. B.
8 اما EL: واما B – صنعته EL: صنعة B.
9 والخطوط EL: الخطوط B.
10 وخط BL: وحنط E.
11 ما EL: فيا B – برج BL: om. E.
12 الكرة EL: in marg. B.
14 بثلاثمائة وستين EL: ٣٦٠ B.

5. Text and translation

In the name of God, the Merciful, the Compassionate.

This book is about knowing the sphere and its use, from the account by Ḥabash al-Ḥāsib. It is in two sections. The first section, on knowing the sphere, is in six chapters: chapter 1 on knowing what the form of the sphere is, what the lines drawn on it are, how many lines they are, and what the name of each of them is; chapter 2 on knowing its poles; chapter 3 on knowing the armilla in which this sphere is mounted; chapter 4 on knowing the armilla that cuts this armilla at right angles; chapter 5 on knowing the stars represented on this sphere; and chapter 6 on knowing how to set it up.

[1] On its form we say that the form of the sphere is the solid round form, which is understood to be produced by beating, filing, turning and paring. The lines drawn on it are the lines of the twelve signs, the belt of the signs and the line of the belt of the *sphaera recta*. The number of lines that are on it is fourteen. The names of the lines are: the lines of the twelve signs – they separate each sign from its neighbour. Each of the lines cuts the sphere in two halves and passes through the north and south poles of the zodiac. In the middle of [*sc.* equidistant between] these two poles is a line that cuts each of these twelve lines into two halves: it is the line of the belt of the signs. It is divided into 360 degrees, so that each sign is divided into 30 degrees. On the sphere is another

عن نطاق البروج في موضعين وهما رأسا الحمل والميزان وهذا الخط يسمى نطاق الفلك المستقيم وهو فلك معدل النهار ⋄

ومحورا فلك البروج هما الموضعان اللذان تقاطعت عليهما خطوط البروج الاثنى عشر من جهة الشمال والجنوب، وأما محورا فلك معدل النهار فهو الموضع الزائل عن هذين المحورين بقدر الميل مما يلى رأس السرطان من جهة الشمال ومما يلى رأس الجدى من جهة الجنوب وهما المحوران اللذان تدور عليهما هذه الكرة ⋄

الحلقة التى تركّب فيها الكرة هى حلقة نصف النهار مشدودة على هذه الكرة بمسارين أحدهما على محور الشمال والآخر على محور الجنوب وهذه الحلقة هى التى تحمل الكرة بكمالها حتى تصيّرها على قدر ما يحتاج إليه من العرض والحلقة مقسومة بثلثمائة وستين درجة ليعرف منها قدر العرض وارتفاع نصف النهار وارتفاع رأس كل برج وارتفاع كل كوكب ⋄

الحلقة القاطعة لهذه الحلقة على أفق الشمال وأفق الجنوب هي حلقة الأفق مقسومة بثلثمائة وستين درجة ليعرف منها خط الاستواء وسعة المشارق والمغارب وبم يطلع الكوكب وبم يغرب وبم تطلع البروج وبم تغرب ومعرفة مقدار مطالع كل واحد من ذلك ⋄

وأما معرفة الكواكب التى فى هذه الكرة فإنها الكواكب الثابتة المضيئة منها تسمى

19 الميل BL: الليل E.
20 المحوران: BE, موضع وسط الممرين L – اللذان: موضع واسط الممرين EL, اللذين B. الذى
21 هى BL: om. E.
23 بثلثمائة وستين EL: ٣٦٠ B.
27 بثلثمائة وستين EL: ٣٦٠ B.
28 مطالع: مطلع MSS.

line, diverging from the belt of the signs in two places, which are the beginning of Aries and of Libra. This line is named the belt of the *sphaera recta*. It is the circle of the equator.

[2] The poles of the zodiac are the two places at which the lines of the twelve signs intersect on the north and south side. The poles of the equator are the place[s] deviating from these poles [*sc.* of the zodiac] by the quantity of the [total] declination, at [the part] near the beginning of Cancer towards the north and at [the part] near the beginning of Capricorn towards the south. They are the two poles on which this sphere rotates.

[3] The armilla in which the sphere is mounted is the meridian-armilla, firmly attached to the sphere by two pins, one at the north pole and the other at the south pole. This armilla is the one that carries the sphere in its entirety so that you may set it to the amount of latitude that is required. The armilla is divided into 360 degrees, so that there may be determined with it the amount of the latitude, the altitude [at] midday, the altitude of the beginning of every sign and the altitude of every star.

[4] The armilla intersecting this armilla at the horizon [at] the north and the horizon [at] the south is the horizon-armilla, divided into 360 degrees so that there may be determined with it the equator-line, the rising- and setting-amplitudes, what a star rises with and what it sets with, what the signs rise and set with, and knowledge of the quantity of the ascension of each of these.

[5] As for knowing the stars that are on this sphere, they are the fixed stars. Of these the bright ones are called *al-biyābānīya* –

البيابانية وهى التى يستعملها أصحاب القياس وما كان دون المضيئة من الكواكب فهى التى تسمى الكواكب الثانية وجميع عددها ١٠٢٢ كوكباً منها فى الشمال ومنها فى الجنوب أما التى فى الشمال فهى ٣٦٠ كوكباً والتى فى الجنوب ٣١٦ كوكباً والشمالية منها هى التى فى الشمال عن نطاق البروج والجنوبية منها هى التى فى الجنوب عن نطاق البروج وأما الكواكب التى فى صور البروج فهى ٣٤٦ وهى قريبة من نطاق البروج وبعضها فى الشمال وبعضها فى الجنوب وفيها منازل القمر وهى ثمان وعشرون منزلة يقطعها القمر فى كل شهر وينزل فى كل ليلة منزلة منها وقد رسمناها على مقاديرها فى الطول والعرض والعظم والصغر لتعرف هيئاتها واسماؤها ويستدل بها على مقادير الليل وجهة القبلة والمواضع التى يسأل عنها أكثر الناس ويستدل بها على السبل والجهات فى البر والبحر ⋄

وأما نصب الكرة فأن تركّب الحلقة التى تحمل الكرة فى حلقة الأفق وتصيّر المحور الشمالى على أفق الشمال والمحور الجنوبى على أفق الجنوب وتصيّر وجه الحلقة المقسومة مما يلى المشرق ليكون ذلك أبين للناظر إليه وأهون فى الاستعمال، تمت المقالة الأولى والحمد لله كثيراً ⋄

31 فهى التى EL: om. B.
32 الثانية: الثابتة MSS – كوكبا :EB كوكبها L.
33 ٣٦٠: ٣٢٠ MSS – الجنوب: الجنوبى MSS.
35 صور :BE صورة L – Post نطاق البروج add. et del. B: واما الكواكب التى فى.
36 ثمان: ثمانية MSS.
37 والعظم :BL العظيم E.
38 Post والمواضع add. et del. B: المتساوية.
39 يسأل :B يسئل EL – والجهات :BL والجهاة E.
40 تركب :E مركب ,L حركة B – فى حلقة: مركب ,EL حركة B.
41 الشمال :BE الشمالى L.
43 كثيرا :B كثير EL.

which are the ones the scientific astronomers [= *aṣḥāb al-qiyās*] use – and those of the stars that are of lesser brightness, which are called secondary [*al-thāniya*] stars. Their total number is 1022, some of them being in the north and some of them in the south. Those in the north are 360 and those in the south are 316. The northerly of them are those that are to the North of the belt of the signs and the southerly are those that are to the South of the belt of the signs. The stars that are in the constellations of the signs are 346. They are near the belt of the signs, some to the North and some to the South. In them are the mansions of the Moon, which are 28 [in number]. The Moon traverses them in every month and stays every night in one of the mansions. We have drawn them [*sc.* the stars] according to their values in longitude, latitude, and magnitude or smallness, so that their forms and names may be recognized and the lengths of the night, the direction of the *qibla* and the topics that most people ask about may be inferred from them and the ways and directions on land and sea may be inferred from them.

[6] To set up the sphere, you mount the armilla that carries the sphere in the horizon-armilla. You put the north pole on the horizon [at] the north and the south pole on the horizon [at] the south. You put the face of the graduated [meridian-]armilla towards the east so that it is clearer to someone looking at it and easier to use. End of the first section. Much praise to God!

المقالة الثانية فى العمل بالكرة وهى ١٤ باباً، الباب الأول فى معرفة حركة الكواكب كلها فى أفلاكها، الباب الثانى فى معرفة الموضع الذى لا عرض له وهو الموضع الذى يكون فيه الليل والنهار مستويين ويسمى الفلك المستقيم، الباب الثالث فى معرفة سعة المشارق لكل بلد، الباب الرابع فى معرفة <سعة> مشرق كل برج أو جزء غير السرطان، الباب الخامس فى معرفة مقدار نهار كل يوم كم يكون من ساعة مستوية ومعرفة أجزاء الساعات الزمانية ومعرفة ما مضى من النهار من ساعة زمانية أو مستوية ومعرفة قوس النهار فى كل يوم لكل بلد، الباب السادس فى معرفة مطالع أى برج من البروج أو جزء من أجزاء البروج، الباب السابع فى معرفة ما يطلع من الدرج لطلوع أى كوكب أردت من الكواكب الثابتة، الباب الثامن فى معرفة ما يتوسط به السماء، الباب التاسع فى معرفة الموضع الذى تكون فيه السنة يوماً وليلة، الباب العاشر فى معرفة الموضع الذى يطلع فيه الثور قبل الحمل، الباب الحادى عشر فى معرفة ارتفاع الشمس والكواكب، الباب الثانى عشر فى معرفة ما تصيب كل درجة، الباب الثالث عشر فى معرفة ما تصيب الدرج، الباب الرابع عشر فى معرفة ميل أى برج ◇

الباب الأول فى حركة الكواكب، فإن الكواكب كلها الثابتة والمتحيرة فإنه يديرها الفلك المستقيم فى كل يوم وليلة دورة واحدة من المشرق إلى المغرب وهى تسير فى البروج

44 فى معرفة BL: om. E.
46 مستويين BL: مشويين E.
48 مقدار نهار BE: مقدارتها L.
49 ما مضى BL: ماضى E.
50 برج BL: ببرج E.
53 فيه السنة EL: السنة فيه B – وليلة EL: وليله B.
54 والكواكب: والكوكب MSS.
55 ما [primum] BL: om. E – الباب BE: om. L.
56 ميل اى BL: اى ميل E.
58 الفلك BL: om. E – تسير BL: om. E.

The second section on the use of the sphere

It has fourteen chapters: chapter 1 on knowing the motion of all the stars in their spheres; chapter 2 on knowing the place that has no latitude, which is the place in which night and day are equal and which is called the *sphaera recta*; chapter 3 on determining the rising-amplitude for every region; chapter 4 on determining the rising-amplitude of every sign or degree except Cancer; chapter 5 on determining the quantity of daylight every day – how much it is in equal hours –, on determining the degrees of the temporal hours, on determining what has elapsed of the day in temporal or equal hours, and determining the arc of day for every day for every region; chapter 6 on determining the ascension of any of the signs or any of the degrees of the signs; chapter 7 on determining what rises from the degrees at the rising of any star you want of the fixed stars; chapter 8 on determining what it mediates with; chapter 9 on knowing the place in which the year consists of a day and a night; chapter 10 on determining the place in which Taurus rises before Aries; chapter 11 on determining the altitude of the Sun and the stars; chapter 12 on determining what every degree attains; chapter 13 on determining what the degrees [of ascension] attain; chapter 14 on determining the declination of any sign.

Chapter 1: on the motion of the stars. All the fixed stars and planets are rotated by the *sphaera recta* once every day and night from east to west. The [planets] move in their spheres in the

فى أفلاكها من المغرب إلى المشرق بخلاف الحركة الأولى وتدور فى مسيرها على محورى فلك البروج ✧

الباب الثانى فى معرفة استواء الليل والنهار، إذا أردت ذلك فصيّر المحور الشمالى على أفق الشمال فإذا فعلت ذلك فقد صار المحور الجنوبى أيضاً على أفق الجنوب وصار رأس الحمل والميزان فى ارتفاع كل واحد منهما إلى وسط السماء على سمت رؤوس أهل ذلك البلد وصارت القوس التى تدور فيها الشمس إذا كانت فى رأس السرطان مساوية للقوس التى تدور فيها الشمس إذا كانت فى رأس الجدى فتكون قوس نهار كل يوم مثل قوس ليله فى البروج الشمالية والجنوبية جميعاً وتكون الشمس زائلة عن سمت رؤوس أهل ذلك البلد ستة أشهر فى الشمال إذا كانت فى البروج الشمالية وستة أشهر فى الجنوب إذا كانت فى البروج الجنوبية وكذلك يكون ظل المقياس فى ذلك الموضع ستة أشهر فى الشمال وستة أشهر فى الجنوب وإذا كانت فى رأس الحمل والميزان لم يكن لشىء من الأشياء ظل نصف النهار إذا صارت الشمس فى وسط السماء ويكون الظل فى هذين اليومين من طلوع الشمس إلى غروبها يمر على خط مستقيم وهو خط الاستواء وتكون سعة مشارق أهل هذا الموضع بقدر الميل ويكون قدر ما يتسع مشرق كل جزء من البروج فى الأفق مثل قدر ميلها فى وسط السماء وتكون الدوائر التى تدور فيها الشمس وسائر الكواكب فى البروج كلها موازية لفلك سمت الرؤوس ومطالع البروج فى هذا البلد هى مطالع الفلك المستقيم وكذا الغروب ✧

59 Post محورى add. E: تسير (v. lin. 58).
62 المحور BL: لمحور E – الجنوب B: الجنوبى EL.
65 فيها L: فيه BE.
66 البروج BL: بروج E.
67 الشمالية . . . اذا EL: om. B.
70 اذا L: واذا BE.
73 ميلها: ميل MSS.

signs from west to east, contrary to the first motion. They rotate in their course[s] about the poles of the zodiac.

Chapter 2: on knowing [the place of] equality of night and day. If you want this put the north pole on the horizon [at] the north. When you do this, the south pole is also placed on the horizon [at] the south and the beginning of Aries and Libra – at the culmination [*irtifāʿ*] of each of them at the meridian – is placed at the zenith of that region. The arc in which the Sun rotates when it is at the beginning of Cancer becomes equal to the arc in which the Sun rotates when it is at the beginning of Capricorn. So the arc-of-day for every day is equal to the arc of its night in all the northern and southern signs. The Sun [at culmination] deviates from the zenith of that region for six months to the North, when it is in the northern signs, and six months to the South, when it is in the southern signs. Similarly, the shadow of a gnomon in that place is six months to the North and six months to the South. When [the Sun] is at the beginning of Aries or Libra, nothing has a midday shadow when the Sun arrives at the meridian. On these two days, from the rising of the Sun to its setting, the shadow passes along a straight line, which is the equator. The rising-amplitude for the people of that place is the amount of the declination: the amount of what the rising-place of each degree of the signs extends to [towards the North or South] at the horizon is equal to the amount of its declination at the meridian. The circles in which the Sun and all the stars in all the signs rotate are all parallel to the circle of [i. e. through] the zenith; the ascension of the signs in this region is the right ascension. Similarly for setting.

الباب الثالث فى معرفة سعة المشارق لكل بلد، إذا أردت ذلك فآرفع المحور الشمالى بقدر عرض ذلك الموضع ثم أدر الكرة حتى يطلع رأس السرطان ثم انظر كم بينه وبين خط الاستواء من الأجزاء فهو سعة المشرق لذلك البلد ◊

الباب الرابع في معرفة <سعة> مشرق كل برج أو جزء غير السرطان، إذا أردت أن تعرف سعة مشرق برج أو جزء غير السرطان فأطلع ذلك البرج أو الجزء من البروج ثم انظر كم بينه وبين خط الاستواء من العد فما كان فهو سعة مشرق ذلك البرج أو الجزء في ذلك البلد وكذلك افعل بالبروج كلها فى كل إقليم أردت معرفة سعة مشرقه ◊

الباب الخامس فى معرفة مقدار نهار كل يوم كم يكون من ساعة مستوية ومعرفة أجزاء الساعات الزمانية ومعرفة ما مضى من النهار من ساعة زمانية أو مستوية ومعرفة قوس النهار فى كل يوم لكل بلد، إذا أردت ذلك فآنظر إلى الجزء الذى فيه الشمس فضعه فى وجه الأفق ثم انظر ما حاذى الأفق من الفلك المستقيم فعلّم عليه ثم أدر الجزء الذى فيه الشمس حتى تضعه على أفق المغرب ثم انظر كم بين العلامة التى علّمت على الفلك المستقيم حين وضعت جزء الشمس على أفق المشرق وبين العلامة حين وضعت درجة الشمس على

76 Post بلد add. B: وكل جزء.

77 Post عرض add. et del. B: بلد – ادر BL: ادرك E – بينه BE: منه L.

79 الباب . . . السرطان EL: om. B – او جزء: وجزء EL.

79–80 اذا . . . السرطان BL: om. E.

79 اذا L: فان B – ان B: in L supra linea.

80 غير L: lacuna in B – فاطلع BL: فطلع E – او الجزء BL: والجزء E.

82 معرفة B: om. EL – مشرقه BL: مشرقة E.

83 الخامس EL: الرابع B – مقدار نهار BE: مقدارتها L.

84 ما مضى BL: ماضى E.

87 حتى B: حين EL – Post على [secundum] add. et del. B: الافق.

88 حين [secundum] BL: حتى E.

Chapter 3: on determining the rising-amplitude for every region. If you want that, raise the north pole by the amount of the latitude of that place. Then turn the sphere until the beginning of Cancer rises. Then see how much there is between it and the equator-line in the degrees [of the horizon]: that is the rising-amplitude for that region.

Chapter 4: on determining the rising-amplitude of every sign or degree except Cancer. If you want to know the rising-amplitude of a sign or a degree except Cancer, let that sign or degree of the sign rise [i. e. put it at the horizon]. Then see how much there is between it numerically and the equator-line. What results is the rising-amplitude of that sign or that degree in that region. Do similarly with all the signs in every climate for which you want to know the rising-amplitude.

Chapter 5: on determining the amount of daylight every day – how much it is in equal hours –, on determining the degrees of the temporal hours, on determining what has elapsed of the day in temporal or equal hours, and determining the arc of day for every day for every region. If you want that [*sc.* the arc of day], look at the degree the Sun is at and put it at the face [upper side?] of the horizon. Then see what [part] of the *sphaera recta* faces the horizon and put a mark on it. Then turn the degree that the Sun is at until you have put it on the horizon [at] the west. [Then make another mark on the part of the equator facing the eastern horizon.] Then see how much there is between the mark you made on the *sphaera recta* when you put the degree of the Sun on the horizon [at] the east and the mark [you made] when you put the degree of the Sun

أفق المغرب فما كان فهو قوس ذلك اليوم فى ذلك الإقليم ، فإن أردت قوس ليله فتمام هذا العدد إلى ٣٦٠ فهو قوس الليل وكذلك تستخرج <قوس> نهار الكوكب ، إذا أردت أن تعلم كم مقدار نهار ذلك اليوم من ساعة مستوية فاّقسم <قوس> النهار على ١٥ فما كان فهو ساعة مستوية وما نقص عن ١٥ فكسر من ساعة مستوية ، إذا أردت أن تعلم أجزاء ساعات نهار ذلك اليوم الزمانية فاّقسم قوس النهار اثنى عشر فما كان فهو أجزاء ساعات النهار فى ذلك البلد ، فإذا أردت أن تعلم ما مضى من ساعات النهار الزمانية وقد عرفت الطالع فاّنظر كم طلع من العدد من الموضع الذى علّمت <عليه> حين وضعت درجة الشمس على الأفق فما كان فهو ما مضى من ساعة زمانية بعد أن تقسمه على أجزاء ساعات ذلك اليوم وما لم يتم فهو كسر من ساعة زمانية ◇

الباب السادس فى معرفة مطالع برج من البروج أو جزء من أجزاء البروج ، فإذا أردت ذلك فأطلع ذلك البرج أو الجزء من البرج ثم انظر ما طلع من العدد من الفلك المستقيم فهو مطالع ذلك البرج أو الجزء من البرج فى ذلك الإقليم وكذلك تفعل فى كل موضع ◇

الباب السابع فى معرفة ما يطلع من الدرج مع طلوع أى كوكب أردت من الكواكب

90 Post الكوكب add. B in marg: الباب الخامس.

92 مستوية [secundum] BL: om. E – Post مستوية add. EL et add. in marg. B: الباب السادس.

93–94 E add. inter lineas 7–8 f. 32r: الشمس . . . والكوكب (separatim).

95 الذى BE: om. L.

96 ما مضى BL: ماضى E.

98 السادس EL: السابع B – او جزء من اجزاء البروج BE: om. L.

99 البرج [secundum] EL: البروج B.

102 السابع EL: الثامن B.

on the horizon [at] the west. What results is the arc of that day in that climate. If you want the arc of its night, then [that is] the complement of this number to 360: it is the arc of night. You similarly find the arc-of-day of a star. If you want to know how great the amount of daylight of that day is in equal hours, divide the arc of day by 15: what results is [the amount in] equal hours, and what falls short of 15 is a fraction of an equal hour. If you want to know the degrees of the temporal hours of daylight of that day, divide the arc of day into 12: what results is the degrees of an hour of daylight for that region. If you want to know what has elapsed of the temporal hours of daylight, and you have already determined the ascendant, see how much has risen numerically from the place that you marked when you put the degree of the Sun on the horizon: what results is the elapse in temporal hours after you have divided it by the degrees of the hours of that day. What is not whole is a fraction of a temporal hour.

Chapter 6: on determining the ascension of any of the signs or any of the degrees of the signs. If you want that, let that sign or degree of the sign rise. Then see what has risen, numerically, from the *sphaera recta*: it is the ascension of that sign or degree of the sign in that climate. You do similarly in every place.

Chapter 7: on determining what rises from the degrees [of the ecliptic] with the rise of any of the fixed stars

الثابتة، فإذا أردت ذلك فأطلع ذلك الكوكب ثم انظر ما طلع معه من جهة الأفق من البروج وأجزائها فهو ما يطلع به الكوكب ✧

الباب الثامن فى معرفة ما يتوسط به السماء، فإذا أردت ذلك ضع الكوكب على خط وسط السماء ثم انظر أى جزء وافى خط وسط السماء من البروج ومن الدرج فهو ما يتوسط به الكوكب السماء وكذلك فآفعل بالغروب ✧

الباب التاسع فى معرفة الموضع الذى تكون فيه السنة الشمسية يوماً وليلة، فإذا أردت ذلك فضع محور الشمال على عرض ص فإذا فعلت ذلك فقد صارت البروج الشمالية ظاهرة فوق الأرض أبداً وصارت البروج الجنوبية غائبة تحت الأرض فلذلك صارت الشمس طالعة على وجه الأرض ستة شهور وغاربة ستة شهور ✧

الباب العاشر فى معرفة الموضع الذى يطلع فيه الثور قبل الحمل، فضع المحور الشمالى على عرض عح فإذا فعلت ذلك فإنك ترى طلوع الثور قبل الحمل ✧

الباب الحادى عشر فى معرفة ارتفاع الشمس أو الكواكب فى وسط السماء، فإذا أردت ذلك فضع جزء الشمس أو أى كوكب أردت على خط وسط السماء ثم انظر ما وافى من العدد فما كان فهو ارتفاعه فى ذلك الإقليم فى الجهة التى هو فيها من الشمال أو الجنوب ✧

الباب الثانى عشر فى معرفة ما تصيب درج السواء من درج المطالع، إذا أردت ذلك

103 معه BE: منه L.

105 الباب الثامن فى EL: واما B – به السماء BE: بالسماء L – ضع BL: موضع E.

108 فيه BL: om. E.

111 ستة (bis) ست MSS.

113 عح: ع ح B, ع ح EL.

114 الكوكب: الكواكب MSS.

116 ارتفاعه BL: in E corr. ex ارتفاع.

117 فى معرفة ما BL: فيما E.

you want. If you want that, let that star rise. Then see what has risen with it at the horizon from the signs and their degrees: it is what the star rises with.

Chapter 8: on determining what it mediates with. If you want that, place the star on the meridian-line. Then see what degree reaches the meridian-line from the signs and from the degrees: it is what the star mediates with. Do similarly with setting.

Chapter 9: on knowing the place in which the solar year is a day and a night. If you want that, put the north pole at latitude 90. When you do that, the northern signs become always visible above the Earth and the southern signs become invisible under the Earth. Because of this the Sun is risen on the face of the Earth for six months and set for six months.

Chapter 10: on determining the place in which Taurus rises before Aries. Put the north pole at latitude 78. When you do this, you will see Taurus rise before Aries.

Chapter 11: on determining the altitude of the Sun or the stars on the meridian. If you want that, put the degree of the Sun or any star you want on the meridian-line. Then see what it reaches numerically: what results is its altitude in that climate on the side it is on, north or south.

Chapter 12: what the degrees of equality [i. e. of the ecliptic] attain in degrees of ascension. If you want to

فأطلع أى برج أو أى جزء أردت معرفة ذلك منه فما وافى من العدد المقسوم للفلك المستقيم فهو <ما> تصيب <من> درج المطالع فى ذلك البلد. ◊

الباب الثالث عشر فى معرفة ما تصيب درج المطالع من <درج> السواء، إذا أردت ذلك فأطلع درج المطالع ثم انظر ما وافى الأفق من درج السواء فهو ما تصيب درج المطالع من درج السواء لذلك البلد ◊

الباب الرابع عشر فى معرفة ميل أى برج من البروج أو جزء من اجزاء البروج، إذا أردت ذلك فانظر ارتفاع رأس الحمل فى ذلك الإقليم فاحفظه ثم ضع ذلك البرج أو الجزء الذى أردت معرفة ميله على خط وسط السماء أيضاً ثم انظر كم بينه وبين ارتفاع رأس الحمل من العدد فما كان فهو ميل ذلك البرج أو الجزء الذى أردت وكذلك تفعل فى البروج كلها فى ذلك الإقليم فافهم ◊

والحمد لله على إتمامه وعلى مزيد آلائه وإنعامه

118 للفلك BL: الفلك E.

119 تصيب: نصيب BL, om. E – Post المطالع scr. et del. B: من السماء.

120 اردت BE: om. L.

122 المطالع EL: المطلع B.

124 فانظر EL: om. B, add. supra alia manus – Post فى ذلك add. et del. B: البلد.

128 Post وانعامه add. L: امين.

know that, let any sign, or any degree of which you want to know that, rise. What it reaches in the subdivided numbering in the *sphaera recta* is what they attain in degrees of ascension for that region.

Chapter 13: on determining what the degree of ascension attains from degrees of equality. If you want that, let the degrees of ascension rise. Then look at what appears at the horizon in degrees of equality: it is what the degrees of ascension attain in degrees of equality for that region.

Chapter 14: on determining the declination of any of the signs or any of the degrees of the signs. If you want that, look at the altitude of the beginning of Aries in that climate and keep it. Then put that sign or degree whose declination you want on the meridian-line also. Then look at how much there is between it and the altitude of the beginning of Aries numerically. What results is the declination of that sign or degree that you wanted. You operate similarly for all the signs in that climate. Understand it well!

Praise be to God for bringing it to an end and for the superabundance of His benefactions and grace!

6. Commentary

I 1–5 The material of all these chapters is in Q 1.

I 5 The term *al-biyābānīya* comes ultimately from ἀπλανής through Middle Persian (Pahlavi)[31].

Qusṭā does not mention the numbers of the stars given by Ḥabash[32].

I 6 Ḥabash seems to have omitted the subsequent raising of the pole to the appropriate height for the locality – a step included by Qusṭā in Q 2.

II 1 Such a general account of the principles of astronomy occurs in other texts on the globe. In Qusṭā (Q 3) it is more definitely geared to showing the correspondence of the globe to the heavens.

II 9, 10 Exotic latitudes were a favourite topic for authors of written descriptions of the sphere. The case in which the north pole is the zenith is to be seen, for instance, in Autolycus' *Moving sphere*, chapter 4. Theodosius treats it in a more sophisticated way in *De habitationibus*, chapter 10. Since the Sun is there supposed still visible when it is up to half a sign on the ecliptic below the horizon, the "day" for this latitude is reckoned at a little more than seven months.

The latitude of 78° is associated by Ḥabash in his treatise on the spherical astrolabe with "Gog and Magog"[33]. The bulk of this treatise is devoted to demonstrating the "wonders" of the various latitudes.

II 12–13 Nallino[34] says that the term *daraj al-sawā'* (and its translation into various languages) occurs in connection with right or oblique ascensions. Al-Battānī[35] explicitly equates them with degrees of the ecliptic. In the same passage he calls the process of finding the degrees of the ecliptic rising with given degrees of ascension *taqwīs al-maṭāli' wa-taḥwīlhā ilā daraj al-sawā'*. Nallino also says that the term occurs often in Ḥabash. Finally, the eighth-

[31] See Kunitzsch.
[32] See e. g. Worrell, 287.
[33] E. g. MS Asghar Mahdavi 503, f 96v lines 13–14.
[34] Battānī, II 339.
[35] Battānī, III 41 line 7.

century astrologer Muḥammad b. ʿUmar b. al-Farrukhān used the term in his *De nativitatibus*. He also wrote a *Kitāb fī ṣūrat al-kura*[36].

Acknowledgements. It is a pleasure to thank the directors of the various libraries for making manuscript material available to us and to thank Professors Sezgin (Frankfurt) and Hamadanizadeh (Tehran) for sending reproductions of manuscripts and Dr. Asghar Mahdavi (Tehran) for allowing us to use a codex in his collection.

Bibliography

Alfonso. *Libros del Saber de Astronomía del Rey D. Alfonso X de Castilla*, compilados por Don Manuel Rico y Sinobas, i–v, Madrid 1863–67.

Autolycus. Autolykos, *Rotierende Kugel und Aufgang und Untergang der Gestirne.* ... übersetzt und mit Anmerkungen versehen von Arthur Czwalina, Leipzig 1931 (Ostwald's Klassiker der Exakten Wissenschaften, Nr. 232).

Battānī. *Al-Battānī sive Albatenii Opus Astronomicum*, ed. C. A. Nallino, i–iii, Milan 1899–1907.

Böker. Aratos, *Sternbilder und Wetterzeichen*, übersetzt und eingeleitet von Albert Schott mit Anmerkungen von Robert Böker, München 1958.

Bossong. *Los Canones de Albateni*, ed. Georg Bossong, Tübingen 1978 (Beihefte zur Zeitschrift für romanische Philologie, Bd. 165).

Celentano. Giuseppe Celentano, „L'epistola di al-Kindī sulla sfera armillare", Istituto Orientale di Napoli, *Annali*, Supplemento n. 33, vol. 42 (1982), fasc. 4.

Cicero. *M. Tulli Ciceronis scripta quae manserunt omnia. De re publica librorum sex quae manserunt.* Recognovit K. Ziegler, Leipzig 1915.

Drechsler. Adolph Drechsler, *Der Arabische Himmelsglobus des Mohammed ben Muyîd el-'Ordhi vom Jahre 1279 im Mathematisch-physikalischen Salon zu Dresden*, 2nd ed., Dresden 1922.

Geminus. *Gemini Elementa Astronomiae*, ed. C. Manitius, Leipzig 1898.

Khāzinī. Richard Lorch, "Al-Khāzinī's 'Sphere That Rotates by Itself' ", *Journal for the History of Arabic Science* 4 (1980), 287–329.

Khwārizmī. Josef Frank, „Die Verwendung des Astrolabs nach al Chwârizmî", *Abhandlungen zur Geschichte der Naturwissenschaften und der Medizin*, Heft 3, Erlangen 1922.

Kunitzsch. Paul Kunitzsch, „Stelle beibenie – *al-kawākib al-biyābānīya.* Ein Nachtrag", *Zeitschrift der Deutschen Morgenländischen Gesellschaft* 131 (1981), 263–67.

Lorch. Richard Lorch, "The *sphera solida* and Related Instruments", *Centaurus* 24 (1980), 153–61.

[36] Sezgin, VI 135.

MILLÁS. J. Millás Vallicrosa, *Assaig d'història de les idees físiques i matemàtiques a la Catalunya medieval*, vol. i, Barcelona 1931 (Estudis Universitaris Catalans, Sèrie monogràfica, 1).

NEUGEBAUER. OTTO NEUGEBAUER, *A History of Ancient Mathematical Astronomy*, I–III, Berlin/Heidelberg/New York 1975.

Philoponus. Jean Philopon, *Traité de l'astrolabe*, ed. A. P. SEGONDS, Paris 1981 (Astrolabica 2).

Qusṭā. Qusṭā b. Lūqā, Treatise on the Sphere. M: Asghar Mahdavi (Tehran) 503, ff 72v–94v, 11c. H; A: Ahmet III (Istanbul) 3475, ff 1v–79r, 10c. H.

SCHLACHTER. ALOIS SCHLACHTER, *Der Globus, seine Entstehung und Verwendung in der Antike*, Leipzig/Berlin 1927 (Stoicheia, VIII).

SCHNELL. HANS SCHNELL, *Die Kugel mit dem Schemel (kurra* [sic] *ḏât al-kursî) nach Alfons X von Kastilien und seinen arabischen Quellen*, Erlangen 1924 (Thesis, typescript).

SEZGIN. FUAT SEZGIN, *Geschichte des arabischen Schrifttums*, I ff., Leiden 1967 ff.

Theodosius. *Theodosii De habitationibus liber, De diebus et noctibus libri duo*, ed. RUDOLF FECHT, Berlin 1927 (Abhandlungen der Gesellschaft der Wissenschaften in Göttingen, Philol.-Hist. Kl., N. F. 19,4).

WIEDEMANN. E. WIEDEMANN, „Beiträge zur Geschichte der Naturwissenschaft III", repr. in his *Aufsätze zur arabischen Wissenschaftsgeschichte*, vol. I, Hildesheim/New York 1970 (originally published 1902).

WORRELL. W. H. WORRELL, "Qusta ibn Luqa on the Use of the Celestial Globe", *Isis* 35 (1944), 285–93.

The *Qibla*-Table Attributed to al-Khāzinī

AL-QAZWĪNĪ INCLUDED in his *Nuzhat al-Qulūb* a *qibla*-table – that is, a table giving the direction of Mecca –, which he says "was drawn up, on the order of the Saljuq Sulṭān Sanjar, by the pious Shaykh ᶜAbd ar-Raḥmān Khāzinī".[1] If this is true, it would put the table later than al-Khāzinī's treatise on "the sphere that moves by itself", which seems to have been written before he was in Sanjar's service.[2] On the other hand it is also possible that al-Khāzinī took the table over from someone else and that al-Qazwīnī confused the table with the *zīj*. The question of dating is mentioned because the "sphere" text gives a value for the *qibla* at Marw which may have been taken from the table.[3]

Since Muslims are required to face Mecca during prayer, tables have been drawn up at various times to give its direction. Like many such tables, the one under consideration is a rectangular array of 20×20 numbers, the values of the difference in latitude (here called $\Delta\varphi$) between the place in question and Mecca being marked horizontally to label the columns from 1° to 20°, and the difference in longitude (ΔL) being similarly marked vertically to label the rows. The entries, here called $q(\Delta\varphi, \Delta L)$, are the angles between the *qibla* and due South.

The entries can scarcely have come from any of the correct methods of calculating the *qibla*,[4] since results computed by such a method, incorporating several different values for φ_M, the latitude of Mecca, differed markedly from the corresponding entries in the table. It will be noticed that, if the value for $q(15,10)$ is set aside, the divergence increases as one goes down Table 1 (below). Further, neither of the two approximate formulae given by King.[5]

$$\tan q = \frac{\sin \Delta L}{\sin \Delta\varphi} \quad (1) \qquad \text{and } \tan q = \frac{\sin \Delta L}{\sin \Delta\varphi} \cos \varphi_M \quad (2)$$

can be the basis of this table. For according to these formulae the diagonal

* Institute for the History of Arabic Science, Aleppo University. I am most grateful to the Alexander von Humboldt-Stiftung for the fellowship that enabled me to carry out the research for this paper at the Institut für Semitistik der Universität München.

1. *The Geographical Part of the Nuzhat-al-Qulūb composed by Ḥamd-allāh Mustawfī of Qazwin in* 740 (1340), edited and translated by G. Le Strange, volume II (translation), (Leiden: 1919), pp. 27-31.
2. R. Lorch, "Al-Khāzinī's 'Sphere That Rotates by Itself'", *Journal for the History of Arabic Science* 4 (1980), p. 288.
3. *Ibid.*, pp. 325-6.
4. D. King, "Ḳibla", *Encyclopaedia of Islam*, second edition, gives several methods.
5. *Ibid.*, p. 84, col. 2. The formulae have been simplified and are given in modern notation.

entries (i. e. those for which $\Delta L = \Delta\varphi$) must be constant, but in the table they increase unsteadily from $41°18'$ to $44°50'$.

Since we can never be certain that any particular value is calculated and not interpolated, and that, even if calculated, it has come down to us as it was written, methods involving averaging were used to analyse the table, in the hope that the accumulated errors would more or less cancel each other out. Individual results obviously at variance with others of the same kind were ignored. To make a start, tests were applied to see if a trigonometrical function of q could be expressed as the product of a function of $\Delta\varphi$ and a function of ΔL. This is the case for formulae (1) and (2), but not for the correct formula, however expressed. Now $f(x,y) = g(x)\,h(y)$ if and only if $f(x,y)/f(x_1,y)$ is a function of x alone. To see if $\sin q(\Delta\varphi, \Delta L)$ is separable in this way, $u(\Delta\varphi, \Delta L) = \sin q(\Delta\varphi, \Delta L) / \sin q(10, \Delta L)$ was computed and tabulated for $\Delta\varphi = 1, 5, 15, 20$ and $\Delta L = 1, 5, 10, 20$. If $\sin q$ had been separable, the rows would have been identical, or, otherwise put, the entries in any one column would have been the same. But this was not so. Sin q was therefore not separable in the above sense. Cos q was likewise found to be inseparable. The result for $\tan q$, which was tested by calculating $\tan q(\Delta\varphi, \Delta L) / \tan q\,(10, \Delta L)$, with $\Delta\varphi = 1, \ldots, 20$ and $L = 1, 5, 10, 16, 20$, was excellent in parts, like the curate's egg, but poor for $\Delta\varphi \leqslant 4$.

None the less, the result was good enough to experiment with the hypothesis that $\tan q = g(\Delta\varphi)\,h(\Delta L)$, for some functions g, h. Now on this assumption g was clearly a decreasing function. Discouraging results were obtained by supposing that g was a simple cosine. So $g(\Delta\varphi) = \text{cosec}\,(\Delta\varphi)$ was tried — a supposition supported by analogy with formulae (1) and (2). Accordingly, $\tan q \sin \Delta\varphi$ was calculated and tabulated for the whole table to see if it was a function of ΔL alone, i. e. to see if the entries along the rows were the same. Obvious errors — including many entries for $\Delta\varphi \leqslant 4$ — were struck out and the rest were averaged, row by row. Only eight values so obtained (including those for $\Delta L = 1, 2, 3, 4$) showed an RMS error greater than 1.5%. The resulting twenty numbers are supposed to be values of some function of ΔL. Plotting them against ΔL produced a graph remarkably like a straight line through the origin. A straight line would mean that $\tan q \sin \Delta\varphi = K\,\Delta L$, for some constant K. But, whatever value of K was taken, the values of a implied by this formula were found to be in the main too large at the top of the table or too small at the bottom, or both. Some other function had to be tried for $\tan q \sin \Delta\varphi$. This function had to be both plausible and not simply a constant multiple of $\sin \Delta L$, since the values on the diagonal from top left to bottom right are not equal on the table.

Accordingly, the twenty numbers were compared with the values of the following function, for which a justification will be given shortly:

$$\frac{\sin \Delta L \ \cos \varphi_M}{\sqrt{1 - \sin^2 \Delta L \cos^2 \varphi_M}} \tag{3}$$

for L=1, ..., 20. Two values of φ_M were tried separately, 21°20′ and 21°40′, the values given by al-Bīrūnī and al-Qazwīnī respectively.[6] In both cases there was good agreement, the average error (no account being taken of sign) being 1.00% and 0.97% respectively. Unfortunately, no very definite result was obtained when formula (3) was used to find φ_M for each of the averaged values of $\tan q \sin \Delta\varphi$, the result varying between 20°9′ ($\Delta L = 6$) and 23°15′ ($\Delta L = 2$), and the average being 21°51′.

Sample results in degrees and minutes are tabulated below (table 1). In each box the top value is from the table, the second and third are calculated with the formula

$$\tan q = \frac{\sin \Delta L}{\sin \Delta\varphi} \cdot \frac{\cos \varphi_M}{\sqrt{1 - \sin^2 \Delta L \cos^2 \varphi_M}} \tag{4}$$

with φ_M = 21°20′ and 21°40′ respectively, and the fourth is the value obtained by the correct formula with φ_M = 21°20′.

ΔL \ Δφ	5	10	15	20
5	43 15 43 4 43 0 43 29	25 36 25 8 25 4 25 17	17 16 17 28 17 26 17 33	13 34 13 24 13 21 13 26
10	62 36 62 0 61 57 63 26	43 39 43 21 43 17 44 12	34 15 32 21 32 17 32 51	25 36 25 36 25 33 25 56
15	70 4 70 39 70 37 73 8	55 4 55 3 54 59 56 54	43 52 43 49 43 45 45 8	35 49 35 59 35 55 36 54
20	75 17 75 27 75 26 78 56	62 17 62 40 62 37 65 37	52 4 52 24 52 20 54 42	44 50 44 30 44 26 46 15

Table 1: Values of q: from top to bottom in any box are the values from the table, those calculated from formula (4) with φ_M = 21°20′ and 21°40′, and correct values.

6. Abū Rayhān Muḥammad b. Aḥmad al-Bīrūnī, *Al-Qanūnu'l-Maṣūdī*, (Hyderabad, 1954-56), volume II, p. 551; Le Strange (see note 1), p. 28.

The values of q from formula (4) with $\varphi_M = 21°20'$ were calculated on the assumption that tangent tables giving values for every degree were used and that intermediate values were found by interpolation. Straightforward calculation (which was used for the case $\varphi_M = 21°40'$) yields exactly the same results except that they are one minute bigger for $(\Delta\varphi, \Delta L) = (5,15), (5,20)$ and $(10,20)$. The value 34°15′ for $q(15,10)$ is probably a mistake for 32°15′. It is suggested here that formula (4) with $\varphi_M = 21°20'$ underlies the table. As far as we know, this formula is not attested in the medieval sources.

Testing the formula against the table

Entries in the table and (beneath them) corresponding values of q calculated from formula (4), with $\varphi_M = 21°20'$, are tabulated below (table 2) in degrees and minutes. The upper entries have been taken from Le Strange's translation of the *Nuzhat al-Qulūb*, except in two cases, $q(15,3)$ and $q(20,7)$, where values exactly equal to the calculated values are taken from the Persian text, which presents the table in *abjad* numerals. (In the translation these two values appear as 10°31′ and 18°23′ respectively). In the fourteen other cases of disagreement the readings are sometimes closer to the table in the translation and sometimes further from it.

The calculated values show 22 exact agreements, 59 values at most 6′ out, a a further 34 out by between 7′ and 10′, 32 (plus 13 that have already been counted as being just 10′ out) which are wrong in only one digit (not the last). Thus over 35% of the table can be reasonably accounted for by the formula. There are further entries that can be justified by supposing copying mistakes.

Three facts about the table may be noted. First, the top row and rightmost column between them include no less than ten values exactly in accord with the formula, three that differ by less than 10′, and seventeen that differ in just one digit (not the last).

Secondly, the following diagonal sequences (in direction, top left to bottom right) are in arithmetic progression: (15,13) to (18,16), with constant difference 20′; (16,15) to (20,19), with difference 10′; (11,11) to (14,14) with difference 2′; (17,17) to (20,20), with difference 9′; (14,15) to (19,20), with difference 5′; (10,12) to (13,15), with difference 10′; (14,16) to (18,20), with difference 10′; (13,16) to (16,19), with difference 10′; (11,16) to (13,18), with difference 10′. There is a further arithmetic progression, again with difference 10′, for the minutes (only) of entries (10,16) to (14,20). But similar differences along the rows, i.e. $q(\Delta\varphi - 1, \Delta L) - q(\Delta\varphi, \Delta L)$, and differences along the columns, i.e. $q(\Delta\varphi, \Delta L) - q(\Delta\varphi, \Delta L - 1)$ show no evident pattern. Indeed, the row-differences for $\Delta L = 14, 15, \ldots, 20$ and column-differences for $\Delta\varphi = 18, 19, 20$, which were examined in detail, were found to be not even monotonic, and seemed totally irregular. It is, of course, possible that the diagonals with constant difference were originally longer. An interesting example is the entry for

(16,20): as it stands, it gives, by interpolation with other entries, excellent approximation to al-Khāzinī's value for the *qibla* of Marw.[7]

Finally, the curious-shaped section enclosed by a line in the bottom left-hand corner of the table contains values with great divergence from those from those calculated by the formula. No explanation is offered for this, apart

ΔL \ Δφ	1	2	3	4	5	6	7	8	9	10	11	12	13	14	15	16	17	18	19	20
1	41 18 42 58	26 59 24 59	17 15 17 15	12 2 13 7	10 14 10 36	8 4 8 50	7 56 7 36	7 40 6 40	7 36 5 56	7 21 5 21	6 32 4 52	6 28 4 28	4 8 4 8	3 51 3 51	3 34 3 36	3 33 3 23	3 31 3 11	3 21 3 1	3 17 2 52	2 28 2 43
2	69 17 61 47	42 5 42 59	31 11 31 51	25 8 25 0	20 40 20 28	15 9 17 17	14 7 14 57	13 11 13 9	11 32 11 45	10 20 10 37	9 44 9 40	6 4 8 53	8 30 8 14	7 20 7 39	7 10 7 10	6 39 6 44	6 15 6 21	6 4 6 1	5 30 5 42	5 26 5 26
3	70 11 70 19	54 9 54 26	43 7 43 0	34 18 34 59	29 30 29 15	26 10 25 2	21 4 21 50	19 20 19 20	16 19 17 20	15 30 15 42	14 34 14 21	14 14 13 13	14 4 12 15	11 14 11 24	10 41 10 41	10 3 10 3	9 40 9 29	8 31 8 59	8 17 8 32	8 7 8 7
4	75 14 75 0	61 37 61 49	51 17 51 12	43 10 43 2	36 20 36 46	31 16 31 55	28 14 28 7	25 41 25 4	22 11 22 36	20 30 20 33	18 14 18 51	17 11 17 23	16 7 16 9	15 21 15 4	15 19 14 7	13 17 13 17	12 36 12 33	11 41 11 54	11 30 11 19	10 17 10 47
5	77 4 77 54	67 18 66 48	56 42 57 16	49 15 49 25	43 15 43 4	36 40 37 56	33 7 33 45	30 4 30 20	26 17 27 30	25 36 25 8	23 47 23 7	21 36 21 24	19 42 19 54	18 24 18 36	17 16 17 28	16 44 16 28	15 14 15 34	14 47 14 46	14 4 14 3	13 34 13 24
6	79 7 79 53	74 18 70 22	60 14 61 51	54 18 54 30	48 16 48 18	43 20 43 6	38 14 38 45	36 16 35 6	32 11 32 1	29 40 29 24	27 43 27 9	25 37 25 12	23 51 23 30	22 46 22 1	22 34 20 42	19 43 19 32	18 19 18 30	17 42 17 34	16 47 16 44	15 50 15 58
7	81 4 81 19	75 34 73 1	72 7 65 23	58 41 58 35	52 14 52 39	46 56 47 33	43 24 43 9	39 22 39 23	36 23 36 9	33 32 33 21	32 28 30 55	28 14 28 47	26 18 26 56	25 34 25 17	23 36 23 49	22 32 22 31	21 48 21 21	20 19 20 18	19 12 19 20	18 28 18 28
8	82 5 82 24	78 14 75 3	74 9 68 11	61 14 61 55	56 9 56 18	51 7 51 21	46 50 47 1	43 31 43 13	39 42 39 53	36 45 36 59	35 4 34 25	32 33 32 10	32 42 30 10	28 36 28 23	26 19 26 48	25 14 25 23	24 23 24 6	21 30 22 56	21 54 21 53	20 19 20 55
9	83 9 83 15	80 19 76 40	76 11 70 24	64 7 64 39	59 49 59 23	54 34 54 38	50 14 50 24	46 45 46 37	43 36 43 17	40 30 40 18	37 30 37 40	35 23 35 19	33 50 33 13	33 33 31 20	29 7 29 39	28 32 28 7	26 14 26 44	25 36 25 29	25 49 24 21	23 38 23 18
10	84 11 83 55	82 23 77 59	77 34 72 17	66 17 66 57	62 36 62 0	56 45 57 28	53 19 53 22	49 35 49 40	46 25 46 20	43 39 43 21	40 41 40 40	38 2 38 15	36 8 36 5	34 17 34 7	34 15 32 21	31 15 30 44	29 32 29 17	27 14 27 57	26 16 26 43	25 36 25 36
11	85 19 84 29	83 42 79 4	79 16 73 50	68 12 68 52	64 14 64 14	59 19 59 56	55 17 55 59	52 12 52 23	49 15 49 6	46 15 46 8	43 43 43 26	40 50 40 59	38 12 38 46	36 18 36 45	35 36 34 55	34 20 33 14	31 45 31 42	31 30 30 18	29 42 29 1	26 50 27 50
12	86 21 84 57	85 51 79 58	79 42 75 9	71 6 70 32	66 9 66 11	62 6 62 6	58 41 58 19	54 49 54 49	51 49 51 38	48 58 48 40	45 54 45 58	43 45 43 31	41 29 41 16	39 4 39 13	36 32 37 20	36 12 35 36	34 24 34 2	32 46 32 34	31 16 31 14	31 28 30 0
13	87 25 85 21	86 8 80 45	82 31 76 16	73 9 71 58	67 17 67 52	64 14 64 0	59 2 60 22	56 17 57 0	53 34 53 52	50 43 50 59	48 48 48 19	45 48 45 52	43 47 43 37	41 39 41 32	39 44 39 37	36 40 37 52	36 20 36 14	34 44 34 44	33 41 33 21	32 4 32 4
14	88 14 85 41	87 11 81 25	82 17 77 14	74 14 73 13	69 40 69 21	65 29 65 41	62 14 62 13	58 18 58 58	55 44 55 56	53 14 53 6	50 23 50 29	48 38 48 3	45 44 45 48	43 49 43 43	41 49 41 47	40 4 40 0	38 36 38 21	36 29 36 49	35 12 35 23	34 4 34 4
15	89 4 85 59	88 16 82 0	82 43 78 6	75 13 74 19	70 4 70 40	68 13 67 11	63 9 63 52	60 42 60 44	56 58 57 48	55 4 55 3	51 46 52 28	50 13 50 4	48 28 47 50	45 40 45 45	43 52 43 49	42 16 42 2	40 24 40 21	38 50 38 48	36 34 37 20	35 49 35 59
16	89 44 86 14	88 32 82 31	82 36 78 51	76 49 75 17	71 36 71 50	68 46 68 31	65 11 65 21	62 17 62 21	59 56 59 30	56 52 56 50	53 33 54 19	51 36 51 57	49 46 49 45	47 45 47 41	45 35 45 45	43 58 43 57	42 26 42 15	40 44 40 41	39 12 39 13	36 50 37 50
17	90 86 28	89 14 82 58	82 4 79 30	76 14 76 9	72 29 72 53	69 17 69 44	66 3 66 42	63 19 63 49	60 17 61 4	58 12 58 28	55 42 56 1	53 23 53 42	51 25 51 31	49 36 49 29	47 35 47 33	45 30 45 45	44 23 44 4	42 36 42 30	41 2 41 0	39 56 39 36
18	90 86 41	90 83 23	83 18 80 7	76 51 76 56	73 42 73 50	70 30 70 49	67 6 67 56	65 30 65 9	62 14 62 30	59 59 59 59	56 44 57 35	55 32 55 20	53 13 53 11	51 5 51 10	49 26 49 16	47 25 47 28	45 25 45 47	44 32 44 12	42 46 42 43	41 19 41 18
19	90 86 52	90 83 45	83 31 80 38	77 7 77 38	74 30 74 41	71 21 71 49	69 4 69 3	66 46 66 23	63 18 63 49	61 14 61 23	59 39 59 3	56 14 56 51	54 22 54 45	52 14 52 46	50 38 50 53	49 16 49 6	47 15 47 25	45 20 45 51	44 41 44 21	42 56 42 56
20	90 87 2	90 84 4	84 9 81 5	83 17 78 16	75 17 75 28	72 34 72 43	70 7 70 4	67 23 67 30	65 29 65 2	62 17 62 40	60 14 60 25	58 7 58 16	56 4 56 12	54 12 54 15	52 4 52 24	50 18 50 39	48 15 48 59	47 5 47 24	45 15 45 55	44 50 44 30

Table 2. Al-Khāzinī's *qibla*-table (upper entries) with calculated values (lower entries).

7. See note 3.

from the suggestion that the entries in this section were taken from another table. For if they are calculated, the underlying formula must be different from the one under consideration, which works noticeably better in the rest of the table; and if they are interpolated, the method of interpolation is obscure and the seven 90s in the corner are difficult to explain. It should be remembered that this part of the table gave wild results for tan $q(\Delta\varphi, \Delta L)/\tan q(10, \Delta L)$ at the beginning of this investigation.

Justification of the formula[8]

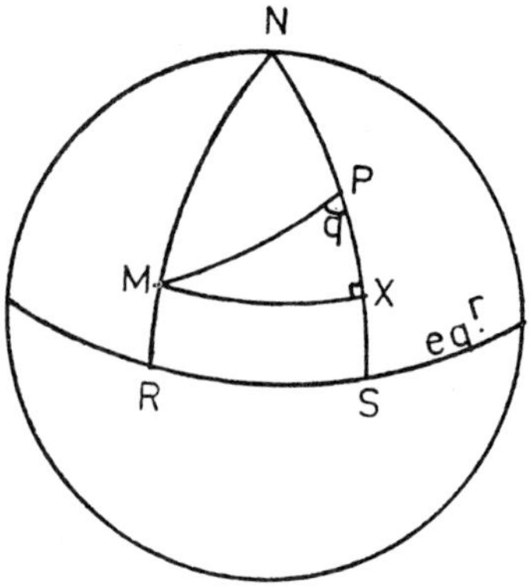

Formula (4) is easy to derive by elementary means. In the diagram N is the North pole of the equator RS, M represents Mecca and P the place in question. $MR = \varphi_M$, $PS = \varphi$ (the latitude of the place in question) and $\angle MNP = \Delta L$. If XM, drawn so that $NM = NX$, is assumed to be at once a great circle and perpendicular to NP — the inconsistency of these conditions characterize the approximation — then $PX = \Delta\varphi$,

$$\sin XM = \cos \varphi_M \sin \Delta L \tag{5}$$

from the sine-theorem in ΔNMX or from the "rule of four quantities", and

$$\tan q = \frac{\tan XM}{\sin \Delta\varphi} \tag{6}$$

by the tangent theorem in ΔMPX. Both results were known to al-Bīrūnī.[9] Since al-Khāzinī borrowed freely from him in other matters,[10] he can be assumed to have known these trigonometrical theorems too. Formula (4) follows immediately from (5) and (6).

Conclusion

In sum, formula (4) with $\varphi_M = 21°20'$ seems to fit too well to be rejected. The grosser irregularities in the table would create difficulties in fitting any smooth function. The diagonal runs mentioned above seem to imply some kind of interpolation. No satisfactory theory is offered here of its nature. Perhaps it was based in some way upon the relatively accurate values on the top row and rightmost column. What is abundantly clear is that the table is corrupt in many places.

8. Other methods are possible. Dr. D. A. King informs me (private communication) that some approximate methnds in use in ninth-century Iraq appear to have been derived by solid geometry.

9. *Qānūn* (see note 6), volume I, pp. 354-60.

10. See, e.g., R.E. Hall, "Al-Khāzinī", *Dictionary of Scientific Biography*, (Charks Scribner's Sons, New York) VII (1973) pp. 335–51.

XV

AL-KHĀZINĪ'S BALANCE-CLOCK AND THE CHINESE STEELYARD CLEPSYDRA*

In the eighth *maqāla* of his *K. Mīzān al-Ḥikma*[1], written 1121–2, 'Abd al-Raḥmān al-Khāzinī describes the construction of a twenty-four-hour water-clock designed for astronomical purposes. The device measures time by letting sand or water flow from a reservoir hung on one arm of a balance and seeing where counterbalancing weights must be placed to restore the arm to the horizontal. Whether or not the clock was a practical proposition, it is clear that some, at least, of the ideas embodied in the device originated in antiquity. For instance, a precious stone with a hole bored in it is used as the sand- or water-outlet[2]. Precious stones were used in this way by pseudo-Archimedes[3] and the idea goes back at least as far as Ctesibius[4] in the third century BC.

In *Science and Civilization in China*, Prof. Needham and his associates[5] confidently calls the public clock at Antioch (described by Chinese travellers of the tenth and eleventh centuries) "the great steelyard clepsydra", and it certainly had something resembling a balance-arm, though its function is far from clear[6]. If Needham's interpretation is correct, it is probable that the idea behind al-Khāzinī's clock goes back to the Byzantine, or more likely Hellenistic, world.

Some doubt should, however, be expressed about the practicability of al-Khāzinī's clock. Assuming a rate of flow of sand that he elsewhere[7] gives as an actual measurement, we may deduce from the calculations he presents in the *Mīzān* that the reservoir would weigh about 76.6 kg (over 1½ cwt) and be about 189 cm (6'2") tall[8]. This must hang on a balance-beam of total length of about 1½ m and approximate cross-section 4 cm × 3.3 cm. If in fact the balance-clock was a theoretical construction, it is fair to assume that the sources that al-Khāzinī relied on (if any) were written descriptions.

* I should like to thank the Alexander von Humboldt-Stiftung for the fellowship granted to me while writing this paper.

1 *Mīzān* 153 et seq. See also Wiedemann, *Stundenwage*; Rozhanskaya; Winter 27; Khanikoff, 17, 24, 105.

2 *Mīzān* 156.

3 See Hill in *Jazarī* 10 and 277. The treatise on clocks attributed to Archimedes was probably written at the end of the eighth century.

4 Drachmann, *DSB Ctesibius*.

5 Needham, *SCC* III 318. The various publications of Needham and associates will be given under the single name, Needham.

6 Wiedemann, *Über Lampen und Uhren* 368.

7 In his description of the self-rotating sphere. See MSS Damascus Ẓāhirīya 4871 f 73r l 27–28, and Oxford Thurston 3 f 118r l 9. A study of the text has been prepared for publication.

8 For the calculations, see the study in the previous note and Wiedemann, *Stundenwage* 59.

Proclus reports from the book *On Water-clocks*, now lost, by Heron[9]

> A vessel is constructed having a hole something like that of a clepsydra, through which the water can flow out evenly, as is usual.

He continues with a description of the use of the clock in measuring the apparent diameter of the Sun. The water that flows out during the rising of the Sun at the equinox is compared with the water that flows out during the remainder of the day till the next sunrise. The ratio of the weights yields the time taken for the Sun to rise, which in turn yields the apparent diameter. In a similar passage, Pappus begins[10]

> The ancients, constructing a vessel with a uniformly flowing stream through a small hole at the bottom and having a supply of some running water, ...

and ends[11]

> How it comes about that the water in the vessel flows uniformly Heron explains in the first book of the *Water-clocks*.

Whatever the construction of this clock[12], it is clearly simpler than al-Khāzinī's – in that there was no continuous weighing –, but probably had a device to ensure uniform flow.

Let us now turn to the "steelyard clepsydras" used by the Chinese. One type sounds remarkably like the clock in the fragment from Heron quoted above[13]:

> Water is placed in a vessel whence it issues by a bronze siphon in shape like a curved hook. Thus the water within is conducted to a silver dragon's mouth which delivers it to the balance-vessel. One *shêng* of water dripping out weighs one *chin*, and the time which has elapsed is one quarter(-hour).

This passage comes from Li Lan's *Clepsydra Technique* (c. 450), which is preserved in an encyclopaedia of c. 700 AD. A more elaborate device, which was used for public and palace clocks throughout the Thang and Sung dynasties (approx. 7–13 c.)[14] is described in two passages from the thirteenth-century encyclopaedia, *Yü Hai*, the first being copied from an account by Wei Cheng and others written in 656[15]:

> The old Taoist technique used a small steelyard clepsydra, with a vessel which weighed the water – thus it was portable. Accordingly (Kêng Hsün and Yüwên K'ai) set up (sun-)shadow measuring (gnomons by which to) graduate the scales of indicator-rods (or steelyard beams), and square vessels to hold the water above.

[9] Proclus, *Hypotyposis* 120–123.
[10] Rome (1931) 87.
[11] *Ibid* 89.
[12] See Proclus, *Hypotyposis* 290–291; Drachmann, *Ktesibios, Philon and Heron* 98–99.
[13] Needham, *HC* 89–90.
[14] Needham, *SCC* III 318.
[15] *Ibid* 327 and *HC* 89.

In these quotations of translations from the Chinese the words in round brackets are added by the translators. The other passage is more detailed[16]:

> The method of this clepsydra was to have a waterbalance with a wooden beam as a steelyard, above which there was an inscription carved, saying *T'ien Ho* (The Celestial River). [footnote: "We suspect that this sentence originally said that graduated scales were inscribed on the beam, but later this was misunderstood ..."] < [gloss to text] The *Chih* says that it was the water that was made to act as a balance. [footnote: "This uncertainty was due to a corruption in the text, the characters for wood and water being easily mistaken for one another if badly written."] > The (lowest vessel) was sufficiently high and wide to hold indicator-rods, of which there were four, each made of wood, 3.5 ft. long, marked with the hours and quarters and night watches, and put in the (lowest vessel of the) Celestial River. [footnote: "This proves that it was the compensating tank that was weighed."] They were used alternately by day and night.

A bronze vessel is hung at one end of the balance and a [movable] weight at the other, both by rings. There is a tank inside the "Water Box", from which a siphon led to the vessel on the balance, and also a square guide for steadying the balance during adjustments. Then

> Whenever (the passage of) time (lit. hours and quarters) made (it necessary to) change (the setting) [footnote: "I.e. when the indicator-rods had to be changed seasonably or the tank's counter-weight moved in position on the beam"], those in charge of the timekeeping put the end of the steelyard into the square ring, and with thick plaited cords laid hold of the large bronze ring above the weight drawing it back and forth (along the steelyard).

Needham assumes that the weights were moved only seasonally and that the time (in seasonal hours) was measured by the height of the water in a further container filled from a hole in the vessel hung on the balance. That is, the function of the balance is to ensure a constant head of water, appropriate for the season. According to this interpretation, the attendants had to move the siphon when the arm diverged from the horizontal.

But a constant head could be marked on the pot. And, besides, the attendants, in the last sentence quoted, are described as shifting the compensating weight after a lapse of time. The instrument is far more probably a piece of apparatus like al-Khāzinī's, but with two essential differences: (1) it was probably devised for seasonal hours, as is implied by the use of different indicator-rods, or scales, at night (and a commentator to the text mentions no fewer than twenty-one indicator-rods); and (2) it measured how much water flowed into the vessel on the balance, whereas al-Khāzinī's measured how much flowed out. This interpretation is confirmed by the mention of hours and quarters on the indicator-rods. If Needham were right, the seasons would be marked. Perhaps he was distracted by the author's calling the beam "The Celestial River". If this curious term, perhaps indicating a connection with astronomy, is kept for the beam, the indicator-rods are put on one arm of the balance, as expected, and not in the "lowest

[16] Needham, *HC* 92.

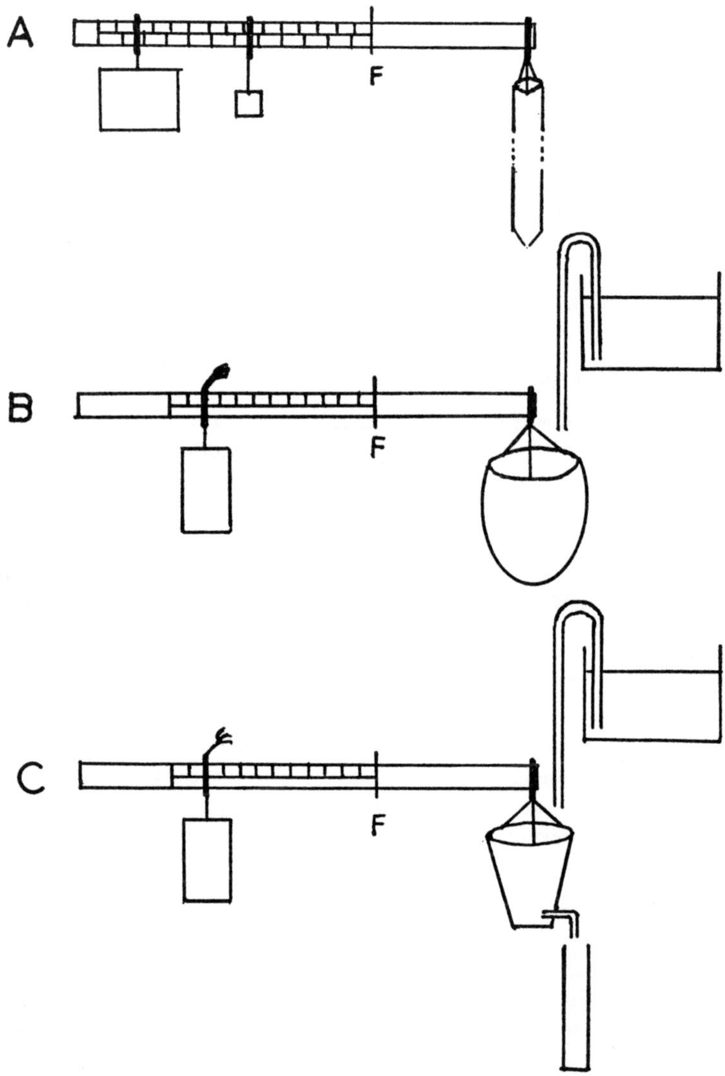

Fig. 1 A: al-Khāzinī's balance-clock with a simple reservoir suitable for sand. The reservoir has been foreshortened. The two scales read directly in equal hours and degrees of celestial motion. *B* and *C:* the Chinese steelyard clepsydra according to the present interpretation and that of Needham respectively. In every case F is the fulcrum.

vessel". The only problem with the present interpretation is the following passage[17], whether or not we accept the additions in brackets:

[17] *Ibid* 93.

> Whenever there was addition or diminution (of the water in the hanging bronze vessel) of the clepsydra, (the beam) moved accordingly and gave an indication to the keepers at the gate.

But if we suppose that the keepers at the gate were different from the attendants, the problem of "addition *and diminution*" vanishes. By counting the movements of the beam the keepers at the gate knew when to perform the various duties that are now described in the text. For on the present interpretation it could be supposed that when the vessel was heavy enough to make the beam tip, the attendants shifted the weight away from the fulcrum, thus causing the beam to tip in the other direction.

A very ingenious interpretation has recently been put forward by Prof. Sleeswyk. The bronze vessel is put inside the tank in the water-box and a siphon joins the two as in fig. 2. The weight has two positions on the other arm of the beam;

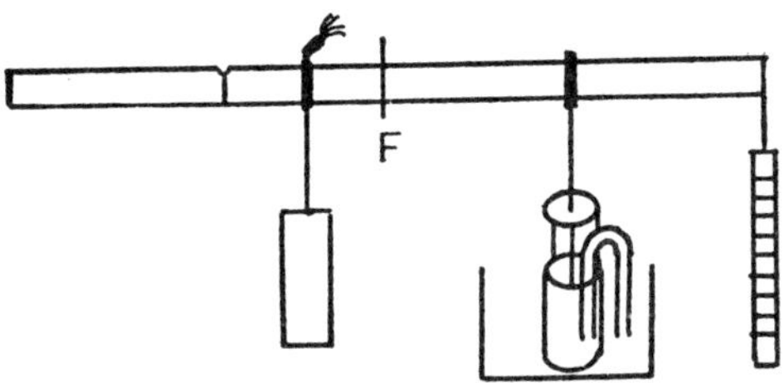

Fig. 2 The Chinese steelyard clepsydra according to Sleeswyk's interpretation.

and the water passes through the siphon, first one way and then the other, to produce an approximately uniform movement of the beam. The time is shown by an indicator-rod (at the end of the beam) seen against some stationary object. A scale model worked well in practice[18]. Unfortunately this interpretation really does not square with the text. Sleeswyk's explanation that *T'ien Ho* is the whole apparatus and it is this, or at least the framework, that is "sufficiently high and wide to hold indicator-rods"[19] is not convincing, for not only would there be no particular need for the frame to be wide (and the height of the *beam* would determine the possible length of the indicator-rod), but it is more natural to suppose that it is the beam that is being described. Again, the interpretation cannot easily accommodate the subsequent modification of the apparatus by adding tanks and siphons[20], though it is suggested[21] that they were added to counter changes in the viscosity of water. But the most powerful objection to the system concerns the position of the tank and water-box. It is incredible that the author would have mentioned the bronze vessel and tank without saying that one was inside the other. Besides, in Wei Cheng's brief account the "square vessels" are described as being *above*. Sleeswyk rightly points out that his interpretation explains the

[18] Sleeswyk 442 and 444 for photographs of the model, and 448 for results of experiments.
[19] *Ibid* 430–431.
[20] Needham, *HC* 93–94.
[21] Sleeswyk 444–445.

author's failure to mention any filling or emptying operations, but this is small compensation for the above objections.

Returning, therefore, to the simpler solution of fig. 1B, we must ask whether the Chinese clock and al-Khāzinī's are really independent. At first sight, the Chinese apparatus seems to be more directly linked to Heron's, for in essence it adds only continuous weighing. On these grounds it might be considered an intermediate step. But it has the advantage over al-Khāzinī's that uniform flow is much easier to bring about; and since he went to such trouble to ensure uniform flow[22], it would be astonishing if he had developed his clock from the elegant Chinese system. Also, his clock, like Heron's, is designed for equal hours, for astronomical purposes. As in the case of the self-rotating sphere[23], a common, probably Hellenistic, source is indicated. Some slight support for this conclusion is to be found in the use of a miniature balance by the banū Mūsā for their trick vase[24]. Oil and water, when poured into this vase, come out at different holes. Inside there is a balance-arm with a small vessel at one end; water is heavy enough, when it fills the vessel, to operate the arm (and connected mechanism), and oil is not. It is well known that the banū Mūsā drew heavily on Hellenistic mechanics.

An obvious candidate for the source of Chinese and Islamic balance-clocks is Heron's lost book *On Water-clocks*. True, Pappus and Proclus do not mention this type of clock, but they may have had an incomplete text and may not have been quoting directly. Whatever the source, it must treat clocks that do not have uniform flow along with those that do.

Since writing this article I have had the benefit of reading in proof Dr. Donald Hill's book, *Arabic Water-Clocks*, which is shortly to be published by the Institute for the History of Arabic Science in Aleppo, and which contains *inter alia* a complete translation of the relevant part of al-Khāzinī's *Mīzān al-Ḥikma*. Dr. Hill points out several things that I had overlooked, all confirming the ideas adumbrated above.

First [Hill, chap. 4], the smaller clock described by al-Khāzinī for temporary use while the big clock is being reloaded worked on the inflow principle, i.e. that of fig. 1B, albeit with two weights on the left. It is remarkable that al-Khāzinī does not use this for the principal clock, for uniform flow of water would be much easier to arrange in this system. Secondly, [Hill, chap. 1, note 24] clocks of this kind – that is, of the inflow type and with two weights – were indeed to be found in China. Thirdly, Dr. Hill explains [chap. 4] the function of al-Khāzinī's "string retainer" for steadying the balance during adjustments: there was a similar device in the Chinese clock. Finally, [Hill, chap. 1, note 29] Philo [or pseudo-Philo – see A. G. Drachmann, *Ktesibius, Philon and Heron* (Copenhagen, 1948) p. 66–70] mentions a tipping-spoon "similar to those used in clocks". The question of transmission remains open, but the most likely origin of the balance-clock still lies in Hellenistic mechanics.

[22] See, e.g., the summary of Rozhanskaya's article in *Referativnyi Zhurnal*. I am most grateful to Dr. Volker Bialas for translating this summary.

[23] See note 7. This device luckily had ancient testimony, Pappus, *Collection* III 10.

[24] Banū Mūsā 190–191.

ABSTRACT

In his *Mīzān al-Ḥikma*, al-Khāzinī (12c.) describes a clock that works by weighing a leaking reservoir of sand or water. It is here suggested that the Chinese "steelyard clepsydra", of which Needham gives the original descriptions in English translation, worked on a similar principle, and that the device goes back to Hellenistic antiquity. This requires an interpretation of the steelyard clepsydra at variance with those offered (1) by Needham, who suggests that the beam was used to confirm that a constant head of water was maintained in the vessel at its end (into and out of which water passes); and (2) by Sleeswyk, according to whom the beam gives a direct measure of time. Needham's interpretation too gives a direct time reading, albeit by means of a separate vessel. My interpretation suggests a cruder instrument that gives only intermittent indications of time.

BIBLIOGRAPHY

(Banū Mūsā.) DONALD R. HILL: *The Book of Ingenious Devices by the Banū Mūsā bin Shākir*. Dordrecht, 1979.

A. G. DRACHMANN: "Ktesibios, Philon and Heron. A study in Ancient Pneumatics", *Acta Historica Scientiarum Naturalium et Medicinalium*, IV, Copenhagen, 1948.

A. G. DRACHMANN: "Ctesibius", *Dictionary of Scientific Biography*, *3* (1971) 491–492.

(Jazarī.) *The Book of knowledge of Ingenious Devices* by Ibn al-Razzāz al-Jazarī, translated by Donald R. Hill, Dordrecht, 1974.

N. KHANIKOFF: "Analysis and Extracts of *Kitāb mīzān al-ḥikma*", *Journal of the American Oriental Society*, *6* (1859) 1–128.

(Mīzān.) 'ABD AL-RAḤMĀN AL-KHĀZINĪ: *Kitāb Mīzān al-Ḥikma*. Hyderabad, 1941.

(Needham *SCC*.) J. NEEDHAM AND WANG LING: *Science and Civilization in China*, volume 3, *Mathematics and the Sciences of the Heavens and of the Earth*. Cambridge, 1959 (repr. 1975).

(Needham, *HC*.) J. NEEDHAM, WANG LING AND D. DE S. PRICE: *Heavenly Clockwork. The Great Astronomical Clocks of Medieval China*. Cambridge, 1960.

Pappi Alexandrini Collectionis Quae Supersunt, ed. F. Hultsch, Berlin, 1876–1878.

Procli Diodachi Hypotyposis Astronomicarum Positionum, edidit C. Manitius, Leipzig, 1909.

Referativnyi Zhurnal, *Astronomija*. Akademija nauk SSSR, Volume for 1979, item 5.51.15.

A. ROME: "Commentaires de Pappus et Théon d'Alexandrie sur l'Almageste", *Studi e Testi*, *54* (1931).

M. M. ROZHANSKAYA: "The Astronomical Clock of al-Khāzinī" [in Russian], *Istoriko-matematičeskia Issledovanija*, *14* (1978) 189–198.

A. W. SLEESWYK: "The Celestial River, a Reconstruction", *Technology and Culture*, *19* (1978) 423–449.

E. WIEDEMANN: "Über Lampen und Uhren", *Aufsätze zur arabischen Wissenschaftsgeschichte*, Hildesheim, 1970, I. 351–376. Originally published in 1909.

E. WIEDEMANN: "Über die Stundenwage", *Aufsätze*, II. 57–68. Originally published in 1914.

H. J. J. WINTER: "Mechanik und Technik des Islams", *Endeavour*, *15* (nr. 57) (1956) 25–8.

The curious term "celestial river" has a counterpart — perhaps coincidentally — in the Arabic word *minjam*. According to Lane's *Arabic-Englisch Lexicon* this can also mean "beam of a balance". Other derivatives of the *NJM* root have meanings to do with the stars.

XVI

The Astronomical Instruments of Jābir ibn Aflaḥ and the Torquetum

I Introduction

The torquetum was an instrument that combined the functions of Ptolemy's astrolabon and the plane astrolabe. It was described in the late thirteenth century by Bernard of Verdun and Franco of Poland[1]. Some authors have supposed that it was invented by Jābir ibn Aflaḥ, who lived in Seville in the first half of the twelfth century, or that it was suggested by the instrument he did invent[2]. There are two instruments attributable to Jābir, and it is the less accessible of these, the one described in the Arabic MSS of his *Iṣlāḥ Almajisṭī*, that has the greater claim in this respect.

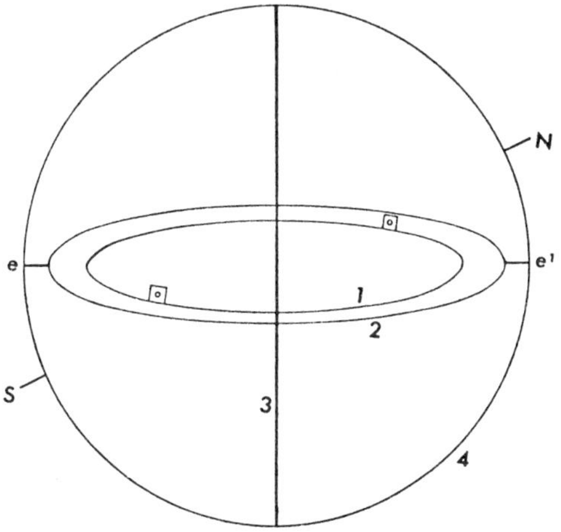

Figure 1

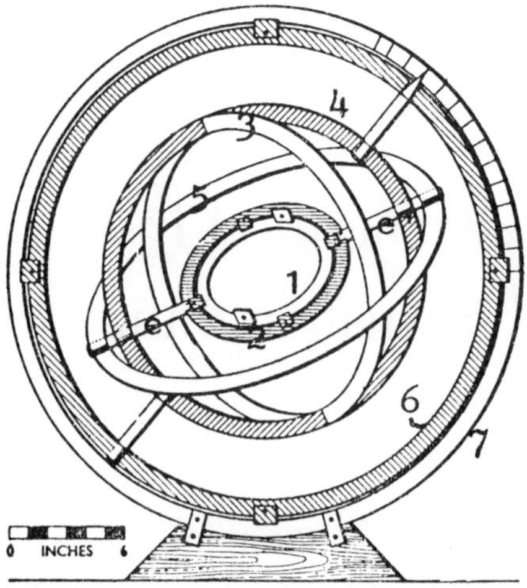

Figure 2

Ptolemy's astrolabon comprised a set of pivoted armillas, of which fig. 2 shows the general arrangement[3]. Fig. 1 shows how the four innermost armillas are fitted together. The pivots N and S are aligned with the poles by mounting them in a meridional armilla, an instrument described in *Almagest* I 12 for finding the arc between the tropics. The movement of the meridional armilla may be used to adjust the astrolabon for local geographical latitude. Armilla 3 represents the ecliptic circle and ee' its poles. The edge of armilla 2, which rotates about ee', indicates longitude on the graduations on armilla 3. Armilla 1 is constrained to move in the same plane as armilla 2 and carries two sights aligned with a pointer, which moves over graduations on armilla 2, thus indicating latitude. There is another armilla, 5, rotating about ee' but outside armilla 4. It is used as an auxiliary in sighting a celestial body. Unlike the circles in the diagrams, the armillas were made so that each just fitted into the next biggest.

In the torquetum, the nested armillas of the astrolabon are replaced by plates piled one on top of the other. Thus the equatorial plate of the instrument is fixed at the appropriate angle to the horizontal plate. Rotating on the equatorial plate, about an axis perpendicular to it, are two circular discs fixed together. One of these (the *basilica*) rotates in the plane of the equator, but the other, representing the ecliptic and

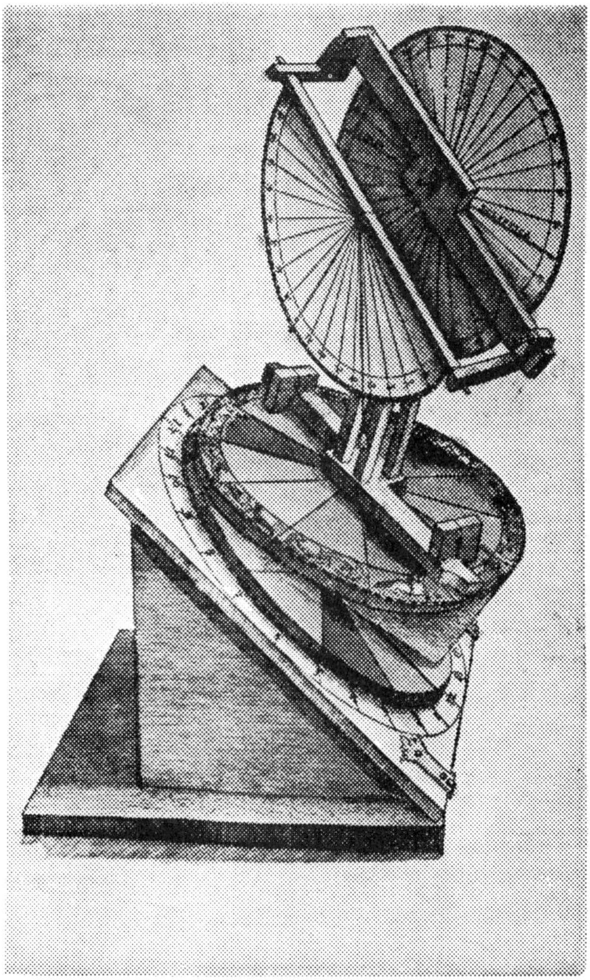

Figure 3

graduated in signs and degrees, is set at an angle to the basilica equal to the obliquity of the ecliptic. Another graduated circular disc (the *crista*) is free to rotate on an axis perpendicular to the ecliptic-disc. Longitude and latitude are respectively measured by means of a pointer attached to the crista and an alidade that rotates about its centre. As described so far, the instrument could be used, like the astrolabon, for finding longitudes

Figure 4

and latitudes. But it was also furnished with a *semissa*, a graduated semicircular disc freely suspended from the alidade, and a plumb-line that hung over this to indicate altitudes (see fig. 3[4]). Together with a scale of degrees or hours on the equatorial plate, these additions made it possible to use the torquetum for the same purposes as the planispheric astrolabe, such as telling the time. Other scales were sometimes added to make the instrument more useful for such purposes.

There were variants of both astrolabon and torquetum. Ptolemy himself is credited with an instrument known by the generic name of *mete-*

ōroskopikos[5]. Others were devised by Muslim writers, and in the fifteenth century Regiomontanus described a simplified astrolabon in a letter to Cardinal Bessarion[6]. A variant of the torquetum was the *rectangulus*, invented by Richard of Wallingford in 1326. It was made entirely of hinged rods, and aptly characterized by Michel as "un véritable *Torquetum* schématisé"[7].

The two instruments described in the *Iṣlāḥ Almajisṭī* and its translations resemble the torquetum in replacing the nested armillas by simpler devices containing the same angles. But no *semissa* nor scale of hours was added: the instruments were not intended for the same purposes as the plane astrolabe. Further discussion of their relation to the torquetum will be left until the instruments themselves have been described.

Jābir claims to have replaced all the instruments in the *Almagest*, that is, not only the astrolabon, but the meridional armilla of *Almagest* I 12 (the quadrant described in the same place does not seem to have been counted in Jābir's total of four), the equinoctial armilla of *Almagest* III 1, and the triquetrum (see fig. 4[8]) of *Almagest* V 12, as well. However, he provides a poor substitute for the triquetrum, and apart from the mountings leaves the simple meridional and equinoctial armillas little changed. His compendious instruments are therefore of interest only when used as substitutes for the astrolabon.

II Texts

1. Arabic

The Arabic text given in figs. 5–8 is transcribed from MSS Escorial 910, ff 117v–119r (E); Escorial 930, ff 59v–61v (F); Berlin 5653, ff 51v–53r (B); Paris Bibliothèque Nationale 1102, ff 12r–14v (P); British Museum Or. 10,725, ff 92v–175v (L). The first two are from the fourteenth century, and the third is dated 626H (1229 A.D.). According to the catalogue, the Paris MS, which is in Hebrew script, may be assigned to the year 1327. The London MS, again in Hebrew script, is not dated.[9] The following is a literal translation.

> There is a way of dispensing with all the instruments mentioned in this book by a single instrument consisting of one armilla and another quadrant and a rule with two sights on it—according to the following description.
>
> We take a brass armilla with parallel surfaces, of diameter about six hands-

وقد يقع الاستغنا عن جميع الآلات المذكورة فى هذا الكتاب بآلة واحدة تشتمل على حلقة واحدة وربع أخرى ومسطرة فيها شظيتان على ما أصف.

نتخذ حلقة نحاس متوازية السطوح ، قطرها نحو ستة أشبار معتدلة الغلظ بحيث لا تعوج محكمة الاستدارة وهى حلقة اب ، ولنقسم محيط دائرة اب ، وهى المخطوط فى سطحها بثلاث مائة وستين جزءا متساوية ونقسم كل واحد من هذه الاجزاء الى ما أمكن ، ولتكن نقطة آ منها رأس السرطان ، وليكن فى ظهرها وعلى نقطتين متناظرتين منها قطعة من حلقة نحاس ، يكون دائرتها أعظم من دائرة اب ، ويكون سطح هذه القطعة قائما على سطح حلقة اب ، على زوايا قائمة ، وهى قطعة اوب ، وليكن عمود دجو من نحاس ، اما من نقطة ج التى هى مركز دائرة اب الى نقطة د فمستدير والى الانخراط قليلا وطوله مساو لنصف قطر دائرة اب ، واما من نقطة ج الى محيط القطعة فطوله نحو الشبر وتحيط به سطوح متوازية . ونتخذ أيضا ربع حلقة شبيهة بحلقة اب ومساوية لها ، وهو ربع زح ، وليكن نصف قطر هذا الربع مسطرة حل متصلة به ، ومركزه نقطة ل ، وليكن فى طرفه حليقة صغيرة مركزها نقطة ل التى هى مركز الربع ، يسع فيها غلظ ج من عمود دج ، ويستدير فيها استدارة سلسة ، ونجعل فى محيط هذه الحليقة وتدا رقيقا تستدير فيه مسطرة من ، ونركب فى طرفى هذه المسطرة شظيتين قائمتين على سطحها ونجعل فيها ثقبين ، وليكن فى طرف ربع زح أيضا فى نقطة ز منه حليقة أخرى صغيرة أصغر من حليقة ل ، يسع فيها طرف عمود دج أعنى الذى عليه نقطة د ، ويستدير فيها . وليكن فى نقطة ب من محدب حلقة اب عمود نحاس مستدير أيضا والى الانخراط قليلا ، وليكن طوله يسامت مركز حلقة اب وهو عمود بس

فاذا أردنا أن نعلم بهذه الآلة مقدار القوس التى بين المنقلبين أركزنا عمود بس فى رخامة قد نصبناها فى منصبة مرتفعة عن الارض وفى موضع مكشوف الشمس ونخرج من وسطها خط نصف النهار ، ونجعل حرف هذه الحلقة مسامتا لهذا الخط ، ونركب ربع زح فى عمود جد ، ونرصد الشمس فى وقت نصف نهار كل يوم بأن نجعل مسطرة هذا الربع لا يفارق سطح حلقة اب وحرف مسطرة من لا يفارق هذا السطح أيضا حتى ترى الشظية السفلى منها مستظلة بأجمعها من الشظية العليا ، فنعلم بذلك فى نصف نهار كل يوم تباعد الشمس من نقطة سمت الرأس ، أعنى نقطة آ . وكذلك نعلم تباعد القمر من هذه النقطة أيضا بأن ننظر من ثقبى الشظيتين حتى نرى القمر من كلى الثقبين فنعلم بذلك بعد موضعه المرئى من نقطة سمت الرأس فى نصف نهار كل يوم .

Figure 5

فاذا علمنا مقدار القوس الذي بين المنقلبين وصلنا حينئذ بعمود دج عمود جه وجعلناهما يحيطان عند نقطة ج بزاوية متساوية لزاوية الميل، أعني أن يكون زاوية هجو مبلغها مبلغ الزاوية التي هي ميل سطح فلك البروج عن سطح معدل النهار فتكون حينئذ نقطة د قطب فلك البروج، واذا جعلنا طول عمود هج مساويا لطول عمود دج صارت نقطة ه التي هي طرف العمود القطب الجنوبي من قطب معدل النهار، وليكن عمود هج متصلا بمحيط قطعة اوب على نقطة ج، وليكن هج منه مستديرا والى الانخراط قليلا كأنه عمود جد.

فاذا أردنا أن نعلم بهذه الآلة نهاية ميل القمر عن فلك البروج على ما رصده هو بالعضادتين الطويلتين أعني أن يكون في انتصاف النهار وهو في نهايته الشمالية من الفلك المائل وموضعه الحقيقي من فلك البروج في نقطة المنقلب الصيفي، فاذا اتفق ذلك أدرنا حينئذ ربع زح حول عمود جد والمسطرة لا تفارق حرفها سطح حلقة اب على ما فعلنا في الشمس وننظر من ثقبي الشظيتين حتى نرى القمر من كلي الثقبين، فنعلم بذلك مقدار بعد موضعه المرئي من سمت الرأس في ذلك الوقت.

واذا أردنا أن نعلم بها وقت كل واحد من الاعتدالين عمدنا الى رخامة وأركزنا في سطحها حلقتين صغيرتين يسع فيها عمود دج، ويكون سطح كل واحدة منهما قائمة على سطح الرخامة، وينصبها في منصبة مرتفعة عن الارض بنحو القامة، ونجعل الخط المار بمركزي الحلقتين يمر بقطبي العالم، فحينئذ يدخل في هاتين الحلقتين عمود دج، فتكون حينئذ حلقة اب في سطح دائرة معدل النهار، فنرصد بها وقت كل واحد من الاعتدالين على ما ذكر.

واذا أردنا أن نعلم موضع كوكب من الكواكب من فلك البروج في الطول العرض وقد تقدم لنا معرفة موضع القمر المرئي من فلك البروج في الطول والعرض وموضع كوكب من الكواكب منه أدخلنا عمود هج في الحلقتين الصغيرتين اللتين في سطح الرخامة وجعلنا وسط مسطرة زح على موضع القمر المرئي أو موضع الكوكب المعلوم موضعه، ثم ندير حينئذ حلقة اب بجملتها حتى نرى القمر أو الكوكب المعلوم موضعه في سطح ربع زح، فاذا كان ذلك لم نحرك بعد حلقة اب وأدرنا حينئذ ربع زح على سطح حلقة اب نحو الكوكب الذي نريد معرفة موضعه حتى نراه في سطح الربع، فنعلم حينئذ موضع الكوكب الحقيقي في فلك البروج بموضع هذا الربع من محيط دائرة اب، ثم ندير مسطرة من حتى نرى الكوكب من كلي ثقبي الشظيتين، فنعلم بذلك عرضه من فلك البروج.

Figure 6

ولما كانت الآلات المستعملة فى الارصاد لا يمكن قسمتها الى أكثر من الدقائق ولا
يمكن القسمة الى الدقائق الا فى حلقة يكون قطرها أزيد من اثنى عشر شبرا، وكان كلما
كبرت الحلقة عسر على الصانع احكامها وقسمتها ، وعسر أيضا على متولى الرصد بها
نصبها على حقيقة ما يريد منها وجب أن يعمل النظر فى حيلة يمكن بها أن نصل بهذه
6v الحلقة التى قطرها نحو الستّة أشبار الى ما نصل بحلقة قطرها مائة شبر أو أزيد لتنقسم
الى الثوانى ويسهل علينا نصبها والتقلب بها لصغرها، وذلك يكون على ما أصف.
وهو أنا نعمد الى لوح أملس طوله نحو أربعة أشبار، ونوتّد فى قريب من طرفه وفى
الوسط من عرضه وتد حديد رقيق، ونثبته فيه حتى لا يتحرّك، ونعمد الى مسطرة
خشب رقيقة، ونثقب فى طرفها ثقبين، يكون بينهما مثل نصف قطر دائرة آب
7r المنقسم محيطها بثلاث مائة وستين جزأ، وندخل الثقب الواحد على عمود الحديد
الرقيق، وندخل فى الثقب الثانى مسمارا حاد الطرف ونثبته فيه حتّى لا يتحرّك،
ونخطّ فى ذلك اللوح قطعة من دائرة، ونجعل ذلك اللوح فى منصبة مرتفعة عن
الارض بمقدار الشبر، ونثقفه فى تلك المنصبة تثقيفا حسنا بحيث لا يتحرّك. ونتخذ
لوحا آخر أملس أيضا ونجعله فى منصبة أخرى ارتفاعها عن الارض كارتفاع الاولى،
7v ونصل مساطر طوالا من خشب حتّى ينتهى طول جميعها اذا وصلت من وتد
الحديد الى هذا اللوح الذى فى المنصبة الثانية، وندخل فى هذه المساطر الموصولة
فى وتد الحديد، وندخل فى طرفها الثانى المسمار ونخطّ به فى هذا اللوح الثانى
قطعة من دائرة عظيمة ونقسم محيط القطعة الاولى بأقسام حلقة آب، ونعمد
الى خيط حرير رقيق مفتول ونعمل فى طرفه عينا يدخل فيه وتد الحديد المركوز فى
8r اللوح الاول ونمدّه على أجزاء القطعة الاولى. فحيث وقع الخيط من قطعة
الدائرة الثانية العظمى التى فى اللوح الثانى رسمنا على تلك النقطة خطّا فى ذلك
اللوح يقطع محيط القطعة، ثمّ نقسم ما بين كلّ خطّين من هذه الخطوط
التى تحدث فى اللوح الثانى الى ما أمكن من الاقسام.
فاذا وقع لنا فى وقت الرصد خطّ وسط مسطرة زح على قسم من أقسام محيط
8v دائرة آب او خطّ وسط مسطرة قن على قسم من أقسام ربع زح أخذنا بخابط
رقيق الطرفين أطول القسمين اللذين انقسمت بها الدرجة على ذلك القسم ونأخذ
من قطعة الدائرة الصغيرة الاولى المساوية لدائرة آب مثل ذلك القسم، ونمدّ الخيط
على آخره الى محيط القطعة العظمة فبدّلنا الخيط حينئذ على كم ثانية هو من
دقيقة تلك الدرجة، وذلك ما أردنا بيانه.

1. يقع الاستعنا: يعاع الاستوا L. الكتاب: الكتب F، كتاب L. 2. تشتمل: وتشتمل L. أخرى: اخرا B. شطبتان PF: شطيتان BE، شطبتين L. اصف: واصف L. 3 تتخذ: فتتخذ B، نتخذ الاعلى L. ستة: من و L. أشبار: اشهار B. 4-3. الغلنا بحيث: الغلوط بحسب L. 4 تعوج: يعوج B، تعاوج L. دائرة آب: دائرة آ B. 5. المحطوط: المخطوطه LP. سطحها: بسطحها L. 6 امكن ولتكن: بمكان ولتكون. السرطان: السرطن P، السارطن L. وليكن:

Figure 7

وليكون LP. في ظهرها : على ظاراها L. 8سطح هذه القطعة : سطحها L. زوايا : زواية L. 9وليكن :
وليكون LP. دجو : جدو B. 10مساو : مساوى P. 11فطوله : فطولها P. وتحيط : ويحيط L. 12ونتخذ :
ولنجد P. شبيهة بحلقة : شبية حلقة L. وهو : وهي F. وليكن : وليكون LP، وليكن في B. 13نقطة F omits
14حليقة : حلقة L. يسع : يساع L. 15سلسة : ثلثة L. ونجعل : ونجعال L. محيط هذه الحليقة وتدا :
محيطها وهذا L. 16تستدير : يستدير L. وتركب : ونركب L. هذه : هذا L. شطبتين : شطيين B(?F). 17ويجعل :
ونجعل L. ثقبين : ثقبتين L. وليكن : وليكون L. طرف ربع : ربع طرف LF (inserted in B) أيضا L omits. منه حليقة :
حلقة L. 18أخرى : اخرا B. 19وليكن : وليكون L. 20نحاس L omits. مستدير : مستديرا B. والى الانخراط قليلا :
الانخراد قللا L. الانخراط : الانحراف B. 21بابس ، بابح L. 23منصبة : مستدة L. عن : من L. مكشوف ، مكشف
24من : في MSS. وسطها : اسط L. 25زح : عمود زح F. دج ، دج LF. وقت L omits. نهار : النهار L. 26هذا الربع :
ربع L. يفارق سطح : تعارق وسطح L. مسطرة : مسترة L. 27يفارق : تفارق L. نرى : يرى L. 27&28 الشطبة :
الشطبية (الشظبية ?) BE(?). مستطيلة : مستنطيلة B(?). 28العليا : الاعلى L. 30ننظر : عظر L. بجموعها : باجمعها L.
نرى : يرى L. كلى ، كلتى L. الثقبين فنعلم : الثقبتين يعلم L. 31المرئي : المر L. 32الذي : التي MSS. وصلنا : وضلعا L.
دج : جد L. 33وجعلناهما : وجعلناها EB. عند : عن L. متساوية : مساوية LPE. 35واذا : اذا EB، واذ F. فاذا L. مساويا :
مساوية L. 36القطب : للقطب L. قطب : قطبين F، قطبي L. 37معدل : معدال L. وليكن : وليكون (twice) . 38
مستديرا : مستدير B. جد : دج F، ص د L. 39فاذا : فانا L. 40بالعضادتين : بهذه العضادتين B. الطويلتين : الطوبلتين L. وهو
L omits. نهايته : العاية L. 41المائل : الميل L. اتقن : اتمن L. 42جد : دج F. تعارق : يفارق L. 43الشطيتين : الشطبتين
B(?) السطحين L. نرى : يرى L. كلى : كلا F، كلتى L. 44مقدار L omits. المرئي : المر L. 45بها : بهذا الآلة L. الى
رجامة : الرجامة L. 46صغيرتين : صغيرين L. دجو B، دج L، دص EFP. 47قائمة : قائما F!. الرخامة ، وينصبها :
الرخمة ، ونصبها L. القامة : القائمة L. 48المار بمركز : مار بمركزي L. فحينئذ : وحينئذ L. يمر : تم P. 49دج LB :
دص EFP. فتكون : فيكون. 50فنرصدها : فرصد بها ما في L. كل واحد L omits. ذكر : ذكرنا F. 51واذا : فاذ B. موضع :
منصاع L. 52معرفة موضع ، موضع P، موضع معرفة F، معرفة منصاع L. المرئي : المرى L. وموضع : ومنصاع L. 53 ج ق :B
د و EFPL. في : من L. الصغيرتين : الصغيرين L. اللتين : التين B، سهما 54 PL وجعلنا : وركبنا ربع زح في عمود دج L adds.
موضع : مواصاع L. المرئي : المرى L. أو موضع : من فلك البروج وموضع L. 55حينئذ : جسد P، L omits. نرى : يرى L.
أو الكوكب : والكوكب L. 56موضعه : في موضعه L. 56&7 فاذا .... زح B omits (there is a compensating marginal note) لم ... اب L omits
تحرك PE : يرحرك F. 57نراه : تراه B، زاده نراه L. 58بموضع : بمقطاع L. 59من : وفي L. نرى : يرى L. كلى L omits.
60من : عن BEF. 61الآلات : الآلة B. الارصاد : ارصاد L. قسمتها : قسمنه L. 62شبار ، وكان : سبار كان L. 63
احكامها : احداما (?) L. متولي الرصد ، متوالو الارصد L. 64يعمل : نعمل LP، نعمل E. في حيلة : فحيالة. بها omits
نصل بهذه : نعمل هذا الآلة اعنى L. 65الستة أشبار F، الستة اسبار L، الستة اشبر P، ستة اشبار B، الستة الاسبار E.
شبر : سبار L. أو أريد لتنقسم : وازيد لتنقسم PL. 66التواني : تواني L. علينا : عليه B، E omits. والتقلب : والانولها L.
67اتا : لنا EF. نعمد : نعمل L. أملس : مماس L. طوله : يكون طوله P. 68رقيق : رقيقة L. الى : اوم L. 69خشب :
حسنة B. وتثقب : نثقب L. طرفها : طرفيها L. يكون L omits. الدائرة عنى دائرة L. 70الثقب : الطرف L.
71مسمارا : مسماكا L. حاد : حد P. حتى : كى L. 72ونجعل : صعد ان نجعل L. عن : على B. 73تلك omit
BF تثقيفا : ثقافا MSS. حسنا : حركة L. 74املس : ملاس L. أخرى ارتفاعها : اخرا رتفعها L. عن : من MSS.
كارتفاع : مثل ارتفاع L. 75مساطر : مساطرا BF. ينتهى : تنتهى P. طوالا من خشب : خشب طوال L.
76الحديد : الحدوك L. في هذه : مذه E، طرف هذه L. المساطر الموصولة : المسطر الموصوبه L. 77وتد :
هذا وتد L. طرفها : طرفوهما L. به : فيه L. 78من L omits. محيط : محيط هذه P، محيطها ومحيط BF.
بأقسام : التي في لوح الاول L. 79خيط حرير : خط رقيق معتدل L، خيط من حرير F. Plus طرفه L adds
في (?) L. المركوز : المركز E، مركزه L. 80القطعة : قطعة الدائرة L. وقع : انتهى وقع P. الخيط : المحيط L.
81الثانية العظمى : العظمى الثانية L. على : في B.3-81 رسمنا .... الثاني L repeats. 84محيط : ربع زح
احركا (?) L. 85خط : الخط الذي من L. 86اللذين : الذين LP. الدرجة : الدارجا L. 87الاولى المساوية : الاوكب
مساوية L. 88آخره : اجزاءه L. فبدلنا : فيبدلنا P. على كم ، كل كم P. 89دقيقة : الدقيقة L. تلك
الدرجة L omits. وذلك Bayan : ان شاء الله عز وجل E adds. بيانه : ان بين B.

Figure 8

breadths, of a moderate thickness so that it does not bend, of a perfect circular shape; and [let it be] the armilla *AB*. We divide the circumference of the circle *AB* and mark its surfaces into 360 equal parts; and each of these parts is divided as far as possible. Let the point *A* in it be the first point of Cancer.

On the back of it let there be a section of a brass armilla whose circle is greater than the circle *AB*, [attached to] the two diametral points, and let the plane of this section be inclined to the plane of the armilla *AB* at right angles; and let it be the section *AUB*.

Let there be a brass rod *DGU*: from point *G*, which is the centre of the circle *AB* to the point *D* [let it be] turned—its length being equal to a semidiameter of circle *AB*—and tapered to a cone; and from point *G* to the circumference of the section [*AUB*], a distance of about a handsbreadth, [let it be] surrounded by parallel surfaces [i.e. let it be of square or polygonal cross-section].

We also procure a quadrant of an armilla, similar to the armilla *AB* and equal to it; and [let it be] the quadrant *ZH*. Let the rule *HL* be a radius of this quadrant, [and let it be held] adjacent to the quadrant (whose centre is *L*). At its extremity let there be a small ring of centre *L*, the centre of the quadrant. The thickness [at] *G* of rod *DG* fits it [the ring] [so that] it rotates easily in it. In the circumference of the ring we put a thin peg on which the rule *MN* turns. Two sights are mounted on the ends of this rule, perpendicular to its surface, and two holes are made in them. At the extremity of quadrant *ZH* and also [sic] at point *Z* let there be another small ring, smaller than the ring *L*. The end of the rod *DG*, i.e. at the point *D*, fits it [so that] it turns on it. At the point *B* of the circumference of armilla *AB*, let there be a rotating brass rod, and let it be tapered to a cone. Let its length point towards the centre of the armilla *AB*. Let it be the rod *BS*.

When we want to know with this instrument the quantity of the arc that is between the tropics, we embed the rod *BS* in a marble slab, which we have set up in an elevated position on the ground and exposed to the Sun. In the middle of it we draw the line of the meridian, and we align the edge of the armilla with this line. We mount the quadrant *ZH* on the rod *GD*, and observe the Sun at noon every day by placing the rule of this quadrant so that it does not leave the plane of the armilla *AB*. The rule *MN* is inclined (so that it does not leave this plane) until the lower sight is seen to be entirely in the shade of the upper sight. By this means we know the distance of the Sun from the zenith, i.e. point *A*, at noon every day.

Similarly we also know the distance of the Moon from this point by looking through the two holes until we see the Moon in the entire hole. By this means we know the distance of its apparent position from the zenith on the meridian every day.

So when we know the quantity of the arc between the tropics we join a rod *GE* to the rod *DG*, and we make them form at point *G* an angle equal to the angle of obliquity—i.e. the extent of angle *EGU* is the extent of the angle which is the obliquity of the plane of the ecliptic with the equatorial plane. Then the point *D* is the pole of the ecliptic. When the length of rod *EG* is made equal to the length of rod *DG*, the point *E*, which is the end of the rod, becomes the southerly of the two poles of the equator. Let the rod *EG* meet the section *AUB* at point *Q*; and let *EQ* of it be turned and tapered to a cone—as was done with rod *GD*.

When we want to know with this instrument the limit of the Moon's inclination to the ecliptic—the same as he [Ptolemy] observed with two long rods [triquetrum], i.e. when it is on the meridian and is at its northern limit from the inclined sphere, and its true position with respect to the ecliptic is at the Tropic of Cancer—, and when that is brought about, we then turn quadrant *ZH* about rod *DG*—while the rule's edge does not leave the plane of the armilla *AB*—just as we do with the Sun. We look through the two holes in the sights until we see the Moon in the entirety of the holes. By this means we know the quantity of the distance of the position of the [object] seen from the zenith at that time.

And when we wish to know by it [instrument] the time of each of the equinoxes, we take a marble slab and in its surface we set up two small armillas, in which the rod *DG* fits. The plane of each of them is perpendicular to the plane of the marble slab, which we set up in a high place about six foot up from the ground. We make the line passing through the centres of the armillas pass through the poles of the world. The rod *DG* then goes into these armillas. The armilla *AB* is then in the plane of the circle of the equator. With it we observe the time of each of the equinoxes according to what was mentioned.

And when we wish to know the position of one of the stars of the zodiac in longitude and latitude if we are previously acquainted with the position of the Moon with respect to the zodiac in longitude and latitude or the position of one of the stars on it, we insert rod *EG* into the small armillas which are in the surface of the marble slab. We position the middle of the rule *LH* according to the apparent position of the Moon or the position of the star of known position. We then continue to turn armilla *AB*—in its entirety—until we see the Moon or star of known position in the plane of quadrant *ZH*. Having done so, we then do not thereafter move the armilla *AB*, but turn the quadrant *ZH* in the plane of the armilla *AB* in the direction of the star whose position we wish to discover until we see it in the plane of the quadrant. We then know the true position of the star in the ecliptic by the position of this quadrant on the circumference of the circle *AB*. Then we turn the rule *MN* until we see the star in the entirety of the two holes of the sights; and by this means we know its latitude from the ecliptic.

Now with the instruments used in the observations it is not possible for the divisions to be finer than minutes, and division cannot be taken into minutes except in an armilla of diameter more than twelve handsbreadths. And when the armilla is increased in size it is accordingly difficult for the artificer [to achieve any] accuracy and [to make] its divisions, difficult also for [the man] responsible for observation with it to set it up correctly as he intends. It is necessary that we pay attention to a device by which we can achieve with this armilla—of diameter about six handsbreadths— what we can achieve with an armilla of diameter 100 handsbreadths or more, so that we can subdivide it to seconds; and it becomes easier for us to erect it and turn it around because of its small size. It is as I shall explain.

This is it. We take up a smooth plate of length about four handsbreadths; near its end and in the middle of its breadth we fix a thin iron peg, and make it firm until it does not move. We take up a thin wooden rule and bore two holes at the ends. Between them is [a distance] the same as half a diameter of circle *AB*, whose circumference is divided into 360 parts. We put one hole on the thin iron rod; and in the second hole we insert a sharp-ended nail, which we make firm till it does not move.

We draw a section of circle on the plate [with the rule and nail], and we put it on a base about a handsbreadth off the ground. We set it in good alignment in this base in such a way that it does not move. We take another smooth plate and put it on another base, whose height from the ground is the height of the first. We join long wooden rules until the length of the whole, when joined, is from the iron peg to this plate which is on the second base. We insert the connected rules in the iron peg; and in its second end we insert a nail and draw with it a section of a large circle in this second plate. We divide the circumference of the first section by the division of the armilla *AB*. We take up a thread of twisted silk, and in its end we make a loop, and insert in it the iron peg fixed in the first plate. We lay it on the parts of the first section, and where the thread falls on the section of the large second circle, which is situated on the second plate, we draw a line from that spot so that it will cut through the circumference of that section. Of the lines just drawn in the second plate we divide [the arc] between each pair as far as possible with divisions.

When at the time of observation we have the central line of the rule *LH* on one of the divisions of the circumference of circle *AB*, or the central line of the rule *MN* on one of the divisions of the quadrant *ZH*, we take with a pair of dividers [bi-ḍābiṭ] with thin ends the span of the two divisions by which the degree is separated on the division [scale?]. On the first (small) section of circle, equal to circle *AB*, we take the equivalent of that division. We then stretch the thread right down to the circumference of the large section. We then re-position the thread according to how many seconds it is of a minute of that degree. And that is what we wanted to show.

For convenience, the uses of this instrument are listed below. The instruments Ptolemy used for the same purposes and references to them in the *Almagest* are put in brackets afterwards.

1. Finding the arc between the tropics (meridional armilla, *Alm.* I 12);
2. finding the distance of the Moon from the zenith (see item 3);
3. finding the limit of the Moon's latitude (triquetrum, *Alm.* V 12);
4. observation of the equinoxes (equinoctial armilla, *Alm.* III 1);
5. finding the longitude and latitude of a star when the longitude and latitude of the Moon or of another star are known (astrolabon, *Alm.* V 1).

In the first three of these uses the armilla is mounted in the meridian, with the rod *BS* inserted into a horizontal marble slab. In the last two, one of the rods *DG* and *EG* is fitted into two brass rings set into a marble slab, so that the rod is aligned with the poles. Fig. 9 shows how the instrument is assembled for finding longitudes and latitudes.

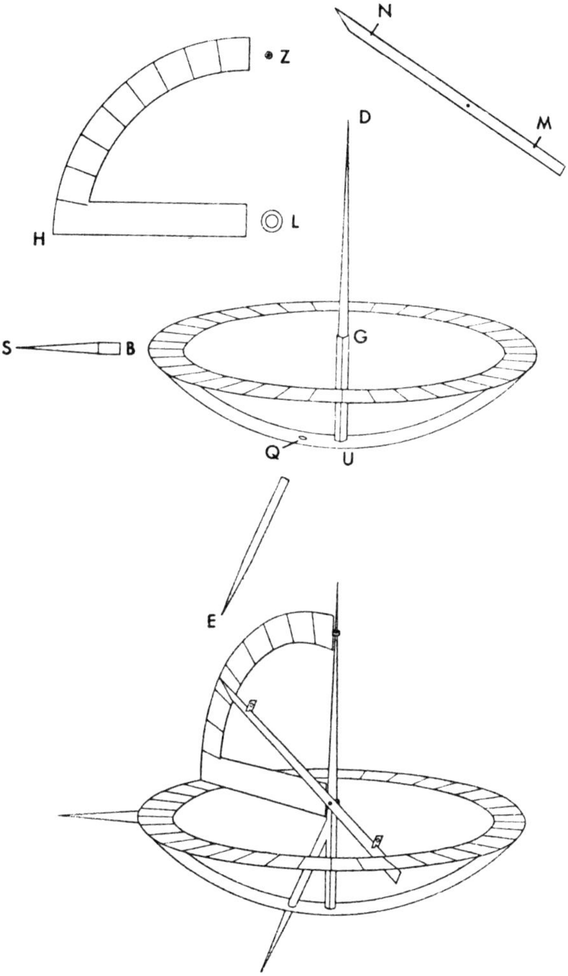

Figure 9

2. Latin

A similar instrument is described in the Latin translation of the *Iṣlāḥ* by Gerard of Cremona. Many of the details, however, are quite different. Here the means of measuring latitude—or angular deviation from the principal armilla, whatever it represents—is similar to the triquetrum. The construction of the instrument is described as follows[10].

Et contingit excusatio ab omnibus instrumentis predictis in hoc libro per armillam unam et duas regulas, secundum quod narrabo. Et illud est quia ego accipiam armillam eris, cuius diameter sit quasi sex palmarum, equalis grossitudinis ita ne torqueatur, decentis rotunditatis, bone rasure; que sit armilla *AB*. Et dividam circumferenciam circuli *AB*, qui est maior circulorum continentium armillam, in 360 partes, et dividam unamquamque illarum partium usque ad illud quod possibile est. Et sit medium grossicie extremitatis armille punctum *A*, quod sit initium signi Cancri, et medium grossicie extremitatis secunde punctum *B*, quod sit initium signi Capricorni; et sit in medio huius diametri, quod est centrum armille, foramen; in quo revolvatur paxillus *GD* rotundus, equalis grossicie, revolutione facili, non currente sicut est revolutio fusorii aque. Et in capite huius paxilli tabella rotunda, amplitudo superficiei eius sit quasi grossitudo 4 digitorum, et sit centrum eius punctum *E*, et grossitudo eius sit medietas grossitudinis paxilli *GD*; et communicet hec tabella cum circulo *ZH* regule *KZT* in meguar transeunte per duo centra duorum circulorum, scilicet duo puncta *E* et *Z*, et sit motus regule *KZT* in circuitu huius meguar motus facilis non currens et in superficie una semper sicut motus duorum crurium circini. Et communicet iterum circulus *T* regule *KZT* cum circulo *M* regule *MN* in meguar transeunte per duo centra duorum circulorum, scilicet per duo puncta *T M*, et sit revolutio circuli *M* in circuitu huius meguar revolutio facilis non currens, et sit in medio regule *KZT* semper sicut revolutio crurium circini iterum. Et sit longitudo que est inter duo puncta, scilicet linea *ZT*, sicut longitudo puncti *Z*, quando componitur perpendicularis *GD* in foramen medii diametri, a circumferencia circuli *AB* divisi; et linea *ZK* sit minor medietate diametri circuli minoris armille parum. Et sit longitudo regule *MN*, scilicet linea *MN*, equalis lateri quadrati cadentis in circulo, medietas diametri eius est linea *ZT*. [f.54r] Et dividam lineam *MN* per divisiones per quas *ZT* est 60 partes equales, et dividam omnem divisionem harum divisionum usque ad illud quod possibile est. Et sint in duabus extremitatibus regule *KZT* due tabelle orthogonaliter super eius superficiem, et sit medium cuiusque earum super lineam *KZT*, et sit in medio latitudinis cuiusque earum foramen super rectitudinem linee *KZT*; et sint due tabelle *p* et *s*. Et sit in dorso diametri *AB* et in medio longitudinis eius paxillus *QF* rotundus, et sit grossitudo eius sicut grossitudo paxilli *GD*; et contineat iste paxillus cum longitudine diametri *BA* angulum addentem super angulum rectum, cuius summa sit 23 partes et 51 minuta, et est angulus maioris declinationis. Et sit in puncto *C* gibbositatis armille *AB*, et est punctum quod est cum duobus punctis *A B* super lineam unam rectam, perpendicularis *CS*; et est super rectitudinem linee *AB*, et sit longitudo eius sicut longitudo paxilli *QF* et grossitudo eius sicut grossitudo illius.

There is a way of dispensing with all the instruments mentioned in this book by one armilla and two rules— according to the following description.

It is this. I take a brass armilla, of diameter about six handsbreadths, of even thickness such that it does not bend, with well-smoothed surfaces; and let it be the armilla *AB*. I divide the circumference of circle *AB*, which is the greater [i.e. outer] of the circles containing the armilla, into 360 parts and divide each of these parts as far as possible. Let *A* be the mid-point of the thickness at [one] end, which is the beginning of the sign of Cancer, and *B* the mid-point of the thickness at the other end, which is the beginning of the sign of Capricorn. In the middle of this diameter, the centre of the armilla, let there be a hole; and let there turn in it an axle *GD*—round, of even thickness, and of easy rotation, but not running free as in the case of the axle tree at the well-head[11]. At the top of this axle let there be a round disc, of breadth of surface the thickness of four fingers; let its centre be the point *E* and its thickness half the thickness of the axle *GD*.

Let this disc be connected to the circle *ZH* of the rule *KZT* by a pivot passing through the centres of the two circles, i.e. the two points *E* and *Z*; and let the motion of the rule *KZT* about this pivot be easy, but not free, and always in one plane like the arms of a pair of dividers. Again, the circle *T* of rule *KZT* is connected to the circle *M* of rule *MN* by a pivot passing through the centres of the two circles, i.e. through the points *T* and *M*; and let the rotation of the circle *M* about this pivot be easy, but not free, and in the joint with [?] rule *KZT* let it be always like the rotation of the arms of a pair of dividers (as before). Let the distance between the two points, i.e. the line *ZT*, be the distance of point *Z* from the circumference of the divided circle *AB* when the column *GD* is fitted into the hole in the middle of the diameter; and let the line *ZK* be a little smaller than the semi-diameter of the smaller [inner] circle of the armilla. Let the length of the rule *MN*, i.e. the line *MN*, be the side of a square inscribed in a circle of semi-diameter *ZT*. I divide the line *MN* into divisions, according to which *ZT* is 60 equal parts, and divide each division as far as possible. At the ends of the rule *KZT* let there be two discs [set] orthogonally to its surface; let the middle of each of them be on the line *KZT*, and in the middle of the width of each let there be a hole in alignment with the line *KZT*. Let the two discs be *p* and *s*.

In the back of diameter *AB* and in the middle of its length let there be a round axle *QF*, its thickness being the thickness of the axle *GD*. Let this axle make with the length of diameter *BA* an angle added to a right angle— the quantity of [this angle] being 23°51′, the angle of greatest declination [i.e. $A\hat{Q}F$ = obliquity of ecliptic + a right angle].

At point *C* of the [outer] curved surface of armilla *AB*, a point that is in a straight line with the two points *A* and *B*, let there be a rod *CS*. Let it be in the direction of the line *AB*; and let its length be the length of the axle *QF* and its thickness the thickness of that [axle].

There must remain some doubt about the mounting of the rule *KZT*. The points *A* and *B* are very carefully introduced: first as defining the armilla, then as defining the circle "maior circulorum continentium armillam", and finally they are described severally as "medium grossicie extremitatis armille". But now the diameter *AB*—not previously mentioned—is called "this diameter" and described as having a hole made at its centre. The omission of a proper description of the diameter *AB* is unfortunate since it is not clear if it carried any device to ensure that the point *Z* of the rule *KZT* (when mounted) was at the centre of the armilla or not. As it stands, the text implies that part of the shank of the axle *GD*, not the disc at the top, is at the centre:

... et sit in medio huius diametri, quod est centrum armille, foramen; in quo revolvatur paxillus *GD* ...

This interpretation is reinforced by the definition of the length *ZT*, which is described as being equal to the distance from *Z* to the outer circumference of the armilla, and not simply as equal to the outer radius of the armilla. At least three reasons may be adduced, however, for placing the centre of *KZT* at the centre of the armilla. First, *ZK* is described without circumlocution as being just less than the inner radius of the armilla. Secondly, rule *MN* is made equal to the chord of a quadrant of which *ZT* is the radius, a length that would just accommodate all

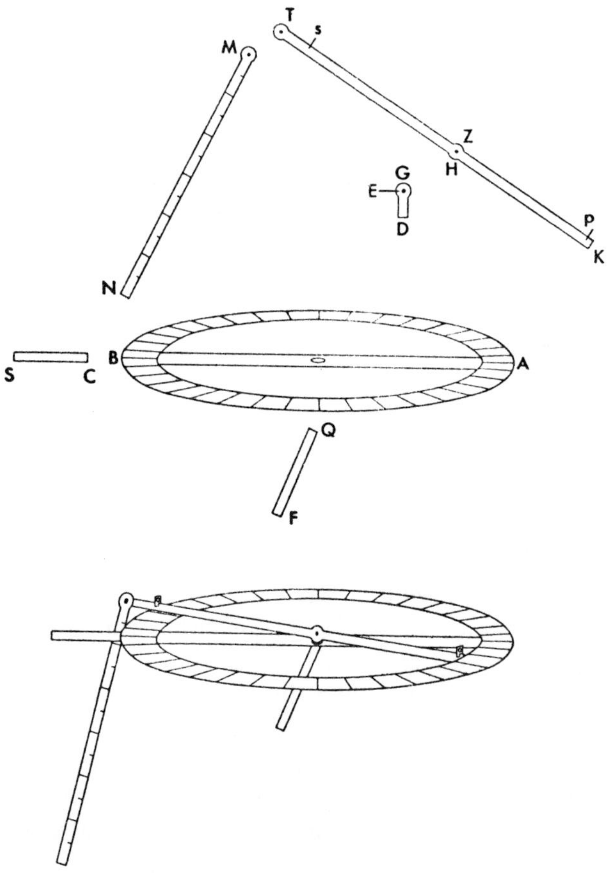

Figure 10

possible latitudes if *Z* were at the centre of the armilla. Thirdly, the graduations on the rule *MN* begin from the pivot *M*. It therefore seems that rule *KZT* is mounted with its centre at the centre of the armilla.

A glance at fig. 10, however, which has deliberately been left vague at the centre of the armilla, brings to mind two further difficulties. First, the axle *QF* cannot reasonably be fixed—as the text says it is—to the middle of the diameter *AB*, since at this place there is a hole for the axle *GD*. Secondly, even if *Z* is not made the exact centre of the armilla, the diameter *AB* will foul the alidade *KZT* whenever they are at all close. This effect could be lessened by suitable shaping of the alidade and diameter, but there is no mention of this in the text. Repsold and Gunther[12] avoided this problem by putting the pivot *Z* a considerable distance from the armilla, but this solution must be rejected for the reasons just given.

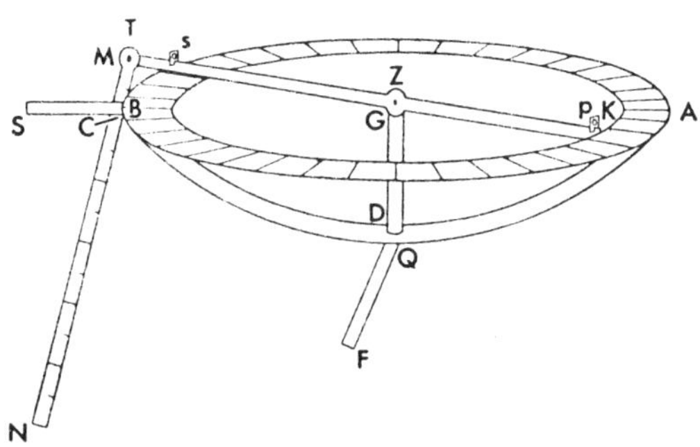

Figure 11

An alternative construction is indicated by fig. 11. Here the "diameter" has been interpreted as a section of armilla attached to the armilla *AB* as in the instrument of the Arabic MSS. The axle *QF* may now be attached to the middle of this section, thus greatly increasing the freedom of the alidade. It is difficult, however, to make this construction tally with the text. The critical passages are:

> et sit in medio huius diametri, quod est centrum armille, foramen;

and

> Et sit in dorso diametri *AB* et in medio longitudinis eius paxillus *QF* rotundus.

In both cases the Latin corresponds word for word to the Hebrew translations of the *Iṣlāḥ* by Moses ibn Tibbon[13] and Jacob ben Maḥir[14] (though the latter has different letters for *QF*). The usual Arabic word corresponding to the Latin *diameter* and Hebrew *quṭr*, which occur in these passages, is *quṭr*. It is true that it can have meanings other than "diameter", but it is not likely in this case where it would obviously be ambiguous. Furthermore, the middle of this *quṭr* has to be the centre of the armilla. Interpreting "diameter" as a mathematical diameter mentioned only to locate the centre, which is to be supported by an unspecified physical construction, is unsatisfactory since the attachment of axle *QF* is still vague and the alidade's freedom of movement must still be restricted. Perhaps fig. 11 is right and the text corrupt, but if the instrument was never constructed—as Delambre[15] claimed—, its inventor may not have considered the problems outlined above.

Another inconsistency might be mentioned here. The angle 23°51′, the obliquity of the ecliptic, is set in the instrument before it has been found with the aid of the instrument. This error is not made in the Arabic text.

Five uses of the instrument are given:

1. finding the arc between the tropics;
2. finding the altitude of the Moon or star;
3. observation of the equinoxes;
4. finding the longitude and latitude of the Moon during the day;
5. finding the longitude and latitude of a star when the longitude and latitude of the Moon or of another star are known.

They are similar to the stated uses of the first instrument. Finding the Moon's position by day is substituted for finding its limits in latitude, and neither operation introduces anything new. As before, the first two uses require a horizontal slab with a vertical hole bored in it, but here two brass washers (*due armille eris*) are embedded in the hole. Plain bearings would suffice if the armilla were to remain in the meridian plane, but the author is concerned with finding altitudes generally, and the continual turning of the axle would require proper bearings.

The mounting for the last three uses is much more elaborate than the one referred to in the Arabic text[16]:

> Et cum voluerimus considerare cum hoc instrumento unamquamque duarum equalitatum, erigemus in loco directo soli duos pedes, altitudo cuiusque quorum sit quasi status unus; et ponemus unum eorum directum ad medium orientis et secundum directum ad medium occidentis. Et extendimus super capita amborum

trabem de ligno, et figemus in medio eius lignum cuius extremitas sit eminens super latitudinem trabis parum; et componam in superficie huius ligni duas armillas parvas, cuiusque quarum superficies sit erecta super superficiem eius orthogonaliter, et amplitudo uniuscuiusque earum sit ut revolvatur in eis paxillus *FQ* revolutione facili non currente. Et intromittam in eis paxillum eris cuius grossitudo sit sicut grossitudo paxilli *FQ*, et sit ex eo eminens ab extremitate ligni quantitas que ingrediatur in foramine quod est in medio diametri armille *AB*. Et extendam filum super dorsum duarum armillarum fixarum in dorso ligni, contingens eas ambas, et ponam illud filum transiens per duos polos mundi. Et tunc constringam duas extremitates trabis in capite duorum pedum constrictione qua non sit possibile ut torqueatur neque ut moveatur. Et extraham tunc paxillum *GD*, cum regulis continuatis cum eo, ex foramine quod est in medio armille, et componam illud medium in illo paxillo qui est in duabus armillis parvis fixis in ligno. Erunt ergo tunc superficies armille *AB* in superficie circuli equatoris diei. Considerabimus ergo tunc solem donec videamus concavitatem armille *AB* obumbrari totam: sciemus ergo quod sol tunc est super circumferenciam circuli equatoris diei.

When we wish to observe each of the equinoxes with this instrument we erect two bases, of height about one status [?] each, in a place open to the Sun, putting one of them due East and the other due West. On top of them we lay a wooden beam and fix at its mid-point a piece of wood [so that] its end projects a little over the breadth of the beam. In the surface of this piece of wood I fit two small rings [*armillas parvas*] [so that] their surfaces are both erected orthogonally to it and the size of each is such that the axle *FQ* rotates in them with an easy, but not free, rotation. I insert a brass axle in them of thickness the thickness of axle *FQ* [so that] the projection over the end of the wood is the length that goes into the hole in the middle of the diameter of armilla *AB*. I stretch a thread over the back of the two rings fixed in the back of the wood, touching them both, and make [the line of the] thread pass through the two poles of the world. I then clamp the ends of the beam on the top of the two bases so tightly that it can be neither twisted nor moved. I then remove the axle *GD*, together with its accompanying rules, from the hole that is in the middle of the armilla, and fit [the vacant hole in] the middle to the axle that is in the two small rings fixed in the wood. The surface of the armilla *AB* will then be in the plane of the equator. We thereupon observe the Sun until we see that the concave surface of the armilla is totally overshadowed: we shall know that the Sun is then on the circumference of the equator.

For the last two uses, finding longitudes and latitudes, the rule *KZT* is put back and the axle *QF* is inserted in the rings in the mounting just described. The instrument is then used like an astrolabon or torquetum. Finding latitudes from the graduation on the rule *MN* is simple: consider the distance from *M* to the graduation concerned as a chord and find the corresponding arc. In the case of the Moon[17]:

... Sciemus, ex partibus linee *MN* que sunt inter punctum *M* et inter partem que est super circumferentiam circuli *AB*, quantitatem partium cordis arcus latitudinis

> lune visibilis in orbe signorum. Arcuabimus ergo illam cordam, et arcus qui fuerit erit latitudo eius visibilis in illa hora.

There are many passages that follow the Arabic, word for word. This is true of almost the entire description of the device to increase the accuracy of the instrument, which starts[18]:

> Proptera autem quod omnium instrumentorum usitatorum in considerationibus divisio non est possibilis in plura nisi in minuta, et divisio non est possibilis in minuta nisi in armilla cuius diameter est maior 12 palmis, ...

3. The Two Instruments

When set up to measure longitudes and latitudes, the instrument of the Arabic MSS is in many ways very similar to that of the torquetum—especially when compared with the astrolabon—and may well have suggested it. Like the torquetum without its horizon-plate Jābir's instrument is essentially the part of the astrolabon shown in fig. 1 with each pair of pivots replaced by a single pivot. It differs from the torquetum in having

(i) a fixed mounting, which was not adjustable for geographical latitude and was not portable;
(ii) no basilica, and hence no means of conversion to the purposes of the plane astrolabe;
(iii) a quadrant set against the ecliptic circle, for measuring latitude rather than a *crista* set well apart from the ecliptic-circle.

The instrument described in the Latin text had the further difference of measuring latitude by hinged rods on the principle of the triquetrum.

The device to increase accuracy, which is common to both instruments, is in fact a device to facilitate the reading of the graduations. It will magnify the error of the instrument—plus the error involved in transferring the reading— and will in no way obviate the principal disadvantage of small instruments, namely that the error of the artificer, who can only work to a certain degree of accuracy, will be more noticeable the smaller the instrument.

4. The Priority Of The Arabic Description

In the aggressive introduction to the *Iṣlāḥ*, Jābir says[19],

Et ipse utitur in eis quibus indiguit de considerationibus instrumentis quatuor, in quibus necessarie sunt armille octo. Nobis autem accidit instrumentum unum, in quo non est necessaria nisi armilla una et quarta alia et regula; et excusamur ab omnibus instrumentis que ipse dixit.

For the observations he needed, he [Ptolemy] used four instruments, which require eight armillas. There has come to us, however, an instrument that requires only one armilla, a further quadrant, and a rule. We dispense with all the instruments he spoke of.

The instrument described here as requiring only one armilla, one quadrant, and one rule, is clearly that described in the Arabic MSS. This is again true when Jābir deals with the observation mentioned in the above extract—i.e. in book V, just after the description of the instrument[20]:

Et inventio quidem finis latitudinis lune est possibilis cum hoc instrumento, quod ostendimus, in omni terra ... Deinde suspendemus filum perpendiculi super centrum armille *AB*. Ergo sciemus per ipsum locum zenith capitis in circumferencia quarte *ZH*; ...

Finding the limit of the Moon's latitude is possible with this instrument, which we have explained, in every country ... Then we hang a plumb-line from the centre of the armilla *AB*. By this means we know the position of the zenith in the circumference of the quadrant *ZH*; ...

It is clear, therefore, that Jābir's original instrument is the one described in Arabic.

It is possible that the other instrument was substituted for it because of a mechanical defect. If fig. 9 represents what the author of the Arabic text intended, the quadrant *ZH* would surely tend to get jammed on the spike *DG*; and for the quadrant to sit squarely in position, the rings at *Z* and *L* would have to be very accurately made. No doubt a collar just below *G* would improve matters, but the system of fig. 10 or fig. 11 avoids the difficulty altogether. If this conjecture is true, it is a pity that the second instrument was so inaccurately described.

5. *The Hebrew Translations*

That the obtrusive passage in Gerard's Latin translation was originally in Arabic is proved by its appearance in Hebrew in Moses ibn Tibbon's translation of 1274[21].

A Hebrew translation of both texts appears in the revision (1335) by Samuel ben Judah of Marseilles of Jacob ben Maḥir's translation. In MS Bibliothèque Nationale heb. 1024, Samuel describes the "Arabic" instrument in the main text, but inserts the "Latin" instrument in smaller writing by splitting four pages (ff. 46r–47v) into two columns. At the beginning of the passage there is an explanatory note in the margin. It appears to be by Samuel and to say that he has been "more precise than he [Jacob]", who did not describe the "instrument that the author of this book invented in accordance with the description that is given here but by another description", which is to be found in the small writing[22]. Samuel put Jābir's subsequent theorem about finding the Moon's distance in the small writing, and Jābir's account of Ptolemy's theorem on this in the main text, whereas the Arabic and Latin versions have both demonstrations, one after the other. Whether this is an indication of the manuscripts he had in front of him, and not of his own critical skill, is a matter for further study. But though the textual history of the *Iṣlāḥ* is complicated, it is fairly certain that the original instrument, i.e. that made with one armilla, one quadrant, and a rule, was not available to Hebrew readers until 1335, the date of Samuel's revision.

III Jābir and the Invention of the Torquetum

Later writers seem to have neglected Jābir's observing-instruments. In his epitome of the *Iṣlāḥ* Quṭb al-Dīn al-Shīrāzī mentions the "instrument which he alleges replaces all the instruments in this book", but unfortunately omits the description because he was working from a poor manuscript[23]. He does not give enough details even to decide which of the two instruments he is referring to. There is no hint of either instrument, or of the torquetum, in al-'Urḍī's list of instruments used in the Marāgha observatory[24], to which al-Shīrāzī was at one time attached. In the West, where the *Iṣlāḥ* was better known, neither of Jābir's instruments seems to have been used, or even mentioned, until in 1469 Regiomontanus included "Jābir's compendious instrument" in a list of instruments that belonged to the "genus Statarium" (i.e. non-portable)[25]:

> Talis quoque est machina collectitia Gebri Hispalensis industrie admodum contexta, quippe quae omnia Ptolemaei instrumenta iocundo quodam compendio mirifice complectitur.

The Latin commentary on the *Iṣlāḥ* in the Bibliothèque Nationale[26] stops in the middle of book IV, and so has nothing to say on the instruments of book V. On the other hand, the torquetum—although, apparently, equally unknown in the East— was used in the West on more than one occasion for observing comets[27], and is mentioned by Richard of Wallingford[28] in a list of standard observing-instruments.

It will be convenient to list the facts concerning the possible connection of Jābir and the torquetum:

1. Description of both instruments attributable to Jābir appeared in the twelfth century;
2. Of the two, the one more like the torquetum was available only to readers of Arabic or (1335 onwards) Hebrew;
3. The torquetum is first mentioned in the late thirteenth century by a Latin writer;
4. It was apparently not known in the East.

If, therefore, Jābir did inspire the invention of the torquetum—an uncertain premiss—it is likely that the inventor lived in the West in the thirteenth (or late twelfth) century, and that he read Arabic. Until new evidence appears, it would perhaps be unwise to speculate further.

I should like to thank the Leverhulme Trust for financial support in preparing this paper. I should also like to thank Dr. H. D. Isaacs for correcting the Arabic text and checking the translation.

NOTES

1. See E. Poulle, "Bernard de Verdun et le Turquet", *Isis* 55 (1964), 200–208; R. T. Gunther, *Early Science in Oxford*, II 35–6 and 370–5, 1923; L. Thorndike, "Franco de Polonia and the Turquet", *Isis* 36 (1945), 6–7. An article on the torquetum is in preparation.
2. E.g. Gunther, *Op. Cit.*, 35; J. Needham, "The Peking Observatory in A.D. 1280 and the Development of the Equatorial Mounting", reprinted from *Vistas in Astronomy*, ed. A. Beer, vol. I, 1955; H. Michel, "Le Rectangulus de Wallingford précédé d'une note sur le Torquetum", *Ciel et Terre* 11–12 (1944), 196. For Jābir b. Aflaḥ, see R. Lorch, "The Astronomy of Jābir ibn Aflaḥ", *Centaurus* 19 (1975), 85–107.
3. Reproduced from D. J. Price's article in *A History of Technology*, III 592, fig. 346, second edition 1964, by kind permission of the editors. The labels on the circles have been added, and correspond to those on the diagram on p. 78 of "L'Astrolabe et le Météoroscope d'après le commentaire de Pappus sur le 5e livre de l'Almageste", by the Abbé Rome, in *Annales de la Société scientifique de Bruxelles*, Série A, Sciences mathématiques, XLVII (1927), 77–102.
4. Reproduced from Peter Apian's *Astronomicum Caesareum*, 1540. There are no page-numbers, but the torquetum is discussed in the last five folios.
5. See Rome's article cited in note 3. On the generality of the instrument's name, see J. D. North, "Werner, Apian, Blagrave and the Meteoroscope", *British Journal of the History of Science* 3 (1966), 57–65 and plate II.

6. Printed appended to Ptolemy's *Cosmographia*, Nuremberg 1514. See North, *Op. Cit.*, 58 and 61.
7. Michel, *Op. Cit.* (see note 2), 202.
8. Reproduced from W. Cuningham, *Cosmographicall Glasse*, 1559.
9. I am grateful to Prof. B. R. Goldstein for pointing this MS out to me. Because it is written in Hebrew script, and is incomplete and misbound, it might be the MS copied in haste by Samuel b. Jehuda of Marseille and his brother David; but more research would be needed to be definite about this. See the note at the end of Samuel's revision of the *Iṣlāḥ*: MS BN 1024ff 107r–108r; and Neubauer and Renan, "Les Écrivains Juifs Français", *Histoire Littéraire* XXXI (1893), 560–3.
10. MS Madrid 10006ff 53v–54r and printed edition (Nuremberg 1534), 61–2. In this and the following Latin quotations I have followed the Madrid MS unless the printed edition offered a better reading.
11. "... sicut est revolutio fusorii aque". Ma 10006 gives a marginal note to this: "fusorii aque, id est spine que fundit aquam ut in loco rio frm [?]; per revolutionem enim huius paxilli inspicitur longitudo stelle". For *fusorii aque* the Hebrew translation by Moses ibn Tibbon and Jacob ben Maḥir (see section II 5) have respectively BRZT BYWRY HMYM (the bung—or tap?—of a water-bottle??) and QPL HMYM (a vessel of water?, a head of water?).
12. J. A. Repsold, *Zur Geschichte der astronomischen Messwerkzeuge von Purbach bis Reichenbach, 1450 bis 1830*, 1908, 11–2 and fig. 5; Gunther, *Op. Cit.* (see note 1), 21.
13. Bodleian MS Opp. Add. fol. 17ff 1–100r. These passages are from ff 41r and v.
14. Paris BN heb. 1024, f 46r. In the second quotation the vocabulary is different from that in the Moses ibn Tibbon translation. See below, section II 5.
15. *Histoire de l'astronomie du moyen âge*, 1819 (reprint 1965) p 183.
16–18. Ma 10006 ff 54v, 55r, 55v, and printed edition pp 63, 63, 64, respectively.
19–20. Ma 10006 ff 1v, 8r, and printed edition pp 2, 65, respectively. In these passages the Latin corresponds well enough with the Arabic in, e.g., MS Berlin 5653 ff 2r, 53v. G. Boffito (*Gli Strumenti della Scienza e la Scienza degli Strumenti*, 1929, 53) thought the quadrant was intended for the device to increase accuracy.
21. See note 13.
22. In spite of help from several Hebraists, the translation is not absolutely certain. But the gist of it seems to be as given.
23. Bodleian MS Thurston 3, f 84r.
24. A. Sayili, *The Observatory in Islam and its place in the general history of the observatory*, Publications of the Turkish Historical Society, series VII no. 38 (1960), 199 foll.
25. *De Torqueto, Astrolabio armillari ...*, 1544, p. 1. See E. Zinner, *Leben und Wirken des Johannes Müller von Königsberg, genannt Regiomontanus*, 2nd edition, 1968, 157.
26. BN 7406 ff 114ra–137vb, 14c.
27. Thorndike, *Op. Cit.* (note 1), and *Latin Treatises on Comets between 1238 and 1368 A.D.*, 1950, 198–204.
28. E. Zinner, *Deutsche und Niederländische Instrumente des 11–18 Jahrhunderts*, second edition 1967, p. 32. The note occurs at the end of Richard's tract on the Rectangulus. See Gunther's edition, *Op. Cit.* (see note 1), 348. A new edition by J. D. North is being published by O. U. P.

XVII

A Note on the Horary Quadrant

THE HORARY QUADRANT WAS OF THE FORM indicated in fig. 1. It is the purpose of this note to enquire about its origin. To tell the time, the instrument was placed in the same vertical plane as the Sun, the sights x and y were aligned with the Sun, and it was observed between (or on) which of the curved hour-lines AC, AD, AE, AF, AG, AH – which severally represent 6, 5, 4, 3, 2, 1 seasonal hours after sunrise or before sunset-the bead M was found. The straight line AB served as the zero hour-line. The seasonal hours thus measured (approximately) were each one twelfth of the time of daylight. To set the bead at the right position for the day, it could be put against the noon-line, the curved line AC, either when the Sun was sighted at noon or when the angle between the thread and AC was made equal to the sum or difference (as appropriate) of the local latitude and the Sun's declinaton for the day. In the latter case the addition or subtraction could be made with tables and calculation[1] or by means of a cursor that slides round BC.[2] We are here not concerned with the cursor, nor with extraneous lines on the quadrant, but only with the hour-lines.

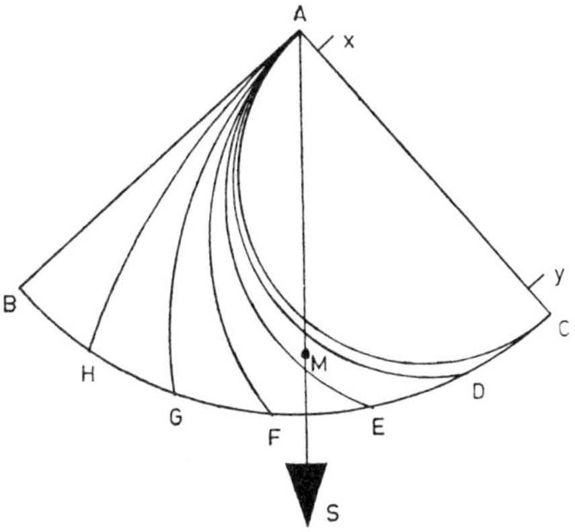

Fig. 1

Examples of the horary quadrant, of Near Eastern provenance, survive from the 4-5/10-11th centuries[3], and there are 3rd-4th/9-10th century treatises

It is a pleasure to thank Professor David A. King, of New York University, for reading though this note, making several pertinent criticisms and supplying much information from his unpublished work (see particularly notes 4, 19 and 20). My thanks, too, go to Professor E. S. Kennedy for his patient advice and help.

1. See Robertus Anglicus, p. 617. Here and elsewhere references are to the bibliography.

2. *Ibid*, p. 615-616. For a photograph of a horary quadrant with cursor, see Poulle, p. 19. Diagrams of such instruments are given by Schmalzl, p. 127 (for Alphonso X's quadrant) and by Tannery in Robertus Anglicus, p. 564.

3. Maddison & Turner, p. 151 (items 70 and 71 in the catalogue). The dates are estimated.

in Arabic describing the instrument.[4] Seasonal-hour diagrams of the same type were inscribed on the backs of astrolabes at least as early as the 4th/10th century, when ᶜAbd al-Raḥmān al-Ṣūfī described them,[5] and examples survive from the 7th/13th century.[6] The Latin treatise on the astrolabe attributed to "Messahala" mentions the lines, but this text has recently been shown to be a western compilation of elements of uncertain date, though the latter part of the treatise (on the use of the astrolabe) appears to be based on astrolabe treatises from the school of Ibn al-Ṣaffār.[7] In the Museum of the History of Science in Oxford there is a splendid western example of an astrolabe carrying these lines – not, as usual, in one or both of the top two quadrants, but occupying the entire back of the instrument.[8] Seasonal hour-lines on the backs of astrolabes appear to have been relatively popular in the medieval West,[9] and only went out of fashion in the mid-seventeenth century.[10]

The hour lines are drawn by dividing the arc *BC* into six equal parts at points *D*, *E*, *F*, *G*, *H*, and joining each of these points to *A* with a circle whose centre lies on *AC* (or *AC* produced). Al-Marrākushī, of the 7th/13th century found the centre of the circle *AF* (to take an example; see figs. 1 and 2) as the intersection of *AC* and the perpendicular bisector of the straight line *AF*.[11] At about the same time Robertus Anglicus described a trial-and-error method to

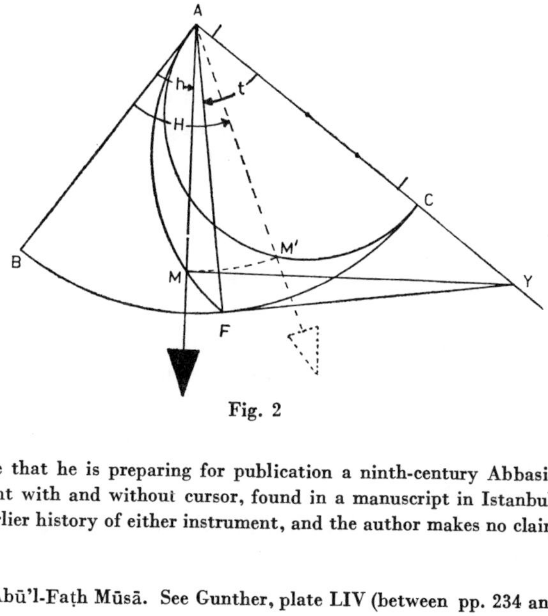

Fig. 2

4. Prof. David King informs me that he is preparing for publication a ninth-century Abbasid Iraqi treatise on the horary quadrant with and without cursor, found in a manuscript in Istanbul. The treatise casts no light on the earlier history of either instrument, and the author makes no claim to have invented either.

5. Al-Ṣūfī, chapter 174, p. 161.

6. E. g. the astrolabe of Sultan Abū'l-Fatḥ Mūsā. See Gunther, plate LIV (between pp. 234 and 235).

7. Kunitzsch, especially pp. 45-46 and 56. The treatises on the astrolabe by John Philoponus and Severus Sebokht (see Gunther, pp. 61-81 and 82-103, for English translations) do not contain a description of these lines.

8. The instrument belongs to Oriel college and dates, perhaps, form 1450. The horary-quadrant lines appear on a *spherical* instrument in the National Museum in Damascus..

9. See previous note by Professor North.

10. Higgins, pp. 109-114. See Cittert, plates XVIII and XXI and p. 35 for late examples.

11. MS Bodleian Hunt. 201 (Uri I, 902), ff. 69v-70r.

find the centre.[12] In 1531 Orontius gave al-Marrākushī's method, which he probably found in his medieval sources.[13]

The lines on the quadrant yield graphical solutions of the formula

$$\sin h = \cos t \sin H \qquad (1)$$

where t (in degrees) is fifteen times the number of seasonal hours before or after noon and is measured by angle CAF (the quadrant is actually graduated directly in hours after sunrise); h is the corresponding solar altitude and is measured by angle BAM; and H is the solar altitude at noon on the same day, measured by angle BAM' ($AM = AM'$). Formula (1) is easily shown to correspond to the hour-lines, since $AM/AF = AM'/AC = \sin H$, and angles M and F being right in semicircle AFY,

$$AM : AF = \sin A\hat{Y}M : \sin A\hat{Y}F = \sin h : \cos t$$

If the instrument were used continously from sunrise to noon, or from noon to sunset, on a day when the solar declination is δ, the bead would trace out the on quadrant an arc of a circle of centre A and radius $AC \sin H = AC \cos(\varphi - \delta)$, where φ is the local latitude. But the quadrant seems not to be directly related to instruments, like the astrolabe, that directly imitate the motion of the Sun about the pole. The quadrant's scale serves not only to measure the time, but also the solar altitudes. The front of the astrolabe which certainly has curved seasonal hour-lines approximated by circular arcs, measures altitudes quite differently. Besides, the radius of a parallel-circle on the astrolabe corresponding to declination δ is $R \cos \delta / (1 + \sin \delta)$, where R is the radius of the circle representing the equator; and if, as is usual, the projection is from the South pole, these parallel-circles are in inverse order to those found on the quadrant, where the smallest circular arc traced out by the bead corresponds to the winter tropic. Furthermore, the hour-lines on the quadrant cannot be projections of the circles of equal azimuth on the celestial sphere, since the hour-lines meet at only one point (A in figs. 1 and 2).

It is conceivable that the curves were formed by joining the appropriate positions of the bead found empirically or by calculation. Such procedures were indeed used by instrument-makers.[14] In fact the hour-lines so plotted for latitude 36° – on the understanding that the length of AM is set at AC

12. Robertus Anglicus, pp. 599-600: ... et alius pes [of the compasses] extendatur ... et queratur punctus in linea AC ... donec pes existens in puncto A fiat mobilis et transeat per puncta A H directe". A and C are as in our diagram; H is our F.

13. Orontius, f. 188v, book II, prop. VIII.

14. Michel, pp. 79-81; Higgins, pp. 109-114.

$\cos(\phi-\delta)$ for the day when the declination is δ – deviate little from those given by formula (1). The errors for the 8 a.m. / 4 p.m. line are the greatest, but even along the summer tropic, where the error is the highest for all hour lines, its deviation is only about 1°49′, equivalent to an error of a little over 8¾ minutes in non-seasonal time.[15] Since the other errors of the instrument – including those of reading the position of the bead and estimating the time when the bead falls between two hour lines – must often exceed a degree or two, such an error would have been acceptable and might even have passed unnoticed. For a contruction of this kind, however, one needs to assume that the bead is set at a distance of $AC \sin H$ from the apex A – a condition hard to justify *a priori*.

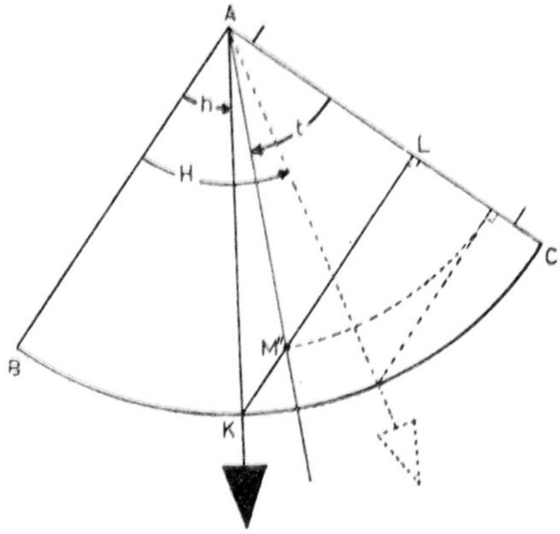

Fig. 3

Easily the most likely origin of the horary quadrant lies in a development of the instrument called by Millás[16] *quadrans vetustissimus*. Instead of curved lines this instrument has a large number of parallel lines perpendicular to the side carrying the sights and reaching from it to the circular arc. The *quadrans vetustissimus* is clearly related to the horary quadrant, since is has a cursor, a thread, and a bead, which is again set at a distance $AC \sin H$ from A (see fig. 3) – albeit by means of the parallel lines – and it works on formula (1). The bead M'' is put against the line KL corresponding to h (see fig. 3); and, since $AL = AC \sin h$ and $AM'' = AC \sin H$, angle CAM'' will measure t according to formula (1) and can be read off on the scale.[17] Since the horary quadrant, with its hour-lines, is easier to use, but is a less obvious embodiment of the formula, we must suppose that it was a development of the *quadrans vetustissimus* and not *vice versa*. Whether the *quadrans vetustissimus* was a specialisation of the sine-quadrant[18] or the sine-quadrant a generalization of the *quadrans vetustissimus* is a matter for speculation. It can be noticed here that the sine-quadrant was usually supplied with at least one semicircle having a side of the quadrant as diameter.

The equivalent of formula (1) is to be found both in al-Khwārizmī's *zīj* (early 3rd/9thcentury) and, in a calculation of the duration of twilight, in a

15. Drecker, p. 86, gives 10, 13, 12, 9, 5 minutes for an instrument for latitude 48°.

16. Millás, p. 235.

17. For four Latin descriptions of the operation of the instrument, see Millàs, pp. 225-226, 227, 239, 239-240. There were other graphical solutions of formula (1) – see, e. g., Sédillot, p. 26.

18. For this instrument see Schmalzl, p. 62 et seq.

recension of Ḥabash's *zīj* (3rd-4th/9-10th century), though both authors gave the correct formula as well.[19] Further, the formula (or equivalent) was used by later Muslim authors and underlies several tables to find the time from the height of the Sun or vice versa. It also occurs in a Byzantine treatise on astronomy. Finally, al-Marrākushî cites it in his treatise on instruments.[20] The formula may be derived from a proof by Abū'l-Wafā' (late 4/10th century) of a formula for the time in terms of solar altitude[21] that Ḥabash had stated and had probably obtained from Brahmagupta[22] (7th century AD). For we find in this proof (see fig. 4), in which $GZTD$ is the day-circle, TM and ZS are perpendiculars from T (the Sun's position) and Z (the Sun's position at noon) to the horizon-plane GDA, that[23]

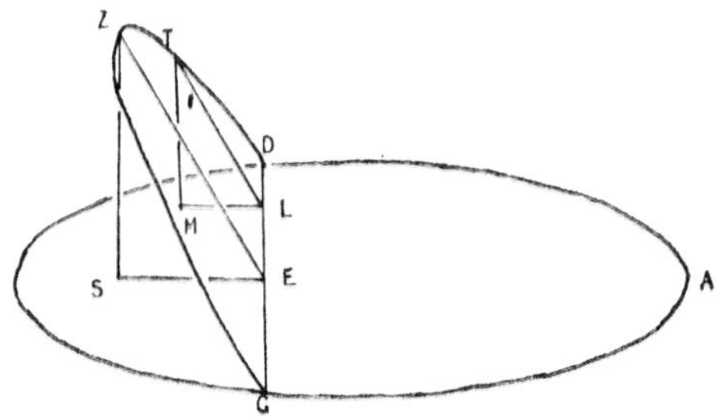

Fig. 4

$$TM : TL = SZ : ZE.$$

Now $TM : SZ = \sin h : \sin H$ and, if only we take $GZTD$ as a semicircle (which, of course, in general it is not), $TL : ZE = \cos t$. Actually, Abū'l-Wafā' does not make this approximation (or mistake), but goes on to prove Ḥabash's correct formula. But the diagram, which in some form must surely have been drawn or thought of to find the correct formula in the first place, is suggestive. That formula (1), or its equivalent in geometrical terms, could have been derived from a mistake is shown by Würschmidt's "derivation"[24] of it by an error similar to that suggested above. Of course, the formula, which is correct for the equinoxes, may have been simply assumed for other times.[25]

In sum, it is suggested here that the horary quadrant was the result of an adaptation – one of great geometrical ingenuity – of an instrumental solution of formula (1), which is an approximation of the true formula. How the formula was arrived at and who invented the instrument remain unknown to us.

19. King, section, 2.5, especially note 21. For al-Khwārizmī, King cites the Hebrew translation of Ibn al-Muthannā's commentary on the *Zīj* (Goldstein, pp. 81-83, 207-208).

20. King, sections 2.5, 4.3 and 4.3.2. On al-Marrākushī, see Sédillot, p. 39.

21. Abū'l-Wafā', first part.

22. Nadir, pp. 460, 462. Al-Khwārizmī (see note 19 above) uses the value 150, common in Indian astronomy, for the radius underlying his treatment of sines.

23. Abū'l-Wafā', p. 4.

24. Würschmidt, pp. 185-186.

25. This is hinted at by Würschmidt, p. 185, and Cittert, p. 45.

Bibligoraphy

Abū'l-Wafā' al-Būzajānī, *Risāla ilā Abī ᶜAlī b. al-Sakr fī iqāmat al-burhān ᶜalā'l-dā'ir min al-falak min qaws al-nahār wa'rtifā' min al-waqt*, (Hyderabad: Osmania Publications Bureau, 1948).

P. H. van Cittert, *Astrolabes. A critical description of the astrolabes, noctilabes and quadrants in the care of the Utrecht University Museum*, (Leiden: E. J. Brill, 1954).

Joseph Drecker, *Theorie der Sonnenuhren.* Band I, Lieferung E of *Die Geschichte der Zeitmessung und der Uhren*, edited by Ernst von Bassermann-Jordan (Berlin & Leipzig: Vereinigung wissenschaftlicher Verleger, 1925).

Goldstein. *Ibn al-Muthannâ's Commentary on the Astronomical Tables of al-Khwârizmî*, Two Hebrew versions, edited and translated, with an astronomical commentary, by Bernard R. Goldstein.

Robert T. Gunther, *The Astrolabes of the World* (London: The Holland Press, 1976).

Kathleen Higgins, "The Development of the Sundial between A. D. 1400 and 1800", unpublished thesis for the degree of B.Sc., Oxford. (Copy in the Museum of the History of Science, Oxford.)

David A. King, *Studies in Astronomical Timekeeping in Medieval Islam* (forthcoming). *Part I: Survey of Medieval Islamic Tables for Reckoning Time by the Sun and Stars.*

Paul Kunitzsch, "On the authenticity of the treatise on the composition and use of the astrolabe ascribed to Messahalla", *Archives internationales d'histoire des sciences*, 31 (1981), 42-62.

Francis Maddison & Anthony Turner, "Catalogue of an exhibition 'Science & Technology in Islam' held at the Science Museum, London, April-August 1976, in association with the Festival of Islam", as yet unpublished.

Henri Michel, *Traité de l'astrolabe*, 2nd edition (Paris: Libraire Alain Brieux, 1976).

J. Millás Vallicrosa, "La introduccion del cuadrante con cursor en Europa", *Isis* 17 (1932), 218-258).

Nadi Nadir, "Abū al-Wafā' on the Solar Altitude", *The Mathematics Teacher*, 3 (1960), 460-463.

Orontii Finei Delphinatis De Solaribus Horologiis et Quadrantibus Libri IIII (Paris: 1531).

Emmanuel Poulle, "Les instruments astronomiques du Moyen Age", *Le Ruban Rouge*, 32 (Mars 1967), 18-29, reprinted as Museum of the History of Science, Oxford, Selected Off-print no. 7.

Robertus Anglicus. Paul Tannery, "Le traité du quadrant de Maître Robert Anglès (Montpellier, XIIIe siècle). Texte latin et ancienne traduction grecque", *Notices et Extraits des Manuscrits de la Bibliothèque Nationale et autres bibliothèques*, 35 (1897) 561-640.

Peter Schmalzl, *Zur Geschichte des Quadranten bei den Arabern* (München: 1929).

J. J. Sédillot, *Traité des Instruments Astronomiques des Arabes* (Paris: 1834).

ᶜAbd al-Raḥmān al-Ṣūfī, *Kitāb al-ᶜAmal bi'l-asṭurlāb* (Hyderabad: Osmania Oriental Publications, 1948).

J. Würschmidt, "Die Bestimmung der krummen Stunden der Deklination und der Gebetszeiten mittels des Astrolabs", *Mitteilungen zur Geschichte der Medizin und Naturwissenschaften*, 18 (1919), 183-190.

XVIII

Al-Ṣaghānī's Treatise on Projecting the Sphere

INTRODUCTION

THE MATHEMATICAL treatise described in this paper is one of the many unstudied texts in the rich tradition of Arabic astrolabe-literature. It is described by the author, Aḥmad ibn Muḥammad ibn al-Ḥusayn al-Ṣaghānī (late A.H. 4th/A.D. 10th century),[1] as his

> Book on how to project the sphere on the plane of the astrolabe so that there are formed in it points, straight lines, circles and the sections of the cone that are known as the parabola, the ellipse and the hyperbola.

As E. S. Kennedy remarked to the present writer a couple of years ago, it is a text well worth investigating.

In a standard astrolabe the stars and ecliptic-circle are represented on a disk that revolves over another disk, on which the horizon coordinates are represented, just as the real stars appear to move across the sky against an imaginary network of almucantars (parallels to the horizon) and azimuths (great circles through the zenith). The upper disk, the rete, is largely cut out, so that the markings on the lower disk, the astrolabe-plate, can be seen. The reduction of sphere to disk is achieved by projecting it, usually from the south pole on to the equatorial (or parallel) plane: each point on the sphere (X in FIG. 1) is connected by a straight line to the south pole, and the intersection of this line with the equatorial plane is the projection (Y) of the point (X). In this type of projection all circles on the sphere become circles on the plane.

This result, which may be easily derived from Apollonius' *Conics* I, 5, and which had been proved a century before by al-Farghānī,[2] was evidently known to al-Ṣaghānī, though he does not mention it explicitly. His own method of projection, in which conic sections are formed by taking as the fixed point, or "pole of projection" (*quṭb al-tasṭīḥ*), not one of the poles but some other point on the line joining them, he claims is new:[3]

> As for the formation of circles, many people have spoken about it. As for the formation of conic sections, no one has spoken about it.

Reprinted from *From Deferent to Equant: A Volume of Studies in the History of Science in the Ancient and Medieval Near East in Honor of E.S. Kennedy*, *Annales of the New York Academy of Sciences* 500, 1987, pp. 237–252. Reprinted by kind permission of The New York Academy of Sciences.

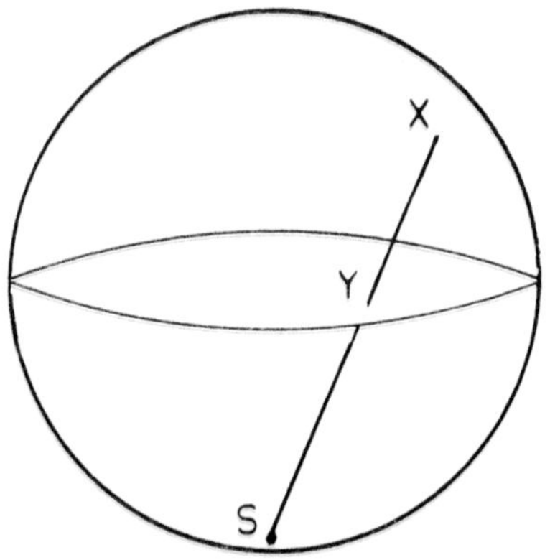

FIGURE 1.

In his book *Istīʿāb al-wujūh al-mumkina fī ṣanʿat al-asṭurlāb*[4] al-Bīrūnī says no-one can deny that al-Ṣaghānī is the inventor of this type of projection.

No astrolabe with the Ṣaghānī projection is known. But al-Bīrūnī may have been encouraged by al-Ṣaghānī's pioneering work to develop the special case of orthographic projection – that is, in modern terms, one with the pole of projection at infinity. It must be said, however, that the orthogonal projection was known before al-Ṣaghānī's time – to al-Farghānī, for instance.[5] Since the orthogonal projection was later used in the West for universal astrolabes,[6] there is the possibility that it was transmitted from the East rather than reinvented. Another special case of the Ṣaghānī projection, in which the pole of projection is a finite distance outside the sphere, was adapted for the universal astrolabe by La Hire at the end of the 17th century.[6]

The twelve sections (*fuṣūl*) of al-Ṣaghānī's book are: 1. mathematical lemmata, *e.g.* one stating, equivalently, that the *latus transversum* (or principal diameter; see below) of an ellipse formed by projection is greater or less than the *latus rectum* (making the north-south diameter greater or less than the east-west diameter) as the point of projection is inside or outside the sphere; 2. on the equator and its parallels; 3. and 4. on the horizon and the almucantars; 5. and 6. on the azimuths; 7. through 9. on the rete, *i.e.* on the ecliptic and on individual stars; 10. lemmata on constructing conics; and 11. and 12. graphical method for almucantars and azimuths. As al-Bīrūnī pointed out,[7] the treatise could do with some reorganization. One subject will appear in different sections, several cases of one theorem could often be treated in a more unified way, and some passages are obscure.

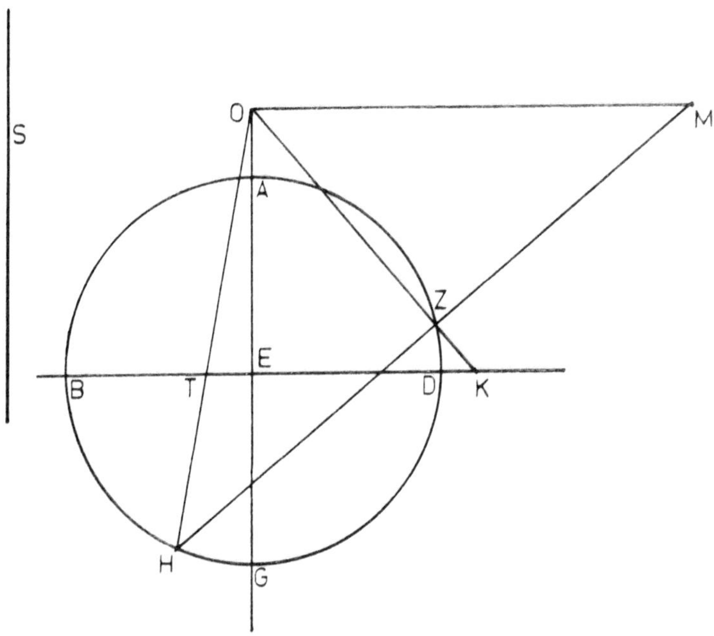

FIGURE 2.

AL-ṢAGHĀNĪ'S TREATMENT OF PROJECTION

It is of the essence of al-Ṣaghānī's style that he considers all the cases produced by the different positions of the pole of projection. For reasons of space, the "southerly" projections, *i.e.*, those produced by putting the pole of projection to the north of the equator, will not be described here. Suffice it to say that the asymmetry between the northerly and southerly projections is occasioned by the author's considering only those almucantars that are above the horizon. He does not mention stars by name, but only assumes that the position of any star to be represented on the rete should be known.

At first, before he deals with the rete, al-Ṣaghānī finds the projection of a circle on a sphere by (1) imagining a cone whose apex is the pole of projection and whose base is the circle to be projected, (2) imagining this cone cut by the "plane of projection", *i.e.*, the equatorial plane, and (3) transferring the figure cut on this plane to the meridian plane by superposition. In the last operation the point directly above *E* (see FIG. 2), the center, is folded on to *A*. So, for instance, in the first case of projecting the almucantars,[8] *ABGD* is the meridian circle with diameters *AG*, *BD* intersecting orthogonally at the

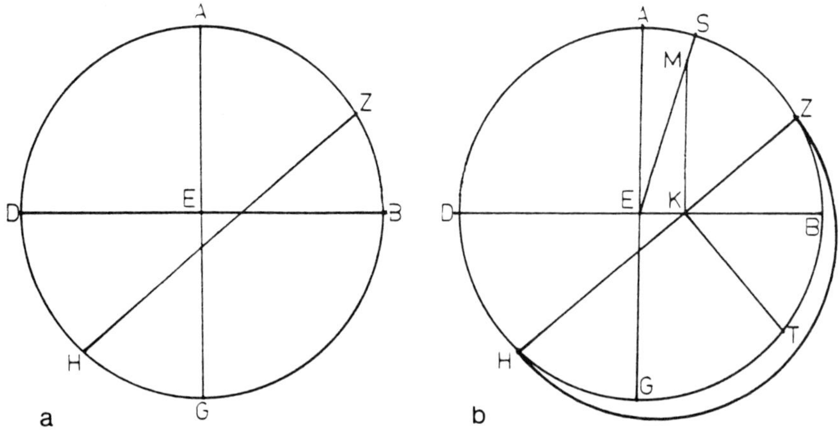

FIGURE 3.

centre E of the sphere, A is the south pole, O (on AG) the pole of projection, ZH the almucantar to be projected; OTH, OZK, HZM are straight lines, MO is perpendicular to EO, and line S is defined by $MO^2 : MH.MZ = KT : S$.

> . . . we imagine a cone whose apex is point O and whose base is the circle whose diameter is ZH—the plane of circle $ABGD$ cuts it and passes through its axis. So their common section is OZH. So if we imagine a plane standing on the plane of circle $ABGD$ at right angles, their common section being BD—*i.e.*, the plane that we call the plane of projection—, that plane cuts the cone, and the common section of that plane and the circle whose diameter is ZH is a line perpendicular to line ZHM. Because triangle OTK is not similar to triangle OZH, the common section of that plane and that cone is an ellipse whose *latus transversum* [*ḍilʿuhu al-māʾil*, here = principal diameter][9] is line TK and whose *latus rectum* [*ḍilʿuhu al-qāʾim*] is line S, . . .
>
> . . . And if we superposed [*aṭbaqnā*] the upright plane on the plane of the astrolabe, the section would coincide [*inṭabaqa*] with the section, and that section is the projection of the circle whose diameter is ZH.

Finding the projection of a circle by forming a cone and cutting it is most startling in the trivial projections. Thus, when the center E is the pole of projection (see FIG. 3a) and ZH is the almucantar to be projected,[10]

> So the cone whose apex is point E and whose base is the circle whose diameter is ZH is cut by the plane of projection, and their common section is a triangle whose apex is point E, as Apollonius showed . . .

However, when projecting the positions of the stars to make the rete (section 7), al-Ṣaghānī naturally enough abandons the intersection of cones and

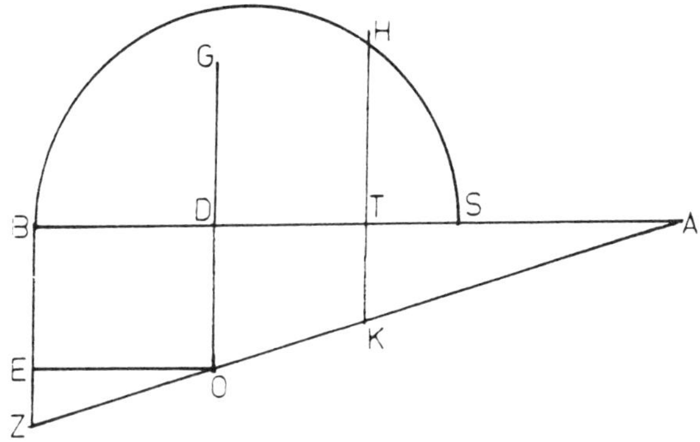

FIGURE 4.

treats projection as a point-to-point operation. At the end of the treatise he gives a point-to-point procedure for almucantars (section 11.4) and azimuths (section 12).

When al-Ṣaghānī describes the conic formed on the plane of projection as the intersection of the plane with a cone, he specifies it either by axis, *latus transversum* and *latus rectum* (axis, vertex and *latus rectum* for a parabola) or by axis, *latus transversum* and one ordinate. In section 10 he gives a graphical construction for finding points on the curve from the latter set of data, though it would do for either, since he finds the *latus rectum* on the way. His procedure for the ellipse and hyperbola may be summarized in the following instructions (FIG. 4):

> *AB* (*latus transversum*) and *GD* (an ordinate) being known,
> apply rectangle *DE*, equal to the
> square of *GD*, to *BD* [so $BD.BE = GD^2$];
> join *AOZ*:
> hence *BZ* is *latus rectum*;
> mark *T* anywhere on *AB*;
> $HTK \perp AB$;
> $TS = TK$;
> semicircle on diameter *BS*:
> hence *H* as point on the conic.

For the parabola it is the same, except that there is no line *AKZ*, and *TS* is put equal to *BE*, in this case the *latus rectum*. Clearly, the procedure rests upon Apollonius' *Conics* I, 11–13, which he quotes, though not by proposition.

In sections 3 and 4 al-Ṣaghānī refers, for the construction of the conics he specifies, to the first book of Apollonius' *Conics*, giving propositions 56, 58,

60 (MS *A* has 55) respectively for the parabola, hyperbola and ellipse. He must be referring to propositions 52, 54, 56 in the modern text,[11] which in this matter agrees with the Arabic of Bodleian MS Marsh 667. For proof that the conics are indeed as he states, he gives, respectively, propositions 32 (MS *A* has 38), 33, 34 of *Conics* I. These are probably propositions 11, 12, 13 in the modern numbering, again the same as in Marsh 667. These citations may shed light on the transmission of the *Conics* in Arabic. It is perhaps worth noting in this context that Omar Khayyām, writing a century or so later, also referred to propositions 52 and 54 of the standard text of Apollonius' book I as propositions 56 and 58 respectively (and also 55 as 59),[12] though he quotes[13] prop. 32 as prop. 33 and prop. 12 as prop. 12. Further, like al-Ṣaghānī, he used the term *al-ḍil' al-mā'il*[14] where one would expect to find *al-quṭr al-mujānib* (or *al-mujānib* or *al-quṭr*), as in Marsh 667, for the *latus transversum*.[15]

The data for the conics – and, in the point-to-point procedures, the projections themselves – are found by graphical methods. Such methods were not uncommon in astrolabe-literature, going back at least as far as Ḥabash al-Ḥāsib in the 3/9th century, as Abū Naṣr Manṣūr (5/11th century) reports.[16] Ḥabash, incidentally, appears to be one of the first to use graphical methods to find the azimuth of Mecca.[17] Al-Ṣaghānī's first such geometrical exercise is also the easiest. The problem is to find the direction of one of the lines forming the trivial projection of almucantar *HZ* from the center (see FIGS. 3a and 3b). The procedure is as follows:

$KT \perp HZ$;
$KM \perp BD$ and $= KT$;
EM is the line required.

It is reminiscent of Ptolemy's procedure in the *Analemma* for finding the azimuth.[18]

THE CHARACTERISTICS OF THE PROJECTIONS

In section 2 al-Ṣaghānī specifies the various configurations of the circles parallel to the equator. Thus (FIG. 5) if the small circle through *H* is the tropic of Capricorn, it is clear that when the pole of projection is at *O*, this tropic will coincide, when projected, with the equator circle. When the pole of projection is beyond *O* or between *O* and *N*, the tropic appears, respectively, inside or outside the equator; when the pole of projection is at *N*, the tropic of Capricorn does not appear on the astrolabe at all; and so on. Meridian circles, *i.e.* circles through the poles, project as straight lines. If *XY* represents an almucantar, then, as is shown in sections 3 and 4, it will become a parabola

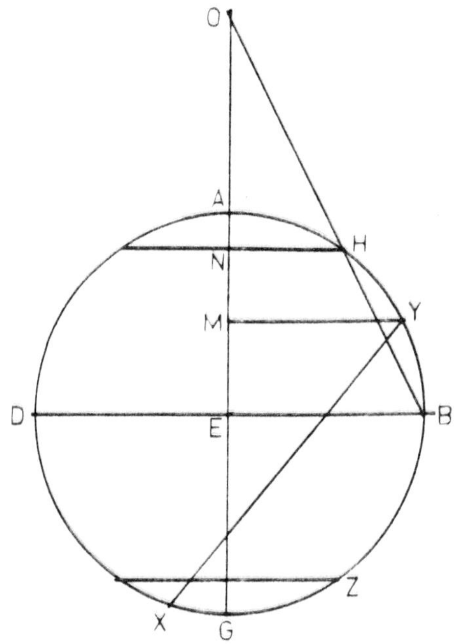

FIGURE 5.

when projected from M (where $YM \perp AG$), a hyperbola when projected from a point between M and E, and an ellipse when projected from a point beyond M. With the azimuths (sections 5 and 6) the situation is a little more complicated, but the principle is the same.

PROJECTING THE ALMUCANTARS AND THE RETE

It is easier to find the projection of the almucantars than of the azimuths, since the axis of the conic representing any almucantar is always BD (see FIG. 2). The method of sections 3 and 4 has already been given. In the case of a parabola (O being such that $ZO \perp AG$ in FIG. 2), the *latus rectum*, C, is given by $ZH^2 : ZO.OH = C : OT$.

In section 11 al-Ṣaghānī gives an entirely different method for almucantars, one that he has given in section 7.3 for the projection of individual stars. As he says in section 7.1, the ecliptic and its parallels may be considered as a horizon (of latitude equal to the complement of the obliquity) and its almucantars. In section 11.1 the position F of one point of the almucantar ZH (see FIG. 6) is found, so that the ellipse formed is determined by one or-

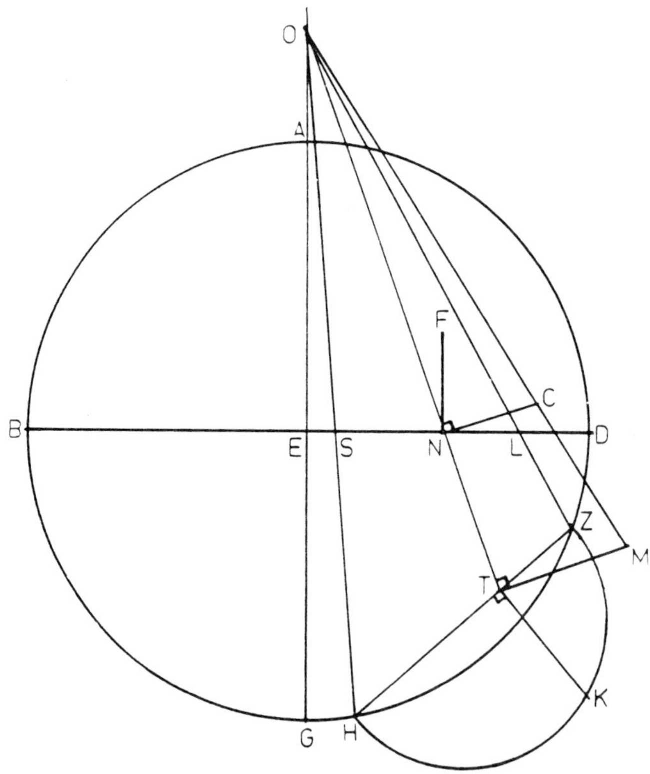

FIGURE 6.

dinate, *FN*, and the *latus transversum SL*. Section 11.2 and 11.3 similarly treat the hyperbola and parabola; and section 11.4 repeats the procedure for the case of the ellipse, but considers it as a point-to-point projection, "without mentioning the conics." Section 11.1 may be summarized:

> *ABGD* is a circle with diameters *AG*, *BD* intersecting orthogonally at *E*; circle *ZH* is to be projected [as in FIG. 2]; *O* is the pole of projection.
> Join *OZ*, *OH* [intersecting *BD* in *L*, *S*];
> *T* is anywhere on *ZH*;
> join *OT* [intersects *BD* at *N*];
> draw semicircle on diameter *ZH*;
> $KT \perp ZH$;
> $TM \perp OT$ and $= TK$;
> $NC \perp OT$ [intersecting *OM* at *C*];
> $NF \perp NS$ and $= NC$.
> Draw ellipse with *latus transversum LS* and ordinate *FN*.

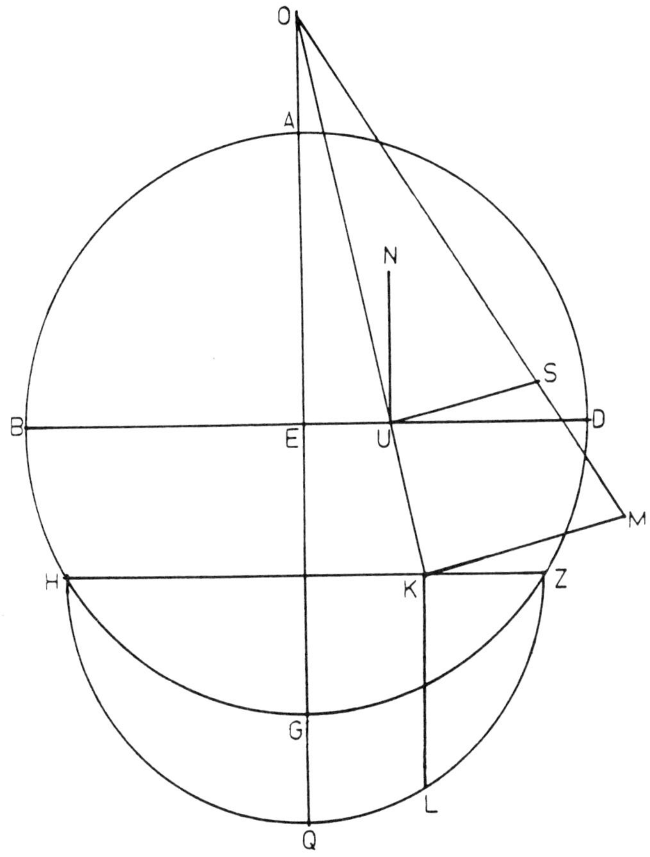

FIGURE 7.

The proof consists of imagining both semicircle *KHZ* and triangle *OMT* as standing upright on plane *ABGD*, so that both *M* and *K* represent the point being projected. In this position *C* will be the projection of *M*, since both *CN* and *MT* are perpendicular to plane *ABGD*.

Two other methods are given for finding the projection of an individual star, both starting with the declination and right ascension. In the first (section 8), *DZ* is put equal to the declination and *QL* to the right ascension (FIG. 7). Then the procedure of section 11.1 is followed, with *ZH* parallel to *BD*. *K* is found by drawing the perpendicular *LK*, and *LK*, thus found, is scaled down by the triangle method of our FIGURE 6; and the result, *US*, is copied on to the perpendicular at *U* to *BD*. In the other method (section 9), it is assumed that declinations are marked, or may be found, directly on the astrolabe-plate – and of course right ascensions may be measured directly.

PROJECTING THE AZIMUTHS

To find the projection of an azimuth, al-Ṣaghānī first (sections 5.1 and 5.2) determines the *ḥāṣil*, the distance along the equator between the meridian and the azimuth, and the *mayl*, the arc measuring the angle the azimuth makes with the equator. Thus he effectively treats the azimuth as a horizon: one whose "meridian" deviates from the east-west line by the *ḥāṣil* and whose "latitude" is the complement of the *mayl*. In FIGURE 8a *ABGD* is the meridian, *BFD* the equator and *EFZ* the horizon (arc *ZB* = arc *ED* = geographical latitude). To find the *ḥāṣil*, *BS*, of azimuth *WSTH*, the procedure (see FIG. 8b) is similar to that associated with FIGURE 6:

ZO = the given arc [*ZT* in FIG. 8a];
OK ⊥ *EZ* [this finds *K*];
join *WK* [this finds *M*];
KT ⊥ *WK* and = *OK*;
join *WT*;
MN ∥ KT;
MC ⊥ *DB* and = *MN*;
join *LC[P]*: hence *BP*, the ḥāṣil [= *BS* of FIG. 8a]

and the proof is also similar.

To find the *mayl*, which controls the characteristics of the resulting conic, al-Ṣaghānī imagines circle *GOVQA*, whose pole is *S*, so that *OV* is the *mayl*. Since circle *HOTSW* goes through the poles of circles *ETZ* and *AQOG*, these two circles must each go through the poles of *HOTSW*. Therefore *Q*, the intersection of circles *ETZ* and *AQOG*, is the pole of circle *HOTSW*. So arc *EQ* = arc *FT* and arc *AQ* = arc *VO*, the *mayl*. This justifies the procedure, which refers to FIGURE 8b, to find the *mayl*:

Put arc *EQ* = arc *FT* of FIGURE 8a [= complement of arc *ZT*];
QX ⊥ *LE*;
join *AX*;
XY ⊥ *AX* and = *XQ*;
join *AY*;
find the arc whose chord is *AY* in circle *ABGD*: this arc = arc *OV* of FIGURE 8a, the *mayl*.

After giving an explanation of the situation (section 5.4) and treating the special case of the azimuth through the east and west points (sections 6.1–6.3), al-Ṣaghānī shows how to determine the characteristics of the conic section produced (sections 6.4 and 6.5). In FIGURE 9:

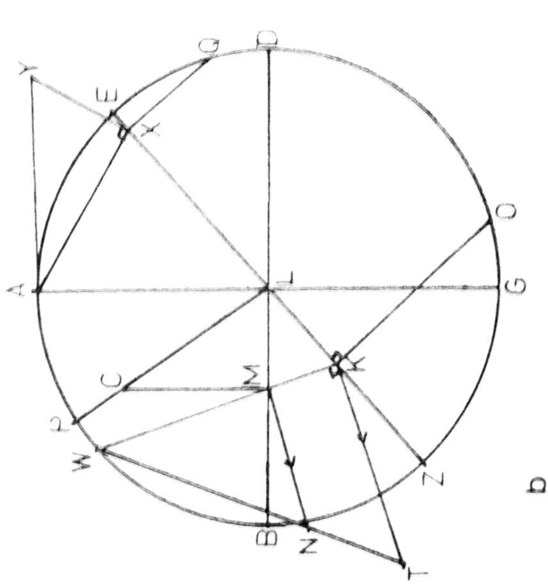

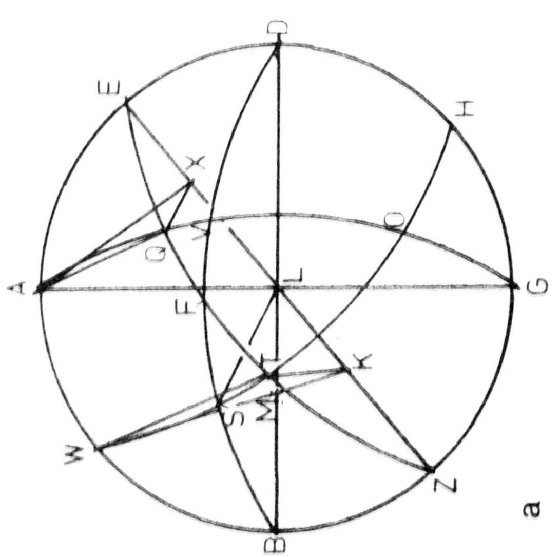

Figure 8.

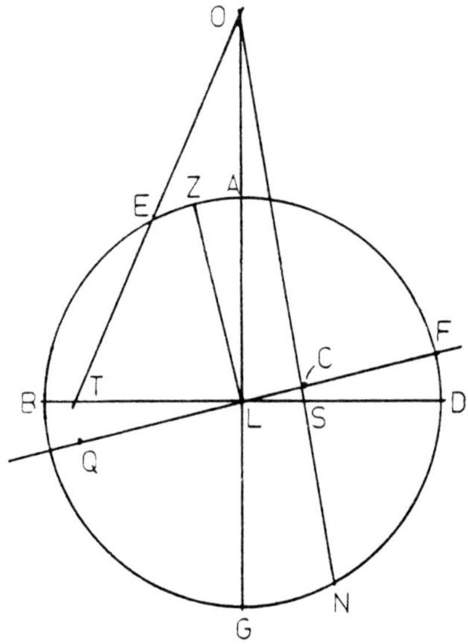

FIGURE 9.

Place *FLQ* [$AF = BZ =$ *ḥāṣil*] by section 5.2;
$ZL \perp FQ$;
$DN = BE =$ *mayl*;
join *ON* and *OE* to find *S* and *T* on *BD*;
$LC = LS$; $LQ = LT$;
draw conic, *latus transversum CQ*, ordinate *LZ*.

Of course, for the parabola the vertex is specified instead of the *latus transversum*.

Finally, in section 12, al-Ṣaghānī applies the method of FIGURE 6 to the projection of the azimuths, using the point-to-point approach. He keeps *HK* equal to the distance along the horizon between the azimuth and the meridian, but varies the position of *ZH*, keeping all the *ZH*'s parallel.

AN EXAMPLE

To give some idea of what one type of Ṣaghānī astrolabe would look like, FIGURE 10 shows the plate for latitude 36° and pole of projection at a distance $r \sin 34°$ from the center of the sphere, where r is the radius of the equator. Almucantars and azimuths are drawn for every 10°; the three concentric circles are the equator and the two tropics. The pole has been so chosen as to make one of the almucantars (20° above the horizon) a parabola. Almucantars of

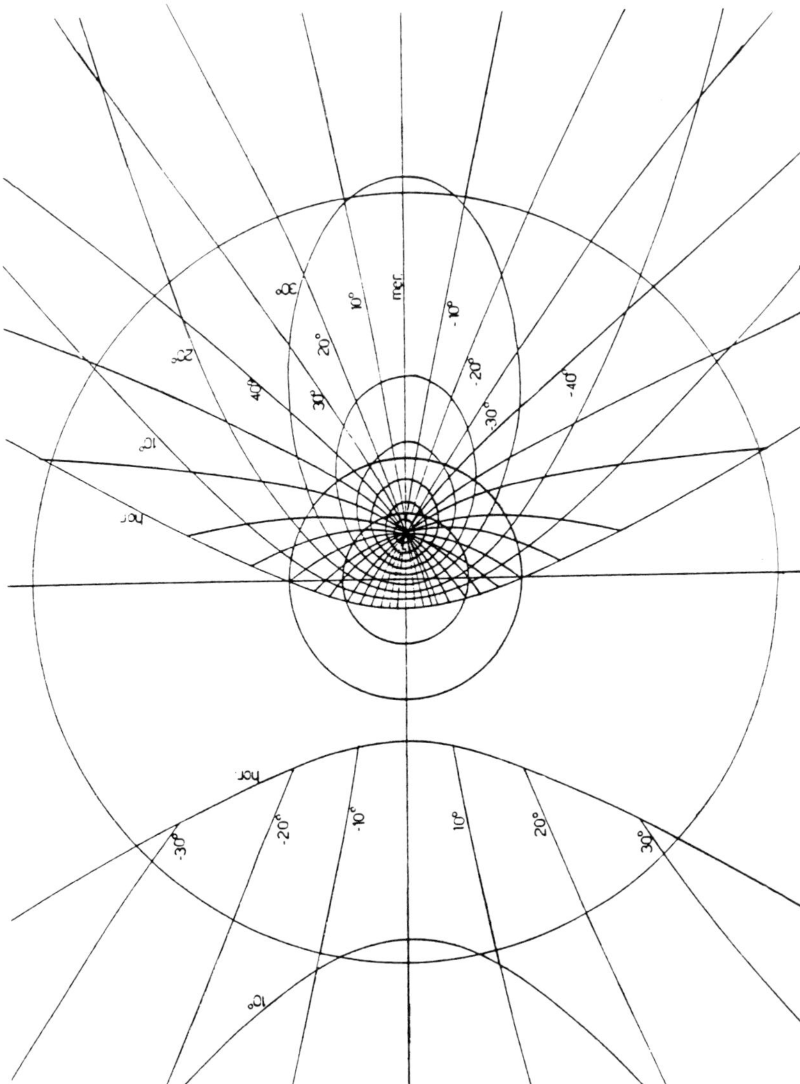

FIGURE 10.

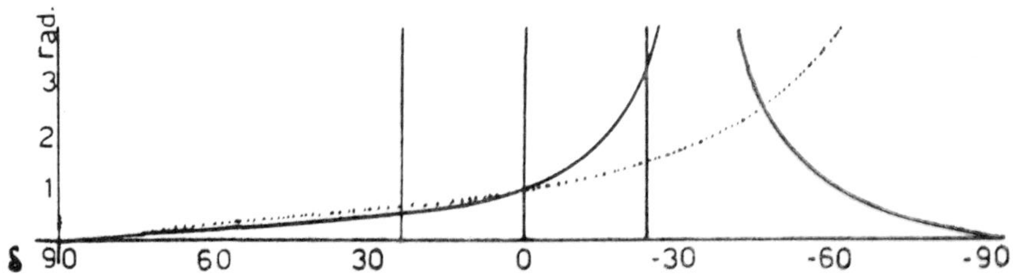

FIGURE 11.

greater altitude are ellipses and those of smaller altitude are hyperbolas, the branches representing the southerly parts of these circles appearing on the left of the diagram. The meridian is, of course, a straight line. Of the other azimuths, which are all hyperbolas, the first four in both directions are labeled with their respective deviations from due south, westerly values being counted positive. The most southerly parts of azimuths for ± 10°, ± 20°, and ± 30° appear on the left of the diagram, each on the opposite side of the meridian to the corresponding branch on the right, which represents the northerly parts of the azimuth.

The rete of such an astrolabe would look much like a normal rete, but with a greater expanse between the equator and the tropic of Capricorn and with the ecliptic in the form of an ellipse, the diameter joining the points of contact with the tropics being about 1.3 times the diameter perpendicular to it.

In FIGURE 10 only the part of the sky above the horizon is furnished with almucantars and azimuths, as is customary with conventional astrolabes. A complete plate would be very confusing. But it will be noticed that one advantage of this type of astrolabe is that the environs of the south pole can be represented, the result being superposed on the projection of the northerly region. In fact, for this latitude and pole of projection all points up to a distance of a little over 46° from the south pole will be projected on to an astrolabe-plate limited by the tropic of Capricorn. FIGURE 11 is a graph of the radii of the projected almucantars against their declinations, north being positive. The radius of the equator is taken as 1. For comparison, the dotted line shows the radii for an astrolabe constructed for the same latitude, but projected from the south pole.

NOTES

1. The opening words of the treatise. It is a pleasure to thank Professor Paul Kunitzsch for checking my translation here and elsewhere and to thank the Institute for the History of Arabic Science, Aleppo, for making the manuscripts available to me.

"Al-Ṣaghānī" is sometimes (e.g., MS *A*, f. 76v, l. 5) spelt "Al-Ṣāghānī." From the dedication to 'Aḍud al-Dawla it is possible to date the text to between 977 and 983. I am grateful to Dr. J. P. Hogendijk of Frankfurt for pointing this out to me.

2. See N. D. Sergeyeva and L. M. Karpova in Thomson, Appendix 3.
3. Ṣaghānī: *P*, p. 3; *A*, f. 76v.
4. Wiedemann and Frank, 535.
5. Frank, 31.
6. Waters, 32–33 and 40–41.
7. Wiedemann and Frank, 535–36. Bīrūnī further says he has made a compendium of the work. The *Istī'āb*, of which MS Leiden Or. 123 (see Sezgin, V, 311 and VI, 218) contains another copy, includes such a compendium.
8. Ṣaghānī: *P*, pp. 16–17; *A*, ff. 79v–80r. "is *OZH* . . . plane of projection –, that plane cuts" is not in *P* (nor in *B*, f. 269v), which substitutes "*BD*, i.e."
9. See Hogendijk, 399 (under no. 84) and 33–34, 37 note 6. See also below and notes 13 and 14.
10. Ṣaghānī: *P*, p. 24; *A*, f. 81v.
11. See, e.g., Heath, 42–52.
12. In his *Algebra*. See Rashed and Djebbar, 32, 37 (French pagination).
13. *Ibid.*, pp. 32 and 39 for prop. 32, and p. 38 for prop. 12.
14. Rashed and Djebbar, p. 32 (Arabic) and pp. 37–38 (French). The use of the term by Omar Khayyām is noted in Hogendijk, 399 (no. 84).
15. *Al-mujānib* where the term is defined (end of book I, props. 12–13); *al-quṭr* and *al-quṭr al-mujānib* appear in the enunciations of book I, props. 54 and 56. A full investigation of this term has not been attempted.
16. Abū Naṣr, 12–16. See Samsó, 49–53 and 89–104 + figs. 14–25.
17. Berggren, 74–77.
18. See, e.g., Gibbs, 112, 116.

BIBLIOGRAPHY

Abū Naṣr — Abū Naṣr Manṣūr ibn 'Alī ibn 'Irāq, *Risāla fī Majāzāt al-sumūt fi 'l-asṭurlāb*, Hyderabad: Osmania Oriental Publications Bureau, 1947. Issued as one of the *Rasā'il Abī Naṣr Manṣūr ibn 'Irāq ila 'l-Bīrūnī*. Hyderabad: Osmania Publications Bureau, 1948.

Berggren, J. L. 1980 — A comparison of four analemmas for determining the qibla. *J. Hist. Arabic Sci* 4: 69–80.

Frank, Joseph 1919 — *Zur Geschichte des Astrolabs*, Erlangen: Universitäts-Buchdruckerei.

Gibbs, Sharon 1976 — *Greek and Roman sundials*. New Haven and London: Yale University Press.

Heath, T. L. 1961 — *Apollonius of Perga*. Cambridge: Heffer. (Reprint of the 1896 Cambridge University Press edition).

Hogendijk, J. P. 1985 — *Ibn al-Haytham's "Completion of the Conics." Sources in the History of Mathematics and Physical Sciences*, no. 7. New York, etc.: Springer Verlag.

Rashed, Roshdi and Ahmad Djebbar 1981 — *L'Oeuvre algébrique d'al-Khayyām*. Aleppo: Institute for the History of Arabic Science.

Al-Ṣaghānī — Aḥmad ibn Muḥammad ibn al-Ḥusayn al-Ṣaghānī, *Kitāb fī Kayfiyat tasṭīḥ al-kura 'alā saṭḥ al-asṭurlāb*: A: MS Istanbul Topkapi, Ahmet III 3342, fols. 76v–91r. B: MS Bankipore 2468/39, fols. 267r–276v. P: Printed text, published as no. 7 of *al-Rasā'il al-mutafarriqa fī 'l-hay'a*, Hyderabad: Osmania Oriental Publications Bureau, 1948.

Samsó Moya, Julio

1969 — *Estudios sobre Abū Naṣr Manṣūr b. 'Alī b. 'Irāq*, Barcelona: Asociación para la historia de la ciencia Española.

Sezgin, Fuat

1967–present — *Geschichte des arabischen Schrifttums*, 9 vols. to date. Leiden: E.J. Brill.

Thomson, Ron B.

1978 — *Jordanus de Nemore and the Mathematics of Astrolabes: De plana sphera*, Toronto: Pontifical Institute of Medieval Studies.

Waters, D. W.

1982 — *The planispheric astrolabe*. Greenwich: National Maritime Museum.

Wiedemann, Eilhard and Josef Frank

1920–21 — Allgemeine Betrachtungen von al-Bīrūnī in einem Werk über die Astrolabien. *Sitzungsberichte der Physikalisch - Medizinischen Sozietät zu Erlangen*. 52–53: 97–121; reprinted in E. Wiedemann, *Aufsätze zur arabischen Wissenschaftsgeschichte*, 2 vols. Hildesheim and New York: Georg Olms, 1970, vol. II, 516–40.

ADDENDA ET CORRIGENDA

The publishers have kindly eliminated several small errors in the reprinted articles. The following *corrigenda* are placed here for reasons of size.

V 430 line 28: for "the lost translation of ROBERT OF CHESTER" **read** "the lost translation of al-Battānī by ROBERT OF CHESTER"

IX 215 line 12: for "*Practica geometriae*" **read** "*Practica geometriae* by Leonardo of Pisa".

XI There is another manuscript of al-Khāzinī's treatise on the "sphere that rotates by itself" in Teheran, University Library 3371, pp. 182–188.

XVIII 246 line 11: for "geographical latitude" **read** "complement of geographical latitude".

General Index

In this index, which is intended to be used in conjunction with the table of contents, most of the personal names (not of modern authors), most of the titles of anonymous writings, and a few of the more important themes have been collected. Personal names have been given in what appears to be their commonest form. In the alphabetic ordering the following have been ignored: the Arabic article "al-", the Arabic "b." (*ibn*) between two names, the Latin preposition "de" and the word "liber" (and similar) in book-titles. "n" following a number means a note on the page indicated by the number; "n" followed by a number means the note of that number (used for articles with notes collected at the end). When a name appears in the text and in a supporting footnote, only the reference to the text is given.

Index of Manuscripts Cited

See also the list of Latin manuscripts of Jābir's astronomy in item VII.

1. Arabic manuscripts

2. Latin manuscripts

3. Hebrew manuscripts